VOLUME ONE HUNDRED AND TWENTY FOUR

ADVANCES IN PROTEIN CHEMISTRY AND STRUCTURAL BIOLOGY

Protein Kinases in Drug Discovery

VOLUME ONE HUNDRED AND TWENTY FOUR

ADVANCES IN PROTEIN CHEMISTRY AND STRUCTURAL BIOLOGY

Protein Kinases in Drug Discovery

Edited by

ROSSEN DONEV

MicroPharm Ltd,
United Kingdom

ELSEVIER

ACADEMIC PRESS

An imprint of Elsevier

Academic Press is an imprint of Elsevier
50 Hampshire Street, 5th Floor, Cambridge, MA 02139, United States
525 B Street, Suite 1650, San Diego, CA 92101, United States
The Boulevard, Langford Lane, Kidlington, Oxford OX5 1GB, United Kingdom
125 London Wall, London, EC2Y 5AS, United Kingdom

First edition 2021

ISBN: 978-0-323-85313-2
ISSN: 1876-1623

For information on all Academic Press publications
visit our website at https://www.elsevier.com/books-and-journals

Publisher: Zoe Kruze
Acquisitions Editor: Ashlie M. Jackman
Editorial Project Manager: Leticia M. Lima
Production Project Manager: Abdulla Sait
Cover Designer: Victoria Pearson

Typeset by SPi Global, India

Contents

Contributors

Carolina Horta Andrade
LabMol—Laboratory for Molecular Modeling and Drug Design, Faculty of Pharmacy, Federal University of Goiás, Goiânia, GO; Laboratory of Tropical Diseases—Prof. Luiz Jacintho da Silva, Department of Genetics, Evolution and Bioagents, University of Campinas, Campinas, SP, Brazil; Department of Infection Biology, London School of Hygiene and Tropical Medicine, London, United Kingdom

Adam Bajinting
St. Louis University School of Medicine, St. Louis, MO, United States

E. Bilsland
Laboratory of Tropical Diseases—Prof. Luiz Jacintho da Silva, Department of Genetics, Evolution and Bioagents, University of Campinas, Campinas, SP; Department of Structural and Functional Biology, Institute of Biology, UNICAMP; Synthetic Biology Laboratory, Institute of Biology, UNICAMP, Campinas, Brazil

Joyce Villa Verde Bastos Borba
LabMol—Laboratory for Molecular Modeling and Drug Design, Faculty of Pharmacy, Federal University of Goiás, Goiânia, GO; Laboratory of Tropical Diseases—Prof. Luiz Jacintho da Silva, Department of Genetics, Evolution and Bioagents, University of Campinas, Campinas, SP, Brazil

Christian Borgo
Department of Biomedical Sciences, University of Padova, Padova, Italy

G.C. Cassiano
Global Health and Tropical Medicine (GHTM), Instituto de Higiene e Medicina Tropical, Universidade Nova de Lisboa, Lisbon, Portugal; Laboratory of Tropical Diseases—Prof. Luiz Jacintho da Silva, Department of Genetics, Evolution and Bioagents, University of Campinas, Campinas, SP, Brazil

Kai Chen
Department of Orthopedics, Changhai Hospital, Naval Military Medical University, Shanghai, China

Fabio Trindade Maranhão Costa
Laboratory of Tropical Diseases—Prof. Luiz Jacintho da Silva, Department of Genetics, Evolution and Bioagents, University of Campinas, Campinas, SP, Brazil

P.V.L. Cravo
Global Health and Tropical Medicine (GHTM), Instituto de Higiene e Medicina Tropical, Universidade Nova de Lisboa, Lisbon, Portugal

Nayana Devang
School of Biotechnology, National Institute of Technology Calicut, Calicut, Kerala, India

Nicholas Furnham
Department of Infection Biology, London School of Hygiene and Tropical Medicine, London, United Kingdom

Preeti Gupta
Centre for Interdisciplinary Research in Basic Sciences, Jamia Millia Islamia, New Delhi, India

Md Imtaiyaz Hassan
Centre for Interdisciplinary Research in Basic Sciences, Jamia Millia Islamia, New Delhi, India

Ritika Kabra
National Centre for Cell Science, SP Pune University Campus, Pune, India

Mingyu Li
Department of Pathophysiology, Key Laboratory of Cell Differentiation and Apoptosis of Chinese Ministry of Education, Shanghai Jiao Tong University, School of Medicine, Shanghai, China

Marilia Nunes Nascimento Lima
LabMol—Laboratory for Molecular Modeling and Drug Design, Faculty of Pharmacy, Federal University of Goiás, Goiânia, GO, Brazil

Yaqin Liu
Medicinal Chemistry and Bioinformatics Centre, Shanghai Jiao Tong University, School of Medicine, Shanghai, China

Shaoyong Lu
Department of Pathophysiology, Key Laboratory of Cell Differentiation and Apoptosis of Chinese Ministry of Education; Medicinal Chemistry and Bioinformatics Centre, Shanghai Jiao Tong University, School of Medicine, Shanghai, China

Sabrina Silva Mendonca
LabMol—Laboratory for Molecular Modeling and Drug Design, Faculty of Pharmacy, Federal University of Goiás, Goiânia, GO, Brazil

M.N. Nascimento
LabMol—Laboratory for Molecular Modeling and Drug Design, Faculty of Pharmacy, Federal University of Goiás, Goiânia, GO, Brazil

Ho Leung Ng
Department of Biochemistry and Molecular Biophysics, Kansas State University, Manhattan, KS, United States

Adyashree Pani
School of Biotechnology, National Institute of Technology Calicut, Calicut, Kerala, India

G.K. Rajanikant
School of Biotechnology, National Institute of Technology Calicut, Calicut, Kerala, India

Ashfaq Ur Rehman
Department of Pathophysiology, Key Laboratory of Cell Differentiation and Apoptosis of Chinese Ministry of Education, Shanghai Jiao Tong University, School of Medicine, Shanghai, China

D.A. Rodrigues
Laboratory of Tropical Diseases—Prof. Luiz Jacintho da Silva, Department of Genetics, Evolution and Bioagents, University of Campinas, Campinas, SP, Brazil

Maria Ruzzene
Department of Biomedical Sciences, University of Padova; CNR Neuroscience Institute, Padova, Italy

Arthur Carvalho Silva
LabMol—Laboratory for Molecular Modeling and Drug Design, Faculty of Pharmacy, Federal University of Goiás, Goiânia, GO, Brazil

Shailza Singh
National Centre for Cell Science, SP Pune University Campus, Pune, India

P. Sunnerhagen
Department of Chemistry and Molecular Biology, Lundberg Laboratory, University of Gothenburg, Gothenburg, Sweden

Aaliya Taiyab
Centre for Interdisciplinary Research in Basic Sciences, Jamia Millia Islamia, New Delhi, India

T.A. Tavella
Laboratory of Tropical Diseases—Prof. Luiz Jacintho da Silva, Department of Genetics, Evolution and Bioagents, University of Campinas, Campinas, SP, Brazil

CHAPTER ONE

Transporter proteins and its implication in human diseases

Ritika Kabra and Shailza Singh*
National Centre for Cell Science, SP Pune University Campus, Pune, India
*Corresponding author: e-mail address: singhs@nccs.res.in

Contents

Abstract

Drug transporters, classified in various ways like efflux transporters and influx transporters; secretory transporters and absorptive transporters; ATP-driven transporters and Solute Linked Carrier (SLC) transporters are of great importance while studying pharmacokinetics. They have impeccable roles in the drug discovery process of infectious diseases. Many of these find a pivotal role in synthetic antimicrobial peptides. The chapter briefly elucidates the varied types and their significance.

1. Drug transport proteins

Drug transporters are membrane proteins that transport drugs either into the cell or out of the cell. In other words, these transporter proteins are responsible for either uptake or efflux of the drug. They are present in various tissues including the ones with a barrier like liver, kidney, intestine, brain and placenta. These drug transporters are of great importance while studying pharmacokinetics and drug-drug interactions as they play crucial role in absorption, distribution, metabolism and elimination (ADME) of drugs. More than 400 membrane transporter proteins present in human genome have been annotated till date. Drug transporters can be classified in various ways like efflux transporters and influx transporters; secretory transporters and absorptive transporters; ATP-driven transporters

Advances in Protein Chemistry and Structural Biology, Volume 124
ISSN 1876-1623
https://doi.org/10.1016/bs.apcsb.2020.08.004

and Solute Linked Carrier (SLC) transporters. Here, we have followed the later classification type (Giacomini et al., 2010; Nigam, 2015; You & Morris, 2014).

2. ATP-driven transporters

Transporters that utilize free energy change obtained from ATP hydrolysis to transport their substrates against chemical gradient are known as primary active transporters or ATP-driven transporters. Since ATP hydrolysis takes place, hence, these transporters are also known as ATPases. They can further be classified as rotary motors (F-, A- and V-type ATPases); P-type ATPases and ABC transporters. Rotary motors and P-type ATPases mostly transport protons or metal ions. While ABC transporters transport a wide variety of small organic and inorganic molecules (like amino acids, sugars, nucleosides, vitamins, metal clusters, small molecules, etc.) as well as large organic molecules (peptides, lipids, oligonucleotides, polysaccharides) (Wilkens, 2015).

2.1 Rotary motors

Rotary motors are molecular ATPases where ATP is either synthesized or hydrolyzed. Most of the ATP synthesis involves proton gradient or sodium ion gradient. Rotary motors can further be categorized as F-type, V-type or A-type ATPases (Stewart, Sobti, Harvey, & Stock, 2013).

F-type ATPases are ATP synthase that transforms energy from a gradient of ions across membrane to synthesize an ATP molecule. Three different variants of F-type ATPase have been seen. The most complex one is found in eukaryotes, in the inner membrane of mitochondria, and helps in synthesis of ATP. Hence, mitochondria named as "Power house of Cell." The other two variants are comparatively simpler ones and are found in chloroplast thylakoid membrane and bacteria. V-type ATPases or Vacuolar ATPase uses ATP hydrolysis energy to pump cations across membrane. They are responsible for acidification of intracellular compartments like lysosomes thereby maintaining its pH. They are also expressed in plasma membrane and neuronal synaptic vesicles as well as central vacuole of yeast cells. They are also trans-membrane proteins and help symporters and antiporters to actively transport ions through membrane by generating electro chemical potential gradient. Similar to F-type ATPase, V-type is also found in eukaryotes, bacteria and archaea. The archaeal V-type ATPase is most commonly called as A-type ATPases (Pharmacology, n.d.-b; Kühlbrandt, 2019; Stewart, Sobti, Harvey, & Stock, 2013).

In general, all rotary ATPases share overall architectural conservedness subdivided in two parts: $F_1/V_1/A_1$ and $F_O/V_O/A_O$. The $F_O/V_O/A_O$ part is a membrane embedded part while the other one is a water soluble motor. The 'o' in $F_O/V_O/A_O$ represents oligomycin which inhibits these pumps. Both parts are connected by a central stalk and peripheral stalks. The number of peripheral stalks may vary as per the motor type. For instance, the F-type ATPase consists of a single peripheral stalk, bacterial V-type and A-type ATPase has two stalks while the eukaryotic V-type comprises of three peripheral stalks. The central stalk is responsible for torque transmission while the peripheral ones stabilize, regulates as well counteracts the rotation (Grüber, Wieczorek, Harvey, & Müller, 2001; Kühlbrandt, 2019; Pharmacology, n.d.-a).

The F-type ATPase, also known as F_1F_O-ATPase, consists of two major subunits, F_1 a spherical head and F_O membrane embedded part. The F1 head consists of 3α, 3β, ϒ, δ and ε subunits. The 3-α and 3-β units are arranged in a symmetrical way alternate to each other with β units exposing the active sites around the central ϒ unit. These active sites of β units are named as empty/open state (O), occupied with Mg-ADP + Pi/loosely bound (L) and bound with ATP/tightly bound (T). The F_0 part consists of 'a' subunit and a rotor ring comprising of 8–17 c-subunits arranged in a ring fashion. The F_O and F_1 are connected to each other through a central stalk encompassing ϒ, δ and ε subunits and a peripheral stalk. The peripheral stalk is further connected with F_1 through another subunit OSCP (Oligomycin Sensitivity Conferral Protein) in mitochondria and δ in chloroplast and bacteria. Along with OSCP, peripheral stalk also includes F6, b and d subunits (Grüber, Wieczorek, Harvey, & Müller, 2001; Kühlbrandt, 2019; Pharmacology, n.d.-a).

The two rotors of F-type ATPase rotate in opposite direction. In comparison with F_1, F_O rotor is stronger. It uses the force generated from the proton gradient to push F_1 part in reverse direction thereby generating ATP from ADP and Pi. The proton gradient is generated by the H^+ ions provided by electron transport chain (ETC) and traverses through F_O c-ring. In response, the central stalk, which is tightly bound to c-ring, rotates 3α3β subunits and initiates the catalytic cycle. As mentioned earlier, the catalytic site of β subunit is divided in three states: L, T and O states. Each rotation is of 120° so the complete cycle is of 360°synthesizing 3 ATP molecules. The 1st state is L-state, where ADP and Pi enters the catalytic site. The torque generated by proton gradient changes the conformational state into the "Tightly bound" state with the newly generated ATP molecule. Finally, the catalytic cycle resets in original state by releasing ATP and attaining O-state and is now ready for the start of next round of the cycle (Nakamoto, Scanlon, & Al-Shawi, 2008) Fig. 1. OSCP subunit plays an important role by maintaining the F_1 catalytic core, i.e. 3α3β, in stationary position against the torque generated by rotating central stalk (Kühlbrandt, 2019; Stewart et al., 2013).

The bacterial F_1F_O-ATPases have been widely used as the model system for understanding the overall structure and general mechanism of rotary motors. The chloroplast F_1F_O-ATPases is also known as cF_1F_O-ATPases and are found in the thylakoid membrane distributed in a random manner. It resembles the bacterial F_1F_O-ATPases in terms of size as well as composition. In cF_1F_O, the proton gradient is generated by the water splitting reaction of PS-II (Photosystem II) and cyclic electron flow driven by light around PS-I (Photosystem I). While the bacterial and chloroplast F-type ATPase exists in monomeric form, the mitochondrial F-type ATPase (mtF_1F_O) forms dimer joined in a 'V' shape form in Fig. 2. (Kühlbrandt, 2019).

The mtF1FO are further classified into four classes. These are Type I, Type II, Type III and Type IV dimers (Kühlbrandt, 2019).

Type I dimers. These are found in animals and fungi and are most widely known as well as most understood type of F-type ATPase. The mammalian and yeast dimers are similar in composition as well as structure at low resolution. The angle between the central stalks of the dimer is ~86°.

Type II dimers. It is the second most well characterized mtF_1F_O found in unicellular and chlorophyll less green algae. Type II ATPase has an extra subunit known as ATP synthase associated (ASA) subunit. The ASA subunits are the characteristic feature of type II dimers and form a massive peripheral stalk. The central stalks of this dimers are 56° apart.

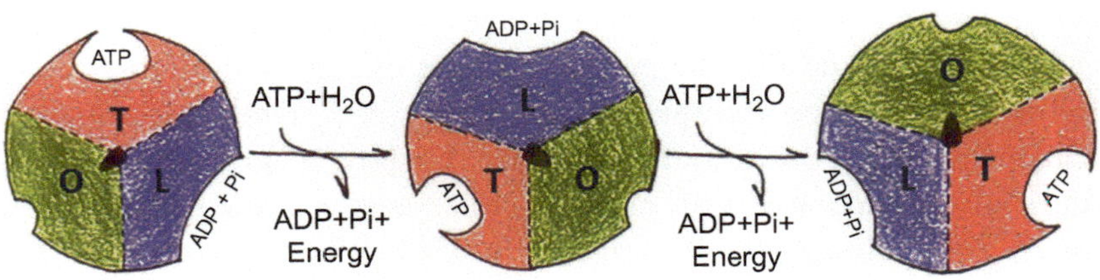

Fig. 1 General catalytic cycle of rotary motors. The three β subunits are represented in three different states namely tightly bound or 'T' state (red color), loosely bound or 'L' state (blue color) and open or 'O' state (green color).

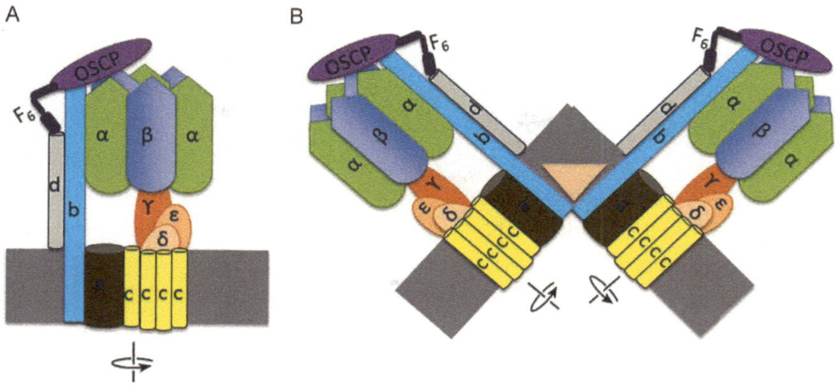

Fig. 2 General architecture of eukaryotic F-type ATPase in monomeric form (A) and dimeric form (B).

Type III dimers. These are found in ciliates like *Paramecium* sp. And the dimers are arranged in a U shape manner rather V shape with dimer angle close to zero. The peripheral stalks are in lateral offsets and the large luminal protein densities forms close interdimer contacts resulting in formation of long helical rows in a regular manner.

Type IV dimers. These dimers are present in *Euglena* and *Trypanosomatids* and share some characteristics with that of Type III dimers in terms of peripheral stalks but with less luminal densities. The F_1 head is pyramid shaped and the dimer angle is ~55°.

Studies by various researchers world-wide have shown that mutations in many subunits of F-type ATPase like subunit A and subunit 8 leads to various severe disorders linked to mitochondria like neuromuscular, neurodevelopmental as well as neurodegenerative diseases. Around 48 disorders have been reported due to mutations in subunit A and 13 diseases by mutations in subunit 8. Most of the serious defects are linked with the proton channel or impairs the stability of F_O subunit (Kühlbrandt, 2019).

The structural composition as well as arrangement of subunits of V-type ATPase is similar to that F-type ATPase. The V_1V_O-ATPase consists of 8 subunits in V_1 part (A-H) and 5 subunits in V_O (a, c, c', c", d and e). The c subunits are arranged in a ring with a subunit on the outer side and the interface of a-c subunits forms the proton channel. Upon ATP hydrolysis, c-rings rotate and a proton is translocated creating a gradient. Thus V-type ATPase is responsible for acidification of various organelles or compartments like Golgi-derived vesicles, clathrin-coated vesicles, synaptic vesicles, liposomes, plant vacuoles, etc. V-type ATPase participates and plays key role in various processes inside cells as in maintaining pH, homeostasis, protein degradation, protein trafficking, receptor mediated endocytosis, release of neurotransmitters, pro-hormone processing, etc. by decreasing the pH of intracellular compartments, it aids in regulating enzymatic activity, dissociation of receptor ligand, transport of substrates across membranes, cytoskeletal tethering, etc. In addition, it assists the entry of various pathogens, viruses or toxins inside the cell. It has been found that mutations in subunit a may lead to osteopetrosis and renal tubular acidosis. The prokaryotic ATPase is similar to that of the eukaryotes and is widely used to study the structure and mechanism of the V-type ATPase (Imamura et al., 2003; Kane, 2006; Maxson & Grinstein, 2014).

The A-ATPase or the Archaeal ATPases also have similar composition as that of F-ATPase and V-ATPase but with less complexity. Since not much have been studied on this type, less is known about A-type ATPases. Another reason for less information is low stability of isolated proteins (Grüber et al., 2001).

2.2 P-type ATPases

P-type ATPases constitutes one of the largest protein family which transports ions and lipids across cell membranes against a concentration gradient by utilizing energy obtained from ATP hydrolysis. They are found in all forms of life irrespective of plants or animals. Based on the sequence homology and phylogenetic analysis, they are classified in five main classes: P1, P2, P3, P4 and P5. Of these, P1 to P3 type ATPases are cation transporters, P4-type ATPases transports phospholipids while P5-type ATPase lie under orphan category as their substrates are still unknown (Andersen, Vestergaard, Mikkelsen, & Mogensen, 2016; Montigny, Lyons, Champeil, Nissen, & Lenoir, 2016). The different classes of P-type ATPases along with their substrates and encoding genes have been summarized in Table 1.

2.2.1 Classification

2.2.1.1 P1-type ATPase

They transports heavy metals mainly monovalent or divalent ions like K^+, Cu^+, Ag^+, Cu^{2+}, Zn^{2+}, etc. they can be further classified as P1A-ATPases and P2A-ATPases. P1A- also known as Type 1A—ATPases are found in bacteria and transport K^+ ions. They are also known as potassium pumps and function as heterotetrameric complex 'KdpFABC'. This unusual ATPase is the smallest known P-type ATPase. Since it is a heterotetrameric

Table 1 P-type ATPases.

Class	Substrate	Disease associated	Genes	Remarks
P1-ATPases				
P1A	K^+	–		Bacterial Kdp-ATPases; smallest P-type ATPase
P1B	Cu^+, Ag^+, Cu^{2+}, Zn^{2+}, Pb^{2+}, Cd^{2+}, Co^{2+}	–	ATP7A, ATP7B	Omnipresent. Helps in metal homeostasis, metal detoxification and metal resistance
P2-ATPases				
P2A / SERCA	Ca^{2+}	Brody's disease, Darier's disease, heart failure, type II diabetes	SERCA1–3/ ATP2A1–3	Located in sarcoplasmic or endoplasmic reticulum; beta subunit: phospholamban and sarcolipin; inhibitors: Thapsigargin and Cyclopiazonic acid
P2A / SPCA	Ca^{2+}, Mn^{2+}	Hailey-Hailey's disease	ATP2C2	Located in secretory vesicles in animals or in vacuolar membrane in fungi; in Golgi membrane transports Mn^{2+}
P2B/PMCA	Ca^{2+}	Hereditary deafness, hearing loss, essential in neuronal Ca^{2+} homeostasis, Familial hypercalcemia	ATP2B1–4, ATP2C1	Located in plasma membrane (animals), plasma membrane, vacuolar membrane, ER, plastid envelope (plants and fungi)
P2C (Na^+/K^+-ATPases)	Na^+, K^+	–	ATP1A1–4, ATP1B1–4	Inhibitor: Ouabain
P2C (H^+/K^+-ATPases)	H^+, K^+	–	gastric: ATP4A, nongastric: ATP12A	–

Continued

Table 1 P-type ATPases.—cont'd

Class	Substrate	Disease associated	Genes	Remarks
P3-ATPases				
P3A	H^+	–	AHA2 (*A. thaliana*); Pma1 (*S. cerevisiae*)	Found in plasma membrane of prokaryotes, protists, fungi and plants
P3B	Mg^{2+}	–		Present in eubacteria and plants
P4-ATPase				
Class 1a	PS, PE	Cerebral Ataxia, Mental Retardation and Disequilibrium Syndrome (CAMRQ)	ATP8A1, ATP8A2	Expression higher in skeletal muscles, spinal cord, brain, retina, thyroid
Class 1b	PC, PS	Intrahepatic cholestasis, Alzheimer's disease, abnormal sperm-egg interaction, defects in sperm capacitation	ATP8B1-B5	Ubiquitous expression but high in small intestine, pancreas, testis
Class 2	–	–	ATP9A, ATP9B	Ubiquitous expression but high in brain, pancreas, testis
Class 5	PC	Obesity, type 2 diabetes, hyperinsulinemia	ATP10A, ATP10B, ATP10D	High expression in brain, pancreas, kidney, lung
Class 6	PS, PE	Marker for metastasis in colorectal cancer, cholestasis, impaired B lymphocyte differentiation	ATP11A, ATP11B, ATP11C	Ubiquitously expressed
P5-ATPases	Unknown	Kufor-Rakeb syndrome; speech-language disorder	ATP13A1-A5	Characteristic motif '**PPxxP**' precedes P-domain

complex, consisting of four different subunits: KdpF, KdpA, KdpB, KdpC; hence it has largest number of subunits. This domain composition is similar but with only seven Transmembrane helices. Also, the N-Domain is small in comparison to other P-type ATPases as well as less homologous (Palmgren & Nissen, 2011).

The second class, P1B-ATPases, transports heavy metal ions like monovalent (Cu^+, Ag^+) and divalent (Cu^{2+}, Zn^{2+}, Pb^{2+}, Cd^{2+}, Co^{2+}). The heavy metals being transported are mostly involved in metal resistance, detoxification or metal homeostasis in most of the organisms. P1B-ATPases are found in all forms of life but are more common in Archaea and Bacteria. The examples includes CadA (Cd^{2+} transporter), ZntA (Zn^{2+} transporter), CopA (Cu^{2+} transporter) in bacteria, HMA4 (Zn^{2+} transporter) in plants, ATP7A, ATP7B (Cu^{2+} transporter) in humans, etc. Amongst them, Cu^{2+} transporters are most abundant and widely studied heavy metal pumps. In order to phosphorylate the enzyme and transport metal ion, it requires metal to first bind at the metal binding sites. They also have a characteristic terminal extensions with either Cys or His residue where metal binds with high affinity. These regions are known as metal binding domains (MBDs). These MBDs are found usually at N terminus with plants being the exception, where they are present at C terminus (Guerrero & Arguello, 2008; Palmgren & Nissen, 2011).

2.2.1.2 P2-type ATPase

P2-ATPase or Type II ATPase is one of the most abundantly studied P-type ATPase for example Ca^{2+} transporting SERCA pumps. They are mostly found in eukaryotes. P2-ATPase is further subdivided in four groups viz. P2A, P2B, P2C and P2D classes (López-Marqués et al., 2015).

P2A-ATPase or Type IIA ATPases are Ca^{2+} transporting ATPases which includes Sarcoendoplasmic reticulum Ca^{2+}-ATPase (SERCA) and Secretory pathway Ca^{2+} – ATPase (SPCA). SERCA pumps are located in sarcoplasmic/endoplasmic reticulum while SPCA pumps are found in secretory vesicles in animals or in vacuolar membrane in fungi. In Golgi membranes, SPCA transports Mn^{2+} ions. SERCA pumps have two Ca^{2+} ion binding sites while SPCA have only one binding site. The bacterial Ca^{2+} ion pumps extrudes Ca^{2+} against transmembrane gradient. In animal cells, SERCA pumps help in pumping back Ca^{2+} ions into Sarcoplasmic reticulum or Endoplasmic reticulum which were released during muscle contractions. In one cycle of ATP hydrolysis, 2 Ca^{2+} ions are transported with counter-transport of H^+ ions. SERCA pumps are encoded by three genes in mammals, SERCA1–3 or ATP2A1–3. Of these the most dominating and abundant one is SERCA1, hence these pumps are commonly known asSERCA1a. SERCA pumps are heterodimeric pumps with their α subunit in accordance with the general structure composition of P-type ATPase (mentioned in detail below). Its β subunit consist of a single transmembrane spanning region, regulating SERCA by its inhibitory action. These accessory subunits are either phospholamban or sarcolipin. Sarcolipin inhibits SERCA1a and SERCA2a in skeletal muscles and is homologs to phospholamban. Apart from regulation, these accessory subunits helps in stabilization of SERCA pumps in E_2 conformational state. Thapsigargin obtained from *Thapsia garganica* and Cyclopiazonic acid obtained from *Aspergillus* and *Penicillum* are potent inhibitors of SERCA pumps (Mandal, Rulli, & Rao, 2003; Olesen et al., 2007; Palmgren & Nissen, 2011; Thever & Saier, 2009).

P2B-ATPase or Type IIB ATPase are also Ca^{2+} transporters. The ions are transported across plasma membrane thus known as plasma membrane Ca^{2+}-ATPase (PMCA). They consists of one Ca^{2+} binding site and are regulated by calmodulin binding and in return assist in establishing Ca^{2+} ion gradient across plasma membrane. P2A and P2B are both Ca^{2+} pumps, but P2B are differentiated from the former based on various characteristic features of P2B-ATPases. P2B contains an auto regulating R-domain present at either C-terminus in animals or at N-terminus in plants. This R-domain provides calmodulin binding site which not only catalysis the activity of pump but also increases the affinity of Ca^{2+} ion binding. As compared to P2A-ATPases, P2B-ATPase have only one Ca^{2+} binding site which corresponds to Ca^{2+} binding site II of P2A-ATPase. The corresponding site I is already occupied with Arg cation in P2B-ATPase, thus, in response of a single Ca^{2+} ion transport only one H^+ ion is counter transported. Another characteristic feature of P2B-ATPase is that in presence of La^{3+} ion, the steady-state phosphorylation levels increases, and while in all other P-type ATPases this phosphorylation decreases. In terms of localization of pumps, as the name suggests, in animals it is ubiquitously found in plasma membrane while in plants and fungi, they are also found in vacuolar membrane, endoplasmic reticulum as well as plastid envelope. In animals, apart from calmodulin, R-domain may also interact with phosphatidylinositol bisphosphate (PIP2) and polyunsaturated fatty acids (PUFA) and increases the affinity of Ca^{2+} ion binding (Brini & Carafoli, 2009; Malmström, Askerlund, & Palmgren, 1997; Palmgren & Nissen, 2011).

Type IIC or P2C ATPases are monovalent ion transporters, i.e. Na^+/K^+-ATPases and H^+/K^+-ATPases. Both these ATPases are closely related to each other with the later found in gastric lumen and 3 NA^+ sites are replaced $2H^+$ sites. In Na^+/K^+-ATPases, each ATP hydrolysis effluxes 3 NA^+ ions and in return imports $2K^+$ ions. This generates an electrochemical gradient across plasma membrane which is important for various processes like secondary transport, signaling, action potential generation, etc. Structurally, the Na^+/K^+-ATPases has a β-subunit and sometimes a ϒ-subunit. The β-subunit is a single transmembrane spanner and is essential for the successful exit of α-subunit from ER. In the outer medulla of kidney, this αβ-complex is accompanied by a ϒ-subunit, belonging to FXYD family with hydrophobic residues and a single transmembrane span. In humans, α-subunit has 4 isoforms, β-subunit has 3 and ϒ-subunit has at least 5 isoforms encoded by ATP1A1–4 and ATP1B1–4 genes. The C′-terminus instead of extruding in cytoplasm, forms a novel regulatory element in a pocket between trans-membranes near Na + ion site III. They regulates Na + affinity as well as binding at site III by blocking the hydrophilic entrance which have been confirmed by exhaustive Molecular Dynamics studies. The ouabain acts as inhibitor of Na^+/K^+-ATPases by interleaving between trans membranes and blocking the entrance route of K+ (Geering, 2001, 2006; Ogawa, Shinoda, Cornelius, & Toyoshima, 2009; Palmgren & Nissen, 2011; Poulsen et al., 2010).

Type IID or P2D ATPases are fungal exporters exporting Na^+ and K^+ ions.

2.2.1.3 P3-type ATPase

P3ATPase or Type III ATPase are found in eubacteria, protists, prokaryotes, fungi and plants. They are further subdivided into two parts, P3A and P3B. P3A-ATPases are also known as proton pumps and transport H^+ while P3B-ATPases transports Mg^{2+} and hence also known as Magnesium pumps.

P3A-ATPases have been most studied in *Arabidopsis thaliana* (plant, encoded by AHA2) and *Saccharomyces cerevisiae* (yeast, encoded by Pma1). These H^+ – exporting ATPases belongs to hydrolases family and creates an electrochemical gradient across plasma membrane with H^+ concentration higher on the extracellular side. The proton pump doesn't counter-transport and ion or small molecule but the electrochemical gradient from it helps in the secondary transport of most metabolites. A single transport cycle uses one ATP molecule to transport a single proton. Structurally, it resembles to P2B—Ca^{2+} ATPases having a C′ terminal autoinhibitory regulatory

(R) domain. Instead of binding with Calmodulin, this R-domain interacts with a variety of proteins of 14–3-3 protein is the most prominent one. Another unique feature of P3A proton pumps is that in E_1 state the central cavity of T-domain is filled with water. This cavity is lined with two important and charged residues Asp and Arg which are essential for transport. The movement of Arg during conformational changes is the determining factor either binding or release of H^+, i.e., whether Asp plays as an acceptor or donor of H + ion. This is possible as pK_a of Asp is regulated by Arg movement (Buch-Pedersen & Palmgren, 2003; Jahn et al., 1997; Palmgren & Nissen, 2011; Pedersen, Buch-Pedersen, Preben Morth, Palmgren, & Nissen, 2007).

2.2.1.4 P4-type ATPase

P4 or Type IV—ATPase is the most unique amongst all P-type ATPases having largest substrate binding site, since, they transport phospholipids such as phosphatidylserine (PS), phosphatidylcholine (PC), phosphatidylethanolamine (PE), etc. Since, they transport inside the cell membrane, they are also known as phospholipid flippases. As for the substrate, it translocases aminophospholipids (PS/PC/PE), lysophospholipids as well as synthetic alkyl phospholipids and their specificity depends on the organisms they are present like in fungi or plants or animals. P4ATPases are involved in many important cellular processes including membrane trafficking and its regulation, lipid metabolism, lipid signaling, and cell division. Different studies have revealed that P4ATPase works in heterodimeric form with its beta subunit CDC50 (Cell Division Cycle 50). This β-subunit, CDC50, is required for the pump to exit ER (Lopez-Marques, Theorin, Palmgren, & Pomorski, 2014). P4ATPase-CDC50 complex performs two main function, firstly, they maintain lipid membrane bilayer asymmetry by flipping phospholipids across the membrane bilayer leaflet. Secondly, they assist in bud- ding of transport vesicles in the endocytic and secretory pathways. This flippase activity is dependent on ATP hydrolysis and can be inhibited by ATPase inhibitor, vanadate (Palmgren & Nissen, 2011; Paulusma & Oude Elferink, 2010).

P4ATPase proteins are expressed only in eukaryotes. In humans, 14 genes are present: ATP8A1, ATP8A2, ATP8B1, ATP8B2, ATP8B3, ATP8B4, ATP9A, ATP9B, ATP10A, ATP10B, ATP10D, ATP11A, ATP11B and ATP11C. These 14 genes are grouped into 5 classes, Class1a (ATP8A1 and ATP8A2); Class 1b (ATP8B1–4); Class 2 (ATP9A, ATP9B); Class 5 (ATP10A, ATP10B, ATP10D) and Class 6 (ATP11A, ATP11B,

ATP11C). In yeast, 5 homologs are found: Drs2p, Dnf1p, Dnf2p, Dnf3p and Neo1p. *C. elegans* expresses 6 homologs TAT1–6 while *A. thaliana* has 12 homologs ALA1–12. As for β-subunit CDC50, in humans three proteins are present (CDC50A, CDC50B and CDC50C), similarly, yeast also has 3 genes (Cdc50p. Lem3p, and Crf1p). Since α-subunits (P4ATPase) are more in number than β-subunits (CDC50), hence, more P4ATPase are associated with same CDC50 protein. Mutations or deficiency in P4ATPase may lead to cholestasis, neurological disorders, liver disorders, immunological issues, type 2 diabetes or obesity (Andersen et al., 2016; Lopez-Marques et al., 2014; van der Mark, Elferink, & Paulusma, 2013).

2.2.1.5 P5-type ATPase

P5-ATPase or Type V ATPase is a large group of ATPases whose substrate specificity is yet to be elucidated. Based on sequence similarity, type V ATPases are further classified in two sub-groups namely, P5A and P5B. The characteristic feature of P5-ATPase is a PPxxP motif preceding P-domain. The two sub-families also differ in this characteristic motif sequence. P5A-ATPase contains motif sequence PP(E/D)xPx(E/D) while that of P5B-ATPase constitutes of PP(A/V)xP(A/V)x motif sequence. The motifs in P5A type has two conserved negatively charged residues while P5B motif has two conserved hydrophobic residues, thus a wide variety of substrate specificity. Structurally, the T-domain of P5A constitutes 12 TMH while P5B has 11 TMH. P5A-ATPases are exclusively found in ER and help in regulating homeostasis. Spf1p of *S. cerevisiae* is the well-characterized P5A-ATPase. In humans, P5B-ATPase is found in lysosomes, in yeast it is present in vacuoles, while in *C. elegans* it is located in plasma membrane. P5B is believed to be associated with neurological diseases. Kufor-Rakeb syndrome, a neurodegenerative disease, is caused due to mutation in a P5ATPase gene, ATP13A2. Similarly, mutations in ATP13A4 may cause disorders related with speech and language (Palmgren & Nissen, 2011; Ramirez et al., 2006; Sørensen, Buch-Pedersen, & Palmgren, 2010).

2.2.2 General architecture

Although different members of P-type ATPase transport different substrates, but they exhibit a common structural fold as well as follows a common catalytic cycle. This protein can be divided in four distinct structural domains: transmembrane domain (M also known as T domain), nucleotide binding domain (N), phosphorylation domain (P) and actuator domain (A) (Fig. 3). The M domain comprises 10 transmembrane helices (TMH)

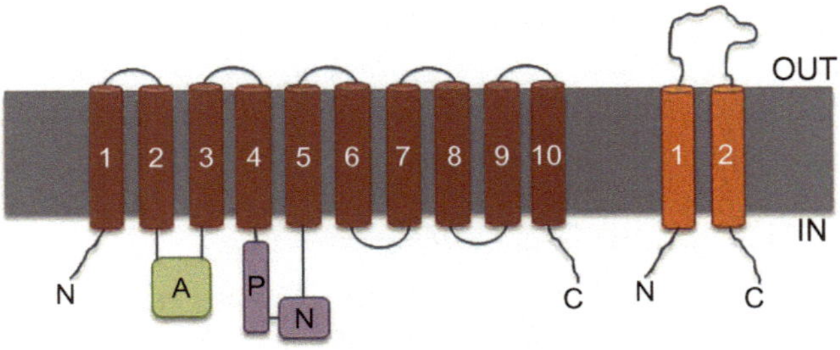

Fig. 3 General architecture of P-type ATPase. P-type ATPase consists of 10 transmembrane helices, actuator domain (A), phosphorylating domain (P), nucleotide binding domain (N) and a β subunit 'CDC50' with two transmembrane helices and a large exocytoplasmic loop.

and forms pore through which substrate is transported from exoplasmic to cytosolic leaflet of cell membrane. It has been observed that M1-M6 forms the main translocating pathway while M7-M10 helps in regulation or supports the transport activity. The N domain or nucleotide binding domain is the most conserved region in P-type ATPases. It provides the binding site for ATP through hydrophobic interaction between adenosine ring of ATP and a conserved phenylalanine residue. It exhibits a protein-kinase activity and phosphorylates the pump, hence permitting an active transport of its substrate. The phosphorylation domain, P domain, gets phosphorylated during the catalysis cycle. The aspartate residue gets phosphorylated which is part of the conserved amino acid sequence DKTG. Along with this, two other conserved sequence motifs present in P domain are TGDN and GDGXND. The actuator domain, A domain, has phosphatase activity and glutamate of conserved sequence DGET (TGES sequence in Ca^{2+} ATPase and Na^{+}/K^{+} ATPase) dephosphorylates the aspartate of P domain and helps in returning the pump in its original conformation (Lopez-Marques et al., 2014).

These regions are connected with amino acid linkers and the complete protein is encoded by a single polypeptide. The β-subunit, CDC50, consists of 2 TMH and a large glycosylated exoplasmic loop. It is believed that heterodimerization of P4ATPase and CDC50 proteins is necessary for release of P4ATPase from ER and its localization at cell membrane (van der Mark et al., 2013).

2.2.3 Catalytic cycle

As mentioned above, all P-type ATPases follow a common catalytic cycle consisting of two main protein conformations, E1 state and E2 state (Fig. 4).

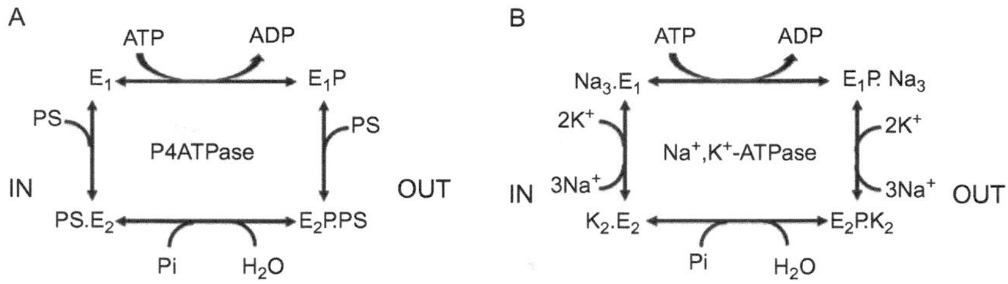

Fig. 4 Catalytic cycle of P4ATPase (A) and Na^+, K^+ – ATPase (B).

When the substrate binds the protein in E1 conformation, structural rearrangement occurs in M domain exposing high binding affinity site for its substrate at the cytosolic face of membrane. This rearrangement leads to phosphorylation of conserved aspartate residue of P domain by ATP bound at N domain and thereby generating E1P conformation. The A domain then reorients itself by 120° and thus forces M domain to reorganize itself. This rotation results in bringing active site of A domain possessing phosphatase activity in close proximity with the phosphorylated aspartyl residue of P domain and transition to E2P conformation. This transition to E2P state exposes lob binding affinity site towards exoplasmic face of membrane and hence release of substrate or in some proteins binding of a counter ion. The phosphatase activity of A domain leads the protein in an unstable E2 conformation which is transitioned to the stable initial E1 conformation by rotating back A domain in its initial orientation. This reorientation of A domain further reorganizes M domain in its initial form subsequently releasing exoplasmic counter ion to cytosolic side of membrane. In case of P4ATPases, the studies have suggested that phospholipids, being bulkier in size, might bound in E2P conformation at exoplasmic site of membrane (analogues to counter ion binding) and is releases to cytosol while transiting back in E1 conformation. It has also been assumed that exoplasmic loop of CDC50 subunit acts as a lid and thus ensuring phospholipid in its place while transporting it through P4ATPase. According to mutational studies performed by various research groups for deciphering the transport pathway, two mechanistic models have been suggested: a canonical pathway, in which phospholipid is transported through the channel formed by P4ATPases as explained above; and a non-canonical pathway also known as 'credit card model', where, only charged phospholipid head group interacts with P4ATPase and hydrophobic tail remains in the lipid bilayer membrane (Andersen et al., 2016; Bublitz, Preben Morth, & Nissen, 2011; Lopez-Marques et al., 2014; van der Mark et al., 2013).

2.3 ABC transporters

The ATP-binding cassette (ABC) transporters constitutes a large superfamily of membrane proteins. These are amongst the most conserved proteins across archaea, eubacteria and eukarya kingdoms. They transport a wide variety of substrates like amino acids, sugars, ions, small organic or inorganic compounds, peptides, proteins, oligonucleotides, polysaccharides, etc. Functionally, ABC transporters consists of four units, two Transmembrane Domains (TMDs: TMD1 and TMD2) and two Nucleotide Binding Domains (NBDs: NBD1 and NBD2) with each TMD consisting of 6–10 Transmembrane Helices (TMH). If all the four units are expressed by a single polypeptide, such ABC transporters are known as 'full transporters' and are mostly found in eukaryotes. In bacteria and some prokaryotes, 'half transporters' are found (two units TMD1 and NBD1 are expressed in a single polypeptide and rest two in another) which assemble in either homo- or hetero-dimeric form to form a complete functional unit. ABC transporters can be broadly classified in two groups as importers and exporters (Seeger & van Veen, 2009; Wilkens, 2015).

2.3.1 ABC importers

The presence of ABC importers is restricted to prokaryotes and they help in uptake of various nutrients and small molecules. In bacteria and other prokaryotic pathogens, they play a key role in maintaining cell viability, virulence as well as pathogenicity. The bacterial importers, apart from four functional unit, requires an accessory subunit required for capturing and transporting the substrate through TMDs to the interior of the cell. In gram negative bacteria, Periplasmic Binding Proteins (PBPs) acts as accessory subunits to capture and transport solutes. While, in gram positive bacteria's, lipoproteins anchored to the outer leaflet of the plasma membrane play this role. Importers can further be classified as Class I Importers, Class II Importers and Class III or Energy Coupling Factor (ECF) transporters. Of these, the Class III Importers are functionally as well structurally diverged from the first two (Locher, 2016; Seeger & van Veen, 2009).

2.3.1.1 Class I importers

Type I Importers facilitate medium affinity uptake of various nutrients like ions, sugars, amino acids, peptides, etc. High resolution experimental structures of only a few importers are available compared to exporters and amongst them too, most of them are Class I Importers. These includes maltose transporter $MalFGK_2$, methionine transporter MetNI, molybdate transporters ModBC, etc. Structurally, they contain less number of transmembrane helices when compared with class II importers. For instance, MetNI has 5 TMH in each TMD and forms a homodimer. Similarly, ModBC have 6 TMH in each TMD. Molybdate transporter is also a homodimeric complex. In case of Maltose transporters, TMD1 and TMD2 are named as MalF and MalG with 8 and 6 TMH respectively. Most of the type I importers (exception is ModBC of *Archaeoglobus fulgidus*) have an additional domain attached with NBDs known as Regulatory Domain (RD). On binding of regulating proteins or substrate with RD, the ATP hydrolysis and hence transport of substrate comes to an abrupt halt. Thus, RDs can provide accessory function along with core transporting function by modulating transport activity. The type I importers follow alternate accessing mechanism to transport its substrate. This model alternates between two conformations, an inward facing open NBDs conformation and an outward facing closed NBDs conformation. This closed NBD dimer is due to binding of ATPs. The alternation of these two conformations results in the movement of substrate across the membrane (Locher, 2016; Rice, Park, & Pinkett, 2014).

2.3.1.2 Class II importers

Class II or Type II Importers includes vitamin B12 transporter BtuCD-BtuF, heme transporter HmuUV and the molybdate transporter MolBC from *Haemophilus influenzae*. The substrates of type II Importers are larger in size, more hydrophobic in nature compared to that of type I, available in low concentration and hence high—affinity uptake of substrates. The mechanism of substrate translocation is well explained with vitamin B12 transporter BtuCD–BtuF (BtuCD–F) of *E. coli* which also serves as model for other type II Importers. Although type I and type II transporters share overall similarity but they still have differences structurally as well mechanistically. Structurally, they consist of large number of TMH with each TMD containing up to 10 helices. Of these, TM 3, 4, 5, 10 and loops between TM 2 and TM 3 forms the main core guiding the majority of conformational changes and forming translocation pathway. The transmembrane helices form three gates, two on the cytoplasmic side (cyto gates 1 and 2) and one on the periplasmic side (peri gate). In the absence of the substrate, the two cyto gates separate the cytoplasm from the central translocation cavity in both the ATP-bound and the nucleotide-free states In these Importers, ATP binding and hydrolysis in itself doesn't generate an inward facing conformation and the open peri gate marks an outward facing conformation (Korkhov, Mireku, Veprintsev, & Locher, 2014; Locher, 2016; Rice et al., 2014).

2.3.1.3 Class III importers

Class III or Type III Importers is sequentially as well as structurally divergent class of Importers and has recently been classified. It is also known as ECF transporters. Folate transporters and the hydroxymethyl pyrimidine transporters from Lactobacillus brevis comes under type III category of importers. Instead of substrate binding protein, ECF have an embedded membrane component that binds and delivers the substrate. ECF transporters consist of 2 NBDs named as EcfA and EcfA'; a TMD as EcfT and a substrate-binding component as EcfS. The NBDs can either be homo- or heterodimeric. EcfS or S component is the unique feature of ECF transporters forming an integral membrane component for substrate binding. Based on EcfS, they can be further classified in two groups. Each S component of Group I ECF transporters forms a complex with a unique energizing module and are located within the same operon. On the other hand, the S components of Group II ECF transporters binds to variety of substrates but shares same energy module to form full transporter for transport of substrate across the membrane. The EcfT domain comprises of 8 helices, of which 5 forms the core domain perpendicular to the membrane bilayer and the remaining ones form an 'L' shape in the cytoplasmic space. TMH 4, 5 and 6 interacts with the substrates and helps in their translocation. The S component lies between 5 core helices and 3 cytoplasmic helices. In ECF transporters, the ATP induced conformational changes leads to substrate translocation and their subsequent release in the cytoplasm (Erkens et al., 2011; Karpowich & Wang, 2013; Locher, 2016; Rice et al., 2014).

References

Andersen, J. P., Vestergaard, A. L., Mikkelsen, S. A., & Mogensen, L. S. (2016). P4-ATPases as phospholipid flippases—Structure, function, and enigmas. *Frontiers in Physiology*, 7(July), 1–23. https://doi.org/10.3389/fphys.2016.00275.

Brini, M., & Carafoli, E. (2009). Calcium pumps in health and disease. *Physiological Reviews*, *89*(4), 1341–1378. https://doi.org/10.1152/physrev.00032.2008.

Bublitz, M., Preben Morth, J., & Nissen, P. (2011). P-type ATPases at a glance. *Journal of Cell Science*, *124*(15), 2515–2519. https://doi.org/10.1242/jcs.088716.

Buch-Pedersen, M. J., & Palmgren, M. G. (2003). Conserved Asp684 in transmembrane segment M6 of the plant plasma membrane P-type proton pump AHA2 is a molecular determinant of proton translocation. *Journal of Biological Chemistry*, *278*(20), 17845–17851. https://doi.org/10.1074/jbc.M212729200.

Erkens, G. B., Berntsson, R. P. A., Fulyani, F., Majsnerowska, M., Vujičić-Žagar, A., Ter Beek, J., et al. (2011). The structural basis of modularity in ECF-type ABC transporters. *Nature Structural and Molecular Biology, 18*, 755–760. https://doi.org/10.1038/nsmb.2073.

Geering, K. (2001). The functional role of β subunits in oligomeric P-type ATPases. *Journal of Bioenergetics and Biomembranes, 33*(5), 425–438. https://doi.org/10.1023/A:1010623724749.

Geering, K. (2006). FXYD proteins: New regulators of Na-K-ATPase. *American Journal of Physiology. Renal Physiology, 290*(2), F241–F250. https://doi.org/10.1152/ajprenal.00126.2005.

Giacomini, K. M., Huang, S.-M., Tweedie, D. J., Benet, L. Z., Brouwer, K. L. R., Chu, X., et al. (2010). Membrane transporters in drug development. *Nature Reviews Drug Discovery, 9*(3), 215–236. https://doi.org/10.1038/nrd3028.

Grüber, G., Wieczorek, H., Harvey, W. R., & Müller, V. (2001). Structure—Function relationships OF A-, F- AND V-ATPases. *The Journal of Experimental Biology, 204*(Pt 15), 2597–2605.

Guerrero, M. G., & Arguello, J. M. (2008). Mechanism of Cu+—transporting ATPases: Soluble Cu+ chaperones directly transfer Cu+ to transmembrane transport sites. *Proceedings of the National Academy of Sciences of the United States of America, 105*(16), 5992–5997. https://doi.org/10.1073/pnas.0711446105.

Imamura, H., Nakano, M., Noji, H., Muneyuki, E., Ohkuma, S., Yoshida, M., et al. (2003). Evidence for rotation of V1-ATPase. *Proceedings of the National Academy of Sciences of the United States of America, 100*(5), 2312–2315.

Jahn, T., Fuglsang, A. T., Olsson, A., Brüntrup, I. M., Collinge, D. B., Volkmann, D., et al. (1997). The 14-3-3 protein interacts directly with the C-terminal region of the plant plasma membrane H(+)-ATPase. *The Plant Cell, 9*(10), 1805–1814. https://doi.org/10.1105/tpc.9.10.1805.

Kane, P. M. (2006). The where, when, and how of organelle acidification by the yeast vacuolar H+-ATPase. *Microbiology and Molecular Biology Reviews, 70*(1), 177–191. https://doi.org/10.1128/MMBR.70.1.177.

Karpowich, N. K., & Wang, D. N. (2013). Assembly and mechanism of a group II ECF transporter. *Proceedings of the National Academy of Sciences of the United States of America, 110*(7), 2534–2539. https://doi.org/10.1073/pnas.1217361110.

Korkhov, V. M., Mireku, S. A., Veprintsev, D. B., & Locher, K. P. (2014). Structure of AMP-PNP-bound BtuCD and mechanism of ATP-powered vitamin B12 transport by BtuCD-F. *Nature Structural and Molecular Biology, 21*(12), 1097–1099. https://doi.org/10.1038/nsmb.2918.

Kühlbrandt, W. (2019). Structure and mechanisms of F-type ATP synthases. *Annual Review of Biochemistry, 88*, 515–549. https://doi.org/10.1146/annurev-biochem-013118-110903.

Locher, K. P. (2016). Mechanistic diversity in ATP-binding cassette (ABC) transporters. *Nature Structural and Molecular Biology, 23*(6), 487–493. https://doi.org/10.1038/nsmb.3216.

López-Marqués, RL, Poulsen, LR, Bailly, A, Geisler, M, Pomorski, TG, & Palmgren, MG. (2015). Structure and mechanism of ATP-dependent phospholipid transporters. *Biochimica et Biophysica Acta, 1850*(3), 461–475. https://doi.org/10.1016/j.bbagen.2014.04.008.

Lopez-Marques, R. L., Theorin, L., Palmgren, M. G., & Pomorski, T. G. (2014). P4-ATPases: Lipid flippases in cell membranes. *Pflügers Archiv / European Journal of Physiology, 466*(7), 1227–1240. https://doi.org/10.1007/s00424-013-1363-4.

Malmström, S., Askerlund, P., & Palmgren, M. G. (1997). A calmodulin-stimulated Ca 2+-ATPase from plant vacuolar membranes with a putative regulatory domain at its N-terminus 1. *FEBS Letters, 400*(3), 324–328. https://doi.org/10.1016/S0014-5793(96)01448-2.

Mandal, D., Rulli, S. J., & Rao, R. (2003). Packing interactions between transmembrane helices alter ion selectivity of the yeast golgi Ca2+/Mn2+-ATPase PMR1. *Journal of Biological Chemistry*, *278*(37), 35292–35298. https://doi.org/10.1074/jbc.M306166200.

Maxson, M. E., & Grinstein, S. (2014). The vacuolar-type H +-ATPase at a glance—More than a proton pump. *Journal of Cell Science*, *127*(Pt 23), 4987–4993. https://doi.org/10.1242/jcs.158550.

Montigny, C., Lyons, J., Champeil, P., Nissen, P., & Lenoir, G. (2016). On the molecular mechanism of flippase- and scramblase-mediated phospholipid transport. *Biochimica et Biophysica Acta*, *1861*(8 Pt B), 767–783. https://doi.org/10.1016/j.bbalip.2015.12.020.

Nakamoto, R. K., Scanlon, J. A. B., & Al-Shawi, M. K. (2008). The rotary mechanism of the ATP synthase. *Archives of Biochemistry and Biophysics*, *476*(1), 43–50. https://doi.org/10.1016/j.abb.2008.05.004.The.

Nigam, S. K. (2015). What do drug transporters really do. *Nature Reviews. Drug Discovery*, *14*(1), 29–44. https://doi.org/10.1038/nrd4461.

Ogawa, H., Shinoda, T., Cornelius, F., & Toyoshima, C. (2009). Crystal structure of the sodium-potassium pump (Na+,K+-ATPase) with bound potassium and ouabain. *Proceedings of the National Academy of Sciences of the United States of America*, *106*(33), 13742–13747. https://doi.org/10.1073/pnas.0907054106.

Olesen, C., Picard, M., Winther, A.-M. L., Gyrup, C., Morth, J. P., Oxvig, C., et al. (2007). The structural basis of calcium transport by the calcium pump. *Nature*, *450*(7172), 1036–1042. https://doi.org/10.1038/nature06418.

Palmgren, M. G., & Nissen, P. (2011). P-type ATPases. *Annual Review of Biophysics*, *40*, 243–266. https://doi.org/10.1146/annurev.biophys.093008.131331.

Paulusma, C. C., & Oude Elferink, R. P. J. (2010). P4 ATPases—The physiological relevance of lipid flipping transporters. *FEBS Letters*, *584*(13), 2708–2716. https://doi.org/10.1016/j.febslet.2010.04.071.

Pharmacology, I. G. (n.d.-a). F—Type ATPases. Retrieved March 28, 2020, from http://www.guidetopharmacology.org/GRAC/FamilyDisplayForward?familyId=156.

Pharmacology, I. G. (n.d.-b). V—Type ATPase. Retrieved March 28, 2020, from http://www.guidetopharmacology.org/GRAC/FamilyDisplayForward?familyId=157.

Pedersen, B. P., Buch-Pedersen, M. J., Preben Morth, J., Palmgren, M. G., & Nissen, P. (2007). Crystal structure of the plasma membrane proton pump. *Nature*, *450*(7172), 1111–1114. https://doi.org/10.1038/nature06417.

Poulsen, H., Khandelia, H., Morth, J. P., Bublitz, M., Mouritsen, O. G., Egebjerg, J., et al. (2010). Neurological disease mutations compromise a C-terminal ion pathway in the Na+/K+-ATPase. *Nature*, *467*(7311), 99–102. https://doi.org/10.1038/nature09309.

Ramirez, A., Heimbach, A., Gründemann, J., Stiller, B., Hampshire, D., Cid, L. P., et al. (2006). Hereditary parkinsonism with dementia is caused by mutations in ATP13A2, encoding a lysosomal type 5 P-type ATPase. *Nature Genetics*, *38*(10), 1184–1191. https://doi.org/10.1038/ng1884.

Rice, A. J., Park, A., & Pinkett, H. W. (2014). Diversity in ABC transporters: Type I, II and III importers. *Critical Reviews in Biochemistry and Molecular Biology*, *49*(5), 426–437. https://doi.org/10.3109/10409238.2014.953626.

Seeger, MA, & van Veen, HW. (2009). Molecular basis of multidrug transport by ABC transporters. *Biochim Biophys Acta*, *1794*(5), 725–737. https://doi.org/10.1016/j.bbapap.2008.12.004.

Sørensen, D. M., Buch-Pedersen, M. J., & Palmgren, M. G. (2010). Structural divergence between the two subgroups of P5 ATPases. *Biochimica et Biophysica Acta, Bioenergetics*, *1797*(6–7), 846–855. https://doi.org/10.1016/j.bbabio.2010.04.010.

Stewart, A. G., Sobti, M., Harvey, R. P., & Stock, D. (2013). Rotary ATPases: Models, machine elements and technical specifications. *Bioarchitecture*, *3*(1), 2–12.

Thever, M. D., & Saier, M. H. (2009). Bioinformatic characterization of P-type ATPases encoded within the fully sequenced genomes of 26 eukaryotes. *Journal of Membrane Biology*, *229*(3), 115–130. https://doi.org/10.1007/s00232-009-9176-2.

van der Mark, VA, Elferink, RP, & Paulusma, CC. (2013). P4 ATPases: flippases in health and disease. *International Journal of Molecular Sciences*, *14*(4), 7897–7922. https://doi.org/10.3390/ijms14047897.

Wilkens, S. (2015). Structure and mechanism of ABC transporters. *F1000Prime Reports*, *9*(February), 1–9. https://doi.org/10.12703/P7-14.

You, G., & Morris, M. E. (2014). Overview of drug transporter families. In *Vol. 9781118489. Drug transporters: Molecular characterization and role in drug disposition* (2nd ed., pp. 1–6). https://doi.org/10.1002/9781118705308.ch1.

CHAPTER TWO

Protein kinase CK2 inhibition as a pharmacological strategy

Christian Borgo[a] and Maria Ruzzene[a,b,*]
[a]Department of Biomedical Sciences, University of Padova, Padova, Italy
[b]CNR Neuroscience Institute, Padova, Italy
*Corresponding author: e-mail address: maria.ruzzene@unipd.it

Contents

Abstract

CK2 is a constitutively active Ser/Thr protein kinase which phosphorylates hundreds of substrates. Since they are primarily related to survival and proliferation pathways, the best-known pathological roles of CK2 are in cancer, where its targeting is currently being considered as a possible therapy. However, CK2 activity has been found instrumental in many other human pathologies, and its inhibition will expectably be extended to different purposes in the near future. Here, after a description of CK2 features and implications in diseases, we analyze the different inhibitors and strategies available to target CK2, and update the results so far obtained by their *in vivo* application.

1. CK2: Structural and functional features

Protein kinase CK2 (formerly known as casein kinase 2, or casein kinase II, or CK-II) is a conserved Ser/Thr kinase, ubiquitously expressed in eukaryotes, originally considered by mistake as the kinase that physiologically phosphorylates casein (hence the name). It is now clear that it

Advances in Protein Chemistry and Structural Biology, Volume 124
ISSN 1876-1623
https://doi.org/10.1016/bs.apcsb.2020.09.003

phosphorylates hundreds of substrates (Meggio & Pinna, 2003), but not casein. Its functions are fundamental for many cellular processes, but also related to different human pathologies.

1.1 Structure

CK2 is usually present in cells as a tetramer, composed of two catalytic and two regulatory subunits. The catalytic subunit could be either α or α', coded by two distinct genes (*CSNK2A1* and *CSNK2A2*, respectively); their relative expression is tissue-specific. In humans, only one regulatory subunit is known, called β, coded by the gene *CSNK2B*. Its regulatory role does not refer to a mechanism of turning the activity on/off, since CK2 is a constitutively active enzyme, as clearly evidenced by the crystal structure of the catalytic subunit (Niefind, Guerra, Pinna, Issinger, & Schomburg, 1998). Instead, the main functions of the β subunit are in driving substrate selection (see below) and maintaining the enzyme stability.

1.2 Regulation

There are not obvious mechanisms of CK2 regulation, its protein level being the major element accounting for the profound variations in activity found in different cells/conditions. However, possible regulatory molecules have been suggested based on *in vitro* observations; for example, polycations are activators (Meggio, Boldyreff, Issinger, & Pińna, 1994) while heparin is inhibitory (Vaglio et al., 1996), but the physiological relevance of these compounds is far from being proven. A crucial aspect in the control of CK2 activity is represented by its subcellular localization: the enzyme is abundant in the nucleus and in the cytosol, but it is also present in organelles and at the plasma membrane, even on the external surface (the so called ecto-CK2 Montenarh & Götz, 2018), and translocation between different compartments has been proposed as related to changes in activity and functions (Filhol & Cochet, 2009; Guo et al., 2001).

Hypothesis for CK2 regulation have been also based on dynamic events which allow different multimolecular assemblies (Filhol, Martiel, & Cochet, 2004), and recent crystal structures of CK2 tetramers aggregated in polymers suggest that CK2 activity might be buffered in a sort of reserve maintaining inactive molecules in supramolecular organization (Lolli, Pinna, & Battistutta, 2012).

All the proposed mechanisms will need further confirmation in cells, and they will likely prove valid only in certain circumstances.

1.3 Specificity and substrates

The number of CK2 substrates, estimated around 300 in 2003 (Meggio & Pinna, 2003), is in fast and continuous increase, growing up to about 400 (Franchin et al., 2017). CK2 is an acidophilic kinase, and its sites are characterized by abundant acidic residues, that may be represented not only by aspartic (D) and glutamic (E) acids, but also by phospho-sites, *i.e.*, other residues (especially S and Y) previously phosphorylated by other kinases (or by CK2 itself). The minimal CK2 consensus sequence is S/T-x-x-D/E/pS/pY, a motif present in a huge amount of proteins. Although a recent study exploiting cells deprived of CK2 by the CRISPR-Cas9 technology concluded that CK2 is not only and always responsible for the phosphorylation of these sites (Franchin et al., 2018), it is undeniable that CK2 targets are present in each and every cellular process of eukaryotic cells. Their phosphorylation, if analyzed *in vitro*, might differ in efficiency if catalyzed by the tetrameric holoenzyme or the monomeric catalytic subunit, and this allowed a classification of the substrates in class I (phosphorylated by both monomer and tetramer), class II (better or exclusively phosphorylated by monomeric CK2), and class III (which are dependent on the presence of the β subunit) (Pinna, 2002). However, the confirmation that this classification has a physiological significance has not been provided, yet.

Many CK2 substrates are proteins related to human diseases. In particular, there are about 40 viral proteins (Meggio & Pinna, 2003) whose phosphorylation by CK2 is essential to the virus life cycle. In many cases, the protein phosphorylated by CK2 is not pathogenic *per se*, but participates to an abnormal signal which is further modified/amplified by CK2. This is especially the case of survival/proliferation signaling proteins, which are deregulated in cancer (such as Akt, NF-κB, PTEN, PML, PAK1, Ikaros, Notch 1, XRCC1) and on which CK2 acts as a "lateral player," potentiating an already aberrant signal (see below).

1.4 Major signaling pathways affected by CK2

Among the substrates of CK2, many components of signal transduction pathways are present, especially those upregulated in cancer cells. CK2, by phosphorylating specific elements in each pathway, has functions in potentiating proliferation, increasing survival and preventing apoptosis, with multiple mechanisms that have been detailed in several reviews. The major pathways regulated by CK2 are PI3K/Akt (Fragoso & Barata, 2015; Ruzzene, Bertacchini, Toker, & Marmiroli, 2017), NF-κB and STAT3

(Manni et al., 2013), Wnt/β-catenin (Dominguez, Sonenshein, & Seldin, 2009), Hedgehog (Jia et al., 2010), TNF-α (Trembley et al., 2010), Androgen Receptor (Götz, Bachmann, & Montenarh, 2007; Trembley et al., 2019), Notch1 (Zhang et al., 2013), Tyr-kinase receptor (de Gooijer, Guillén Navarro, Bernards, Wurdinger, & van Tellingen, 2018; Trembley et al., 2010). Moreover, CK2 counteracts caspase actions (Duncan et al., 2010), promotes epidermal-mesenchymal transition (Filhol, Giacosa, Wallez, & Cochet, 2015), enhances HIF-1 activity (Mottet, Ruys, Demazy, Raes, & Michiels, 2005), potentiates DNA repair (Trembley et al., 2010), protects from unfolded protein response (Buontempo et al., 2014; Manni et al., 2012), controls chaperone machineries (Miyata, 2009), impairs Ikaros functions (Gowda et al., 2020), inhibits tumor suppressors (Scaglioni et al., 2008; Vazquez et al., 2001).

2. CK2 and human diseases

CK2 has been found implicated in a plethora of human diseases (Guerra & Issinger, 2008). Its best documented role is in cancer, where it counteracts apoptosis and promotes survival and proliferation (Duncan et al., 2010; Gowda et al., 2017; Trembley et al., 2010; Trembley, Wang, Unger, Slaton, & Ahmed, 2009; Unger, Davis, Slaton, & Ahmed, 2004), migration (Benavent Acero et al., 2017; Di Maira et al., 2019; Im, Cheong, Lee, Oh, & Yang, 2019; Niechi et al., 2015; Siddiqui et al., 2017; Zou et al., 2011), drug resistance (Borgo & Ruzzene, 2019), angiogenesis (Montenarh, 2014), senescence escape (Kalathur et al., 2015), metabolic rewiring (Silva-Pavez & Tapia, 2020 and Zonta et al., manuscript in preparation). Prominent roles for CK2 have been reported for practically all types of cancers, either blood malignancies or solid tumors (Buontempo et al., 2018; Chua et al., 2017; Lian et al., 2019; Piazza et al., 2012).

CK2 functions in bacterial, parasites and viral infections are also well-known: especially in the latter, several viruses have been shown to exploit host cell CK2 for their life cycle (reviewed in St-Denis & Litchfield, 2009). Very interestingly, also the SARS-CoV-2 virus, responsible of the Covid-19 pandemic, has been found to promote the phosphorylation of several CK2-dependent sites in host and viral proteins, and apparently CK2 activity increases in the infected cells (Bouhaddou et al., 2020; Gordon et al., 2020).

CK2 is also considered an emerging target for neurological diseases, and psychiatric disorders due to CK2 mutations have been recently identified (Castello, Ragnauth, Friedman, & Rebholz, 2017).

Diverse ophthalmic diseases have been identified where CK2 might represent a drug target, such as proliferative retinopathy (Kramerov et al., 2008; Ljubimov et al., 2004) and macular degeneration (Morooka et al., 2015); in a rat model of optic nerve injury, CK2 inhibition promotes axonal regeneration (Cen et al., 2018).

CK2 has been proposed as target in inflammatory diseases (Singh & Ramji, 2008) and autoimmune disorders (Gibson & Benveniste, 2018), and a specific mechanism has been detailed for Crohn's disease (Yang et al., 2020).

The hypothesis of considering CK2 inhibition for cystic fibrosis therapies has been entertained for many years (Mehta, 2008); although recent findings tend to exclude this approach to increase the plasma membrane level of F508delCFTR (pathological mutant of the transmembrane-conductance regulator) (D'Amore, Borgo, et al., 2020), the possibility still exists that combined treatments with CK2 inhibitors and the proteostasis regulator cysteamine are advantageous (Stefano et al., 2014).

CK2 modulates also insulin signaling and obesity (Borgo, Milan, et al., 2017) and has been proposed as target in diabetes (Ampofo, Nalbach, Menger, Montenarh, & Götz, 2019).

As far as cardiovascular diseases is concerned, CK2 is associated to pathogenesis of cardiac ischemia reperfusion injury (Zhou et al., 2018), atherosclerosis (Wadey et al., 2017), cardiac hypertrophy (Eom et al., 2011; Hauck et al., 2008). However, in contradiction, CK2 has been also defined as an anti-hypertrophic pathway (Murtaza et al., 2008).

Fig. 1 summarizes human pathologies where CK2 is involved and is (or may be) considered a drug target.

3. CK2 inhibitors

A plethora of compounds is available to target CK2 (Cozza, 2017; Cozza, Pinna, & Moro, 2012; Qiao et al., 2019), either natural or chemically synthesized, and many of them are commercially available. They differ in efficacy, selectivity, cell permeability, mode of binding; for some of them the crystal structure of the complex with the kinase has been solved, some are patented. They will be here classified according to their mechanism of action, with a brief summary of their features.

3.1 CK2 inhibitors classes

3.1.1 ATP-competitive CK2 inhibitors

This is by far the most abundant class of CK2 inhibitors, and probably the most studied one. Their strong potential is due to the unusual features of the

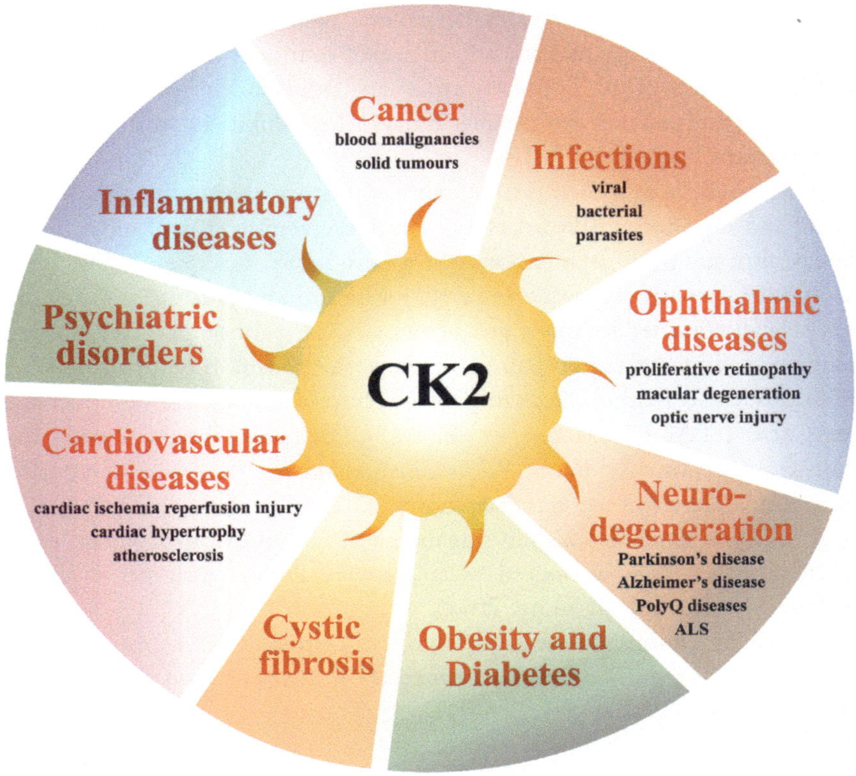

Fig. 1 Human pathologies where CK2 implication has been demonstrated.

CK2 ATP pocket, which is narrower than that of the majority of the other kinases (Battistutta, De Moliner, Sarno, Zanotti, & Pinna, 2001). For this reason, CK2 is able to strongly bind small compounds, which instead fit too loosely in the ATP site of most other kinases, and this account for the high selectivity of ATP-competitive CK2 inhibitors.

In this class, we find **polyhalogenated benzimidazole and benzotriazole** derivatives: DRB (Zandomeni, 1989) (actually a dual-site inhibitor) can be considered the precursor of this lucky series of compounds, where we find TBB, which is still frequently used in cellular studies. Several TBB derivatives have been developed, allowing to improve selectivity and efficiency. Among them, we can mention TBI (also called K17), DMAT (also called K25), K137, and TDB. This latter is a dual inhibitor, being able to significantly inhibit Pim1, besides CK2 (Cozza et al., 2014); since both kinases are related to cancer, their simultaneous targeting might offer better therapeutic potential, according to the principle of selectively-nonselective targeting (Morphy, 2010). Some polyhalogenated inhibitors have been proposed as possibly useful to differentially inhibit CK2α and CK2α' (Bollacke, Nienberg, Borgne, & Jose, 2016), which might become relevant in case isoform-dependent diseases will be identified, as possibly occurring in glioblastoma (Villamañan et al., 2019).

Other ATP-competitive CK2 inhibitors are natural **flavonoid** compounds: in this group we find apigenin and quercetin, but also several other natural molecules with the common property of inhibiting CK2 (McCarty, Assanga, & Lujan, 2020).

Coumarin (as ellagic acid) (Cozza et al., 2011) and **anthraquinone** (as emodin and quinalizarin) derivatives are other ATP-competitive CK2 inhibitors, either natural or derivatives. Despite the fact that usually these molecules tend to be more promiscuous, the anthraquinone quinalizarin is one of the most efficient and specific CK2 inhibitors (Cozza, Venerando, Sarno, & Pinna, 2015). Interestingly, the anthraquinones MNA and MNX (Meggio et al., 2004), which are structurally very similar, display different inhibitory capacity against monomeric and tetrameric CK2 (Salvi et al., 2006).

Other potent inhibitors are **pyrazolotriazine derivatives**, developed by Polaris Pharmaceuticals, and including the most potent compounds, with very low Ki values (Nie et al., 2007), the **SRPIN803** derivatives, with an unusual binding mode to the CK2 ATP pocket (Dalle Vedove et al., 2020). On the contrary, curcumin, which was claimed as CK2 inhibitor, in our hand is not able to inhibit the recombinant enzyme, and its cellular effects are due to degradation products, mainly ferulic acid (Cozza et al., 2020).

A very interesting class, defined on the bases of the presence of a carboxy group, is that of **carboxyl acid derivatives**. TBCA (Pagano et al., 2007) and IQA (Sarno et al., 2003) belongs to this class, as well as the compounds initially produced by Cylene Pharmaceuticals (tricyclic quinoline derivatives) which includes the very promising CX-4945 (Pierre et al., 2011) and CX-5011 (Battistutta et al., 2011). CX-4945 (commercially known as silmitasertib), very potent and specific, is presently the only ATP-competitive CK2 inhibitor in clinical trials, and it is demonstrating favorable safety, pharmacokinetic, and pharmacodynamic characteristics (Marschke et al., 2011; Padgett et al., 2010). It is able to cross the blood brain barrier, and a recent work evaluates its concentrations in human plasma and cerebrospinal fluid from treated patients, and in primate or mouse brain (Zhong, Campagne, Salloum, Purzner, & Stewart, 2020). FDA has recently granted Orphan Drug Designation to CX-4945 for the treatment of cholangiocarcinoma (CX-4945 Granted Orphan Drug Designation, 2017). It is worth to mention, however, that, in cholangiocarcinoma cells, also effects on methuosis have been reported for CX-4945, which are not-mediated by CK2 inhibition, especially when the inhibitor is used at high doses (Lertsuwan et al., 2018). Similarly, also the CX-4945 derivative CX-5011 was demonstrated to be a CK2-independent inducer of methuosis, even more potent than CX-4945 (D'Amore, Moro, et al., 2020). Moreover, CK2-independent effects of CX-4945 as a splicing regulator have been found (Kim et al., 2014, p. 4945).

Very recently, a new compound, **GO289**, has been identified as very potent and selective for CK2 (Oshima et al., 2019); it consists of a triazole structure linked to bromoguaiacol, methyl thioether, and phenyl groups; it appears very promising for future studies.

Table 1 summarizes the classes of ATP-competitive CK2 inhibitors and reports the Ki (or IC_{50}) values of some relevant compounds. Where available, the table also shows the indicative number of kinases composing the panel used to assess specificity.

Table 1 ATP-competitive CK2 inhibitors.

	Ki (or $^{d}IC_{50}$) (nM)	Number of kinases present in the specificity panel	References
Polyhalogenated benzimidazole and benzotriazole compounds			
DRB	7000[a]	n.a.	Zandomeni (1989)
TBB	400	70	Andrzejewska, Pagano, Meggio, Brunati, and Kazimierczuk (2003), Pagano et al. (2008)
TBI (K17)	300	n.a.	Pagano et al. (2004)
DMAT (K25)	40	70	Pagano et al. (2004, 2008)
K137[d]	130	140	Cozza, Zanin, et al. (2015)
TDB	15	124	Cozza et al. (2014)
TMCB	31/19[b]	n.a.	Bollacke et al. (2016)
Flavonoids			
Quercetin	1180	44	Sarno et al. (2003)
Apigenin	740	44	Sarno et al. (2003)

Table 1 ATP-competitive CK2 inhibitors.—cont'd

	Ki (or [d]IC_{50}) (nM)	Number of kinases present in the specificity panel	References
Coumarins			
Ellagic acid	20	n.a.	Cozza et al. (2011)
DBC	60	n.a.	Chilin et al. (2008)
Anthraquinones			
Emodin	1850	44	Sarno et al. (2003)
MNX[d]	340	33	Meggio et al. (2004)
MNA[d]	300/2800[c]	33	Meggio et al. (2004)
Quinalizarin	58	140	Cozza, Venerando, et al. (2015)
Pyrazolotriazine derivatives			
Compound 9e	0.35	n.a.	Nie et al. (2007)
Carboxyl acid derivatives			
IQA	170	44	Sarno et al. (2003)
TBCA	77	72	Cozza et al. (2009)
CX-5011	0.22	235	Battistutta et al. (2011)
CX-4945	0.17	235	Battistutta et al. (2011)
Other compounds			
SRPIN803 derivative compound 4[d]	280	320	Dalle Vedove et al. (2020)
GO289[d]	7	59	Oshima et al. (2019)

[a]different values are reported throughout the literature, probably due to the mixt type of inhibition.
[b]Ki values for α or α' containing tetramer, respectively, are reported.
[c]Ki values for tetrameric or monomeric CK2, respectively, are reported.
[d]IC_{50} is reported instead of Ki.
n.a., not applicable (specificity not assessed).

3.1.2 Peptide-competitive CK2 inhibitors

The most important compound of this class is a peptide, originally termed P15, developed by screening a cyclic peptide-library for interference with phosphorylation by CK2 of the human papillomavirus (HPV-16) E7 protein (Perea et al., 2004). The active peptide was fused to the cell-penetrating peptide Tat to confer cell permeability. The chimera peptide, called CIGB-300, is very effective in preventing CK2-dependent substrate phosphorylation and displays anti-tumor properties (Perea, Baladrón, Valenzuela, & Perera, 2018). Although its mechanism of action is not completely clear and also off-targets seem to be involved in mediating cellular effects (Zanin et al., 2015), it binds to CK2 subunits and impair the catalytic activity (Perera et al., 2020). It is very effective against tumor cell proliferation *in vitro* and in mouse tumor models, and is presently in clinical trials for cervical cancers, alone or in combination with chemoradiotherapy (see below).

3.1.3 Allosteric compounds

These inhibitors do not compete with ATP or protein substrates, but affect the catalytic activity by binding to different sites. For CK2, compounds have been identified that target three major allosteric sites, named site 1 (the α/β interface), site 2 (the αD pocket), and site 3 (the interface between the αC helix and the glycine-rich loop), as recently reviewed (Chen, Li, Wang, Chen, & Zhang, 2020).

The most promising approach is probably that of targeting site 1, thus interfering with the CK2 holoenzyme formation (Prudent, Sautel, & Cochet, 2010), since this would ideally preserve the catalytic activity of monomeric CK2, but, preventing the association (or promoting the dissociation) of the β subunit, would affect the phosphorylation of selected substrates (only those of class III, requiring the tetrameric CK2, see above). The ATP-competitive inhibitor DRB mentioned above, indeed, also targets the α/β interface (Raaf, Brunstein, Issinger, & Niefind, 2008), but one of the first compound specifically developed for this purpose is the 11-mer peptide named Pc (Laudet et al., 2007), which mimics the C-terminal α binding motif of β. Further modifications were later proposed, to improve the binding affinity (Hochscherf et al., 2015). Recently a not-peptide compound was also developed, with pro-apoptotic functions in breast cancer cells (Kufareva et al., 2019), and a bi-functional inhibitor (see below) was found to bind also to the α/β interface (Pietsch et al., 2020).

Another interesting class of CK2 allosteric inhibitors is represented by polyoxometalate (POM) derivatives, inorganic compounds with antiviral, antitumoral, and antibiotic activities. They do not bind to the CK2 catalytic site nor to the α/β interface, and their mechanism of action is still poorly known, but they inhibit CK2 in the nanomolar range and displayed specificity for CK2 when assayed on a panel of 29 protein kinases (Prudent et al., 2008).

Several other allosteric inhibitors have been proposed (Chen, Li, et al., 2020; Cozza et al., 2012), but their actual interest in still to be clarified.

3.1.4 Bi-functional inhibitors

Under this term, we include all those inhibitors composed of two different structural moieties, thus targeting two sites. The most immediate application of this concept is represented by CK2 inhibitors binding to both the ATP and the peptide sites (bi-substrate inhibitors). Several examples have been reported (Cozza, Zanin, et al., 2015; Pietsch et al., 2020; Rahnel et al., 2017; Viht et al., 2015). The inhibitors contain a classical ATP-competitive head, fused through a linker to a short acidic peptide, which optimally fits onto the substrate cleft of the CK2 catalytic subunit. Their undoubtful advantage is in term of specificity, and they are usually better than the parent ATP-competitive molecules also in term of Ki. A problem is instead represented by their cell permeability. However, this issue has been bypassed by a pro-drug formulation (for example, as acetoxymethyl ester) which can cross plasma membranes and is cleaved inside the cell to release the active compound (Viht et al., 2015). In alternative, these cell-impermeable compounds have been found useful to selectively inhibit ecto-CK2 (Cozza, Zanin, et al., 2015), the form of CK2 present at the outer side of plasma membrane, which, despite quite neglected so far, is expected to be implicated in human pathologies, as Alzheimer's disease (Montenarh & Götz, 2018).

Apart from bi-substrate inhibitors, a different strategy (multi-target drugs) exploits the combination of two drugs on the same molecule, to exert anti-tumor effects by simultaneously hitting different targets (and avoiding problems of drug cocktail administration). Very recently, multi-target drugs inhibiting CK2 and histone deacetylase (HDAC) have been developed (Martínez et al., 2020), by fusing TBB or DMAT (to inhibit CK2) and a zinc binding group (to inhibit HDAC). Preliminary results on different cancer cells, also with drug-resistance phenotype, are encouraging, and this approach is expected to be further developed and applied in next future.

Another strategy exploited the conjugation of cisplatin (or derivatives) with a CK2 inhibitor (CX-4945), to produce a compound with good antitumor activity in mouse xenograft models, able to overcome chemo-immune-resistance (Chen, Pei, Wang, Zhu, & Gou, 2020).

In summary, several potential drugs based on inhibiting CK2 are already available for clinical purposes, and more are expected to be developed. A part from the clinical trials already ongoing (see below), we think that potentially good results can be obtained by new inhibitors (such as

GO289 and dual HDAC/CK2 inhibitors), or by the identification of compounds able to selectively target CK2 substrates specifically related to a certain pathology (as in the case of CIGB-300).

Moreover, apart from chemical compounds altering CK2 catalytic activity, the possibility exists of regulating CK2 protein amount. This can be achieved either by genetic manipulations, or by selective delivery of anti-CK2 RNAi through nanocapsules (Trembley et al., 2017), or by the PROTAC strategy. This latter has been recently developed with chimeras of a CK2 inhibitor (CX-4945) and a pomalidomide for CK2 targeting to the ubiquitin-proteasome system (Chen, Chen, Liu, Wang, & Gou, 2018).

3.2 *In vivo* studies with CK2 inhibitors: Past, present and future

The first hypothesis of CK2 targeting to kill cancer cells dates back to 2000, when Faust and colleagues demonstrated that antisense oligonucleotides against CK2α inhibited growth of squamous cell carcinoma of the head and neck (Faust, Tawfic, Davis, Bubash, & Ahmed, 2000). In 2002, we proposed the usage of CK2 inhibitors in tumor cell cultures, and demonstrated the consequent occurrence of apoptosis (Ruzzene, Penzo, & Pinna, 2002). This was followed by a huge number of studies exploiting CK2 inhibitors to produce apoptosis in cancer cell cultures, until the research moved to animal treatments. To our knowledge, the first *in vivo* study with CK2 inhibitors was performed in a mouse model of oxygen-induced retinopathy, aimed at investigating the CK2 function in retinal vascularization (Ljubimov et al., 2004). In that study, the CK2 inhibitors emodin and TBB were found to decrease the extent of retinal neovascularization by up to 70%, with an effect primarily concerning the actively proliferating endothelial cells concentrated in neovascular tufts. The authors concluded that CK2 inhibitors could be useful for treatment of proliferative retinopathies. The inhibitors were systemically administered (intraperitoneal injection), and this was also the first evidence for the lack of major side effects and toxicity for the animals.

The anti-angiogenic function identified in this model paved the way for subsequent studies on the effects of CK2 inhibition on neovascularization in other diseases (Montenarh, 2014); consistently, most of *in vivo* studies with CK2 inhibitors were then performed in animal cancer models. However, the benefit of CK2 targeting in ophthalmic diseases has been considered again recently: a clinical potential for the CK2 inhibitor SRPIN803 (which also inhibits SRPK1) has been suggested for macular degeneration, since its topical administration as eye ointment significantly inhibited choroidal neovascularization in mice (Morooka et al., 2015), and TBB and DMAT were found beneficial for optic nerve injury in rats (Cen et al., 2018).

Most animal studies exploiting CK2 inhibitors were performed in mouse cancer models; among the first studies, Ahmed and his group found that the administration of anti-CK2 oligonucleotides were effective in xenograft models of prostate cancer (Slaton, Unger, Sloper, Davis, & Ahmed, 2004). A multitude of xenograft studies were then performed showing the anti-tumor effects of CK2 inhibitors in different cancers. Since a detailed list is beyond the purpose of this review, we invite the readers to refer to reviews on the specific type of cancer they are interested on. We just want to point out that promising results were so far obtained in blood cancers (Buontempo et al., 2018; Gowda et al., 2017; Piazza et al., 2012) but also in many solid tumors (Chua et al., 2017; Lian et al., 2019). Among them, glioblastoma deserves special attention. Studies on this tumor give us the opportunity to mention a critical emerging problem, concerning the suitability of xenografts for investigating the anti-tumor potential of CK2 inhibitors. In fact, conflicting results were obtained in a glioblastoma syngeneic model in immunocompetent mice (Ferrer-Font et al., 2017) compared to xenograft (Zheng et al., 2013) or other immunocompromised mouse models (Pencheva et al., 2017). Although there might be several reasons for the discrepancy, which are under investigation, we should remind that nude mice employed in xenograft studies (or other immunocompromised models) do not allow to appreciate the eventual effects on the immune system in mediating the animal response to the treatments. Since mounting evidences are accumulating for the CK2 participation to immune cell functions (Gibson & Benveniste, 2018), we cannot exclude that some conclusions obtained from xenograft studies are to be reappraised.

The view of CK2 as a valuable target for cancer therapy is relatively recent for an enzyme known for over 60 years. Its ubiquitous expression and its involvement in many cellular processes has long prevented its targeting. Initially, strategies were explored to selectively reduce CK2 functions in tumor cells. For example, an approach based on tenascin or tenfibgen nanocapsules for the specific delivery of anti-CK2 oligonucleotides in a squamous cell carcinoma of head and neck tumor model was developed (Brown et al., 2010). But it is becoming increasingly evident that worries about side effects induced by CK2 inhibitors might have been overestimated since, in most cases, cancer cells are sensitive to lower doses of inhibitors compared to normal cells. This is essentially explained by the addiction to CK2 of cancer cells, which in turn is due to the CK2-dependent potentiation of specific oncogenic pathways, not active in normal cells.

A further confirmation has come from genetic manipulation by CRISPR-Cas9 techniques, which allowed to produce double knock out (KO) of both CK2 catalytic subunit α an α' only in normal cells (Borgo, Franchin, et al., 2017), while no double KO tumor clone has been found to survive so far (unpublished). An additional interesting hypothesis has been recently suggested, which considers the extracellular glucuronidation frequently occurring on inhibitors, at least for flavonoid compounds, and the higher β-glucuronidase activity usually displayed at cancer cell surface: since conjugated compounds are not cell permeable, only cancer cells would become accessible to the inhibitors (McCarty et al., 2020).

It is expectable that the tolerability and the effects in not-tumor cells are different in dependence of inhibitor chemistry, administration protocols, animals, *etc.*, and further studies will be necessary. Nevertheless, human experimentation has started in the meantime, with two major lines of research, one based on CX-4945 and the other on CIGB-300.

3.2.1 Clinical trials with CX-4945

There are several studies ongoing with CX-4945 (https://www.cancer.gov/about-cancer/treatment/clinical-trials/intervention/silmitasertib-sodium).

CX-4945 was initially produced by Cylene Pharmaceutical, and the same company started human clinical trials in 2010 for patients with relapsed or refractory Multiple Myeloma (ClinicalTrials.gov Identifier: NCT01199718).

Then Cylene was taken over by Senhwa Biosciences, who now produces CX-4945, and human experimentations are continuing on:

- patients with cholangiocarcinoma (ClinicalTrials.gov identifier: NCT02128282), phase I/II, estimated completion date on November 2021;
- patients with recurrent SHH (sonic hedgehog) medulloblastoma (ClinicalTrials.gov Identifier: NCT03904862), phase I/II, estimated completion date on May 2022; very interestingly, CK2 affects terminal step of SHH signaling (Purzner et al., 2018), and is therefore likely to overcome resistance to other inhibitors (like those targeting SMO protein) due to mutations within this pathway.
- patients with metastatic basal cell carcinoma (BCC) (ClinicalTrials.gov Identifier: NCT03897036), phase I, estimated completion date on December 2022.

In all these studies CX-4945 is administered orally.

Moreover, a combination of CX-4945 and an inhibitor of ATM kinase (Ku 60,019) is under evaluation on organoid cultures of human renal tumors (ClinicalTrials.gov Identifier: NCT03571438), estimated completion date on September 2024.

The communication of the results of these on-going trials will give the actual perception of the feasibility of CK2 targeting against cancer.

3.2.2 Clinical trials with CIGB-300

This is not a *sensu stricto* CK2 inhibitor, since it is expected to block the phosphorylation of specific substrates, and not the CK2 global activity. It was developed by Perea and collaborators, who applied the compounds by local administration for treatment of cervical malignancies. The results of clinical investigations with CIGB-300 have been recently summarized by the same authors (Perea et al., 2018). Since the beginning they found significant improvement in women with high-grade squamous intraepithelial lesion, with reduced lesion in 90% of patients (28/31), and good tolerability (Solares et al., 2009). Then a phase 1 trial was conducted on women with locally advanced cervical cancer. Although only minor or mild side effects were produced, an important observation was that CIGB-300 induced histamine release, and it is suggested that premedication with antihistamines improves tolerability. The mechanism by which CIGB-300 causes histamine release is not yet understood, and it is possible that it might be related not to CK2 but to the multiple mechanisms of action of CIGB-300 (Zanin et al., 2015). Thus, the results should not be exported to what expected with other CK2 inhibitors. The authors moved to a phase 2 trial, combining CIGB-300 local administration with conventional chemoradiotherapy (Perea et al., 2018). Initial results are encouraging, suggesting a best response in comparison to chemoradiotherapy alone. The authors are collecting the 5-year overall survival data.

4. Conclusions and perspectives

In conclusion, CK2 appears as a multi-purpose target exploitable for different human diseases. Its inhibition is already carried out at clinical grade for cancer, but it will be expectably extended to other pathologies in the near future, possibly including Covid-19, for which it warrants further testing. It is also possible that future studies, by illuminating the mechanism of other diseases, will disclose novel opportunities of CK2 targeting.

As far as the available tools, different kinds of inhibitors have been developed and characterized, and, in some cases, they are endowed with both high efficacy and selectivity. However, even the best compound, CX-4945, is not devoid of unspecific effects. In a short time, the outcome of its employment in human trials will be known, providing information on the real feasibility of its clinical application. If, on one side, small molecule drug research will continue and hopefully improved compounds will be developed, on the other side it will be important to confirm that CK2, a protein expressed and active also in normal tissues, is really a convenient drug target. In case, the possibility of selective delivery of the compound at the disease site is under investigation. Moreover, apart from CK2 inhibitors, new approaches are attainable. In particular, strategies based on targeting specific CK2 substrates or CK2-dependent pathways which are upregulated and crucial in pathological circumstances, as in the promising case of CIGB-300, could be a possible solution for a more disease-tailored CK2 blockage.

Acknowledgments

The authors would like to thank all the members of their laboratories for their contributions. The work performed in our laboratory was supported by grants from Associazione Italiana per la Ricerca sul Cancro (AIRC, grant IG 18756) and Institutional grants from University of Padova to M.R.

References

Ampofo, E., Nalbach, L., Menger, M. D., Montenarh, M., & Götz, C. (2019). Protein kinase CK2-A putative target for the therapy of diabetes mellitus? *International Journal of Molecular Sciences*, *20*, 4398. https://doi.org/10.3390/ijms20184398.

Andrzejewska, M., Pagano, M. A., Meggio, F., Brunati, A. M., & Kazimierczuk, Z. (2003). Polyhalogenobenzimidazoles: Synthesis and their inhibitory activity against casein kinases. *Bioorganic & Medicinal Chemistry*, *11*, 3997–4002.

Battistutta, R., Cozza, G., Pierre, F., Papinutto, E., Lolli, G., Sarno, S., et al. (2011). Unprecedented selectivity and structural determinants of a new class of protein kinase CK2 inhibitors in clinical trials for the treatment of cancer. *Biochemistry*, *50*, 8478–8488.

Battistutta, R., De Moliner, E., Sarno, S., Zanotti, G., & Pinna, L. A. (2001). Structural features underlying selective inhibition of protein kinase CK2 by ATP site-directed tetrabromo-2-benzotriazole. *Protein Science: A Publication of the Protein Society*, *10*, 2200–2206.

Benavent Acero, F., Capobianco, C. S., Garona, J., Cirigliano, S. M., Perera, Y., Urtreger, A. J., et al. (2017). CIGB-300, an anti-CK2 peptide, inhibits angiogenesis, tumor cell invasion and metastasis in lung cancer models. *Lung Cancer (Amsterdam, Netherlands)*, *107*, 14–21.

Bollacke, A., Nienberg, C., Borgne, M. L., & Jose, J. (2016). Toward selective CK2alpha and CK2alpha' inhibitors: Development of a novel whole-cell kinase assay by autodisplay of catalytic CK2alpha'. *Journal of Pharmaceutical and Biomedical Analysis*, *121*, 253–260.

Borgo, C., Franchin, C., Scalco, S., Bosello-Travain, V., Donella-Deana, A., Arrigoni, G., et al. (2017). Generation and quantitative proteomics analysis of CK2α/α'(-/-) cells. *Scientific Reports*, 7, 42409.

Borgo, C., Milan, G., Favaretto, F., Stasi, F., Fabris, R., Salizzato, V., et al. (2017). CK2 modulates adipocyte insulin-signaling and is up-regulated in human obesity. *Scientific Reports*, 7, 17569.

Borgo, C., & Ruzzene, M. (2019). Role of protein kinase CK2 in antitumor drug resistance. *Journal of Experimental & Clinical Cancer Research*, *38*, 287.

Bouhaddou, M., Memon, D., Meyer, B., White, K. M., Rezelj, V. V., Marrero, M. C., et al. (2020). The global phosphorylation landscape of SARS-CoV-2 infection. *Cell, 182*, 685–712.e19. https://doi.org/10.1016/j.cell.2020.06.034.

Brown, M. S., Diallo, O. T., Hu, M., Ehsanian, R., Yang, X., Arun, P., et al. (2010). CK2 modulation of NF-kappaB, TP53, and the malignant phenotype in head and neck cancer by anti-CK2 oligonucleotides in vitro or in vivo via sub-50-nm nanocapsules. *Clinical Cancer Research: An Official Journal of the American Association for Cancer Research, 16*, 2295–2307.

Buontempo, F., McCubrey, J. A., Orsini, E., Ruzzene, M., Cappellini, A., Lonetti, A., et al. (2018). Therapeutic targeting of CK2 in acute and chronic leukemias. *Leukemia, 32*, 1–10.

Buontempo, F., Orsini, E., Martins, L. R., Antunes, I., Lonetti, A., Chiarini, F., et al. (2014). Cytotoxic activity of the casein kinase 2 inhibitor CX-4945 against T-cell acute lymphoblastic leukemia: Targeting the unfolded protein response signaling. *Leukemia, 28*, 543–553.

Castello, J., Ragnauth, A., Friedman, E., & Rebholz, H. (2017). CK2-an emerging target for neurological and psychiatric disorders. *Pharmaceuticals (Basel, Switzerland), 10*, 7. https://doi.org/10.3390/ph10010007.

Cen, L.-P., Liu, Y.-F., Ng, T. K., Luo, J.-M., van Rooijen, N., Zhang, M., et al. (2018). Casein kinase-II inhibition promotes retinal ganglion cell survival and axonal regeneration. *Experimental Eye Research, 177*, 153–159.

Chen, H., Chen, F., Liu, N., Wang, X., & Gou, S. (2018). Chemically induced degradation of CK2 by proteolysis targeting chimeras based on a ubiquitin-proteasome pathway. *Bioorganic Chemistry, 81*, 536–544.

Chen, X., Li, C., Wang, D., Chen, Y., & Zhang, N. (2020). Recent advances in the discovery of CK2 allosteric inhibitors: From traditional screening to structure-based design. *Molecules (Basel, Switzerland), 25*, 870. https://doi.org/10.3390/molecules25040870.

Chen, F., Pei, S., Wang, X., Zhu, Q., & Gou, S. (2020). Emerging JWA-targeted Pt(IV) prodrugs conjugated with CX-4945 to overcome chemo-immune-resistance. *Biochemical and Biophysical Research Communications, 521*, 753–761.

Chilin, A., Battistutta, R., Bortolato, A., Cozza, G., Zanatta, S., Poletto, G., et al. (2008). Coumarin as attractive casein kinase 2 (CK2) inhibitor scaffold: An integrate approach to elucidate the putative binding motif and explain structure-activity relationships. *Journal of Medicinal Chemistry, 51*, 752–759.

Chua, M. M. J., Ortega, C. E., Sheikh, A., Lee, M., Abdul-Rassoul, H., Hartshorn, K. L., et al. (2017). CK2 in cancer: Cellular and biochemical mechanisms and potential therapeutic target. *Pharmaceuticals (Basel, Switzerland), 10*, 18. https://doi.org/10.3390/ph10010018.

Cozza, G. (2017). The development of CK2 inhibitors: From traditional pharmacology to in silico rational drug design. *Pharmaceuticals (Basel, Switzerland), 10*, 26. https://doi.org/10.3390/ph10010026.

Cozza, G., Gianoncelli, A., Bonvini, P., Zorzi, E., Pasquale, R., Rosolen, A., et al. (2011). Urolithin as a converging scaffold linking ellagic acid and coumarin analogues: Design of potent protein kinase CK2 inhibitors. *ChemMedChem, 6*, 2273–2286.

Cozza, G., Girardi, C., Ranchio, A., Lolli, G., Sarno, S., Orzeszko, A., et al. (2014). Cell-permeable dual inhibitors of protein kinases CK2 and PIM-1: Structural features and pharmacological potential. *Cellular and Molecular Life Sciences: CMLS, 71*, 3173–3185.

Cozza, G., Mazzorana, M., Papinutto, E., Bain, J., Elliott, M., di Maira, G., et al. (2009). Quinalizarin as a potent, selective and cell-permeable inhibitor of protein kinase CK2. *The Biochemical Journal, 421*, 387–395.

Cozza, G., Pinna, L. A., & Moro, S. (2012). Protein kinase CK2 inhibitors: A patent review. *Expert Opinion on Therapeutic Patents, 22*, 1081–1097.

Cozza, G., Venerando, A., Sarno, S., & Pinna, L. A. (2015). The selectivity of CK2 inhibitor quinalizarin: A reevaluation. *BioMed Research International, 2015*, 734127.

Cozza, G., Zanin, S., Sarno, S., Costa, E., Girardi, C., Ribaudo, G., et al. (2015). Design, validation and efficacy of bisubstrate inhibitors specifically affecting ecto-CK2 kinase activity. *The Biochemical Journal*, *471*, 415–430.

Cozza, G., Zonta, F., Dalle Vedove, A., Venerando, A., Dall'Acqua, S., Battistutta, R., et al. (2020). Biochemical and cellular mechanism of protein kinase CK2 inhibition by deceptive curcumin. *The FEBS Journal*, *287*, 1850–1864.

CX-4945 Granted Orphan Drug Designation. (2017). *Oncology times*. *Vol. 39* (p. 23). Lippincott Williams & Wilkins (LWW). https://doi.org/10.1097/01.COT.0000514203.35081.69.

Dalle Vedove, A., Zonta, F., Zanforlin, E., Demitri, N., Ribaudo, G., Cazzanelli, G., et al. (2020). A novel class of selective CK2 inhibitors targeting its open hinge conformation. *European Journal of Medicinal Chemistry*, *195*, 112267.

D'Amore, C., Borgo, C., Bosello-Travain, V., Vilardell, J., Salizzato, V., Pinna, L. A., et al. (2020). Deciphering the role of protein kinase CK2 in the maturation/stability of F508del-CFTR. *Biochimica Et Biophysica Acta. Molecular Basis of Disease*, *1866*, 165611.

D'Amore, C., Moro, E., Borgo, C., Kenichiro, I., Tsuyoshi, H., Pinna, L. A., et al. (2020). "Janus" efficacy for CX-5011: CK2 inhibition and methuosis induction by independent mechanisms. *Biochimica Et Biophysica Acta. Molecular Cell Research*, *1867*, 118807.

de Gooijer, M. C., Guillén Navarro, M., Bernards, R., Wurdinger, T., & van Tellingen, O. (2018). An experimenter's guide to glioblastoma invasion pathways. *Trends in Molecular Medicine*, *24*, 763–780.

Di Maira, G., Gentilini, A., Pastore, M., Caligiuri, A., Piombanti, B., Raggi, C., et al. (2019). The protein kinase CK2 contributes to the malignant phenotype of cholangiocarcinoma cells. *Oncogenesis*, *8*, 61.

Dominguez, I., Sonenshein, G. E., & Seldin, D. C. (2009). Protein kinase CK2 in health and disease: CK2 and its role in Wnt and NF-kappaB signaling: Linking development and cancer. *Cellular and Molecular Life Sciences: CMLS*, *66*, 1850–1857.

Duncan, J. S., Turowec, J. P., Vilk, G., Li, S. S. C., Gloor, G. B., & Litchfield, D. W. (2010). Regulation of cell proliferation and survival: Convergence of protein kinases and caspases. *Biochimica et Biophysica Acta*, *1804*, 505–510.

Eom, G. H., Cho, Y. K., Ko, J.-H., Shin, S., Choe, N., Kim, Y., et al. (2011). Casein kinase-2α1 induces hypertrophic response by phosphorylation of histone deacetylase 2 S394 and its activation in the heart. *Circulation*, *123*, 2392–2403.

Faust, R. A., Tawfic, S., Davis, A. T., Bubash, L. A., & Ahmed, K. (2000). Antisense oligonucleotides against protein kinase CK2-alpha inhibit growth of squamous cell carcinoma of the head and neck in vitro. *Head & Neck*, *22*, 341–346.

Ferrer-Font, L., Villamañan, L., Arias-Ramos, N., Vilardell, J., Plana, M., Ruzzene, M., et al. (2017). Targeting protein kinase CK2: Evaluating CX-4945 potential for GL261 glioblastoma therapy in immunocompetent mice. *Pharmaceuticals (Basel, Switzerland)*, *10*, 24. https://doi.org/10.3390/ph10010024.

Filhol, O., & Cochet, C. (2009). Protein kinase CK2 in health and disease: Cellular functions of protein kinase CK2: A dynamic affair. *Cellular and Molecular Life Sciences: CMLS*, *66*, 1830–1839.

Filhol, O., Giacosa, S., Wallez, Y., & Cochet, C. (2015). Protein kinase CK2 in breast cancer: The CK2β regulatory subunit takes center stage in epithelial plasticity. *Cellular and Molecular Life Sciences: CMLS*, *72*, 3305–3322.

Filhol, O., Martiel, J.-L., & Cochet, C. (2004). Protein kinase CK2: A new view of an old molecular complex. *EMBO Reports*, *5*, 351–355.

Fragoso, R., & Barata, J. T. (2015). Kinases, tails and more: Regulation of PTEN function by phosphorylation. *Methods (San Diego, Calif.)*, *77–78*, 75–81.

Franchin, C., Borgo, C., Cesaro, L., Zaramella, S., Vilardell, J., Salvi, M., et al. (2018). Re-evaluation of protein kinase CK2 pleiotropy: New insights provided by a phosphoproteomics analysis of CK2 knockout cells. *Cellular and Molecular Life Sciences: CMLS*, *75*, 2011–2026.

Franchin, C., Borgo, C., Zaramella, S., Cesaro, L., Arrigoni, G., Salvi, M., et al. (2017). Exploring the CK2 paradox: Restless, dangerous, dispensable. *Pharmaceuticals (Basel, Switzerland)*, *10*, 11. https://doi.org/10.3390/ph10010011.

Gibson, S. A., & Benveniste, E. N. (2018). Protein kinase CK2: An emerging regulator of immunity. *Trends in Immunology*, *39*, 82–85.

Gordon, D. E., Jang, G. M., Bouhaddou, M., Xu, J., Obernier, K., White, K. M., et al. (2020). A SARS-CoV-2 protein interaction map reveals targets for drug repurposing. *Nature*, *583*, 459–468.

Götz, C., Bachmann, C., & Montenarh, M. (2007). Inhibition of protein kinase CK2 leads to a modulation of androgen receptor dependent transcription in prostate cancer cells. *The Prostate*, *67*, 125–134.

Gowda, C., Sachdev, M., Muthusami, S., Kapadia, M., Petrovic-Dovat, L., Hartman, M., et al. (2017). Casein kinase II (CK2) as a therapeutic target for hematological malignancies. *Current Pharmaceutical Design*, *23*, 95–107.

Gowda, C., Song, C., Ding, Y., Iyer, S., Dhanyamraju, P. K., McGrath, M., et al. (2020). Cellular signaling and epigenetic regulation of gene expression in leukemia. *Advances in Biological Regulation*, *75*, 100665.

Guerra, B., & Issinger, O.-G. (2008). Protein kinase CK2 in human diseases. *Current Medicinal Chemistry*, *15*, 1870–1886.

Guo, C., Yu, S., Davis, A. T., Wang, H., Green, J. E., & Ahmed, K. (2001). A potential role of nuclear matrix-associated protein kinase CK2 in protection against drug-induced apoptosis in cancer cells. *The Journal of Biological Chemistry*, *276*, 5992–5999.

Hauck, L., Harms, C., An, J., Rohne, J., Gertz, K., Dietz, R., et al. (2008). Protein kinase CK2 links extracellular growth factor signaling with the control of p27 Kip1 stability in the heart. *Nature Medicine*, *14*, 315–324.

Hochscherf, J., Lindenblatt, D., Steinkrüger, M., Yoo, E., Ulucan, Ö., Herzig, S., et al. (2015). Development of a high-throughput screening-compatible assay to identify inhibitors of the CK2α/CK2β interaction. *Analytical Biochemistry*, *468*, 4–14.

Im, D.-K., Cheong, H., Lee, J. S., Oh, M.-K., & Yang, K. M. (2019). Protein kinase CK2-dependent aerobic glycolysis-induced lactate dehydrogenase A enhances the migration and invasion of cancer cells. *Scientific Reports*, *9*, 5337.

Jia, H., Liu, Y., Xia, R., Tong, C., Yue, T., Jiang, J., et al. (2010). Casein kinase 2 promotes Hedgehog signaling by regulating both smoothened and Cubitus interruptus. *The Journal of Biological Chemistry*, *285*, 37218–37226.

Kalathur, M., Toso, A., Chen, J., Revandkar, A., Danzer-Baltzer, C., Guccini, I., et al. (2015). A chemogenomic screening identifies CK2 as a target for pro-senescence therapy in PTEN-deficient tumours. *Nature Communications*, *6*, 7227.

Kim, H., Choi, K., Kang, H., Lee, S.-Y., Chi, S.-W., Lee, M.-S., et al. (2014). Identification of a novel function of CX-4945 as a splicing regulator. *PLoS One*, *9*, e94978.

Kramerov, A. A., Saghizadeh, M., Caballero, S., Shaw, L. C., Li Calzi, S., Bretner, M., et al. (2008). Inhibition of protein kinase CK2 suppresses angiogenesis and hematopoietic stem cell recruitment to retinal neovascularization sites. *Molecular and Cellular Biochemistry*, *316*, 177–186.

Kufareva, I., Bestgen, B., Brear, P., Prudent, R., Laudet, B., Moucadel, V., et al. (2019). Discovery of holoenzyme-disrupting chemicals as substrate-selective CK2 inhibitors. *Scientific Reports*, *9*, 15893. https://doi.org/10.1038/s41598-019-52141-5.

Laudet, B., Barette, C., Dulery, V., Renaudet, O., Dumy, P., Metz, A., et al. (2007). Structure-based design of small peptide inhibitors of protein kinase CK2 subunit interaction. *The Biochemical Journal*, *408*, 363–373.

Lertsuwan, J., Lertsuwan, K., Sawasdichai, A., Tasnawijitwong, N., Lee, K. Y., Kitchen, P., et al. (2018). CX-4945 induces methuosis in cholangiocarcinoma cell lines by a CK2-independent mechanism. *Cancers*, *10*, 283. https://doi.org/10.3390/cancers10090283.

Lian, H., Su, M., Zhu, Y., Zhou, Y., Soomro, S. H., & Fu, H. (2019). Protein kinase CK2, a potential therapeutic target in carcinoma management. *Asian Pacific Journal of Cancer Prevention: APJCP*, *20*, 23–32.

Ljubimov, A. V., Caballero, S., Aoki, A. M., Pinna, L. A., Grant, M. B., & Castellon, R. (2004). Involvement of protein kinase CK2 in angiogenesis and retinal neovascularization. *Investigative Ophthalmology & Visual Science*, *45*, 4583–4591.

Lolli, G., Pinna, L. A., & Battistutta, R. (2012). Structural determinants of protein kinase CK2 regulation by autoinhibitory polymerization. *ACS Chemical Biology*, 7, 1158–1163.

Manni, S., Brancalion, A., Mandato, E., Tubi, L. Q., Colpo, A., Pizzi, M., et al. (2013). Protein kinase CK2 inhibition down modulates the NF-κB and STAT3 survival pathways, enhances the cellular proteotoxic stress and synergistically boosts the cytotoxic effect of bortezomib on multiple myeloma and mantle cell lymphoma cells. *PLoS One*, *8*, e75280.

Manni, S., Brancalion, A., Tubi, L. Q., Colpo, A., Pavan, L., Cabrelle, A., et al. (2012). Protein kinase CK2 protects multiple myeloma cells from ER stress-induced apoptosis and from the cytotoxic effect of HSP90 inhibition through regulation of the unfolded protein response. *Clinical Cancer Research: An Official Journal of the American Association for Cancer Research*, *18*, 1888–1900.

Marschke, R. F., Borad, M. J., McFarland, R. W., Alvarez, R. H., Lim, J. K., Padgett, C. S., et al. (2011). Findings from the phase I clinical trials of CX-4945, an orally available inhibitor of CK2. *Journal of Clinical Oncology*, *29*, 3087.

Martínez, R., Di Geronimo, B., Pastor, M., Zapico, J. M., Coderch, C., Panchuk, R., et al. (2020). Multitarget anticancer agents based on histone deacetylase and protein kinase CK2 inhibitors. *Molecules*, *25*, 1497. https://doi.org/10.3390/molecules25071497.

McCarty, M. F., Assanga, S. I., & Lujan, L. L. (2020). Flavones and flavonols may have clinical potential as CK2 inhibitors in cancer therapy. *Medical Hypotheses*, *141*, 109723.

Meggio, F., Boldyreff, B., Issinger, O. G., & Piṅna, L. A. (1994). Casein kinase 2 down-regulation and activation by polybasic peptides are mediated by acidic residues in the 55-64 region of the beta-subunit. A study with calmodulin as phosphorylatable substrate. *Biochemistry*, *33*, 4336–4342.

Meggio, F., Pagano, M. A., Moro, S., Zagotto, G., Ruzzene, M., Sarno, S., et al. (2004). Inhibition of protein kinase CK2 by condensed polyphenolic derivatives. An in vitro and in vivo study. *Biochemistry*, *43*, 12931–12936.

Meggio, F., & Pinna, L. A. (2003). One-thousand-and-one substrates of protein kinase CK2? *FASEB Journal: Official Publication of the Federation of American Societies for Experimental Biology*, *17*, 349–368.

Mehta, A. (2008). Cystic fibrosis as a bowel cancer syndrome and the potential role of CK2. *Molecular and Cellular Biochemistry*, *316*, 169–175.

Miyata, Y. (2009). Protein kinase CK2 in health and disease: CK2: The kinase controlling the Hsp90 chaperone machinery. *Cellular and Molecular Life Sciences: CMLS*, *66*, 1840–1849.

Montenarh, M. (2014). Protein kinase CK2 and angiogenesis. *Advances in Clinical and Experimental Medicine: Official Organ Wroclaw Medical University*, *23*, 153–158.

Montenarh, M., & Götz, C. (2018). Ecto-protein kinase CK2, the neglected form of CK2. *Biomedical Reports*, *8*, 307–313.

Morooka, S., Hoshina, M., Kii, I., Okabe, T., Kojima, H., Inoue, N., et al. (2015). Identification of a dual inhibitor of SRPK1 and CK2 that attenuates pathological angiogenesis of macular degeneration in mice. *Molecular Pharmacology*, *88*, 316–325.

Morphy, R. (2010). Selectively nonselective kinase inhibition: Striking the right balance. *Journal of Medicinal Chemistry*, *53*, 1413–1437.

Mottet, D., Ruys, S. P. D., Demazy, C., Raes, M., & Michiels, C. (2005). Role for casein kinase 2 in the regulation of HIF-1 activity. *International Journal of Cancer*, *117*, 764–774.

Murtaza, I., Wang, H.-X., Feng, X., Alenina, N., Bader, M., Prabhakar, B. S., et al. (2008). Down-regulation of catalase and oxidative modification of protein kinase CK2 lead to the failure of apoptosis repressor with caspase recruitment domain to inhibit cardiomyocyte hypertrophy. *The Journal of Biological Chemistry*, *283*, 5996–6004.

Nie, Z., Perretta, C., Erickson, P., Margosiak, S., Almassy, R., Lu, J., et al. (2007). Structure-based design, synthesis, and study of pyrazolo[1,5-a][1,3,5]triazine derivatives as potent inhibitors of protein kinase CK2. *Bioorganic & Medicinal Chemistry Letters*, *17*, 4191–4195.

Niechi, I., Silva, E., Cabello, P., Huerta, H., Carrasco, V., Villar, P., et al. (2015). Colon cancer cell invasion is promoted by protein kinase CK2 through increase of endothelin-converting enzyme-1c protein stability. *Oncotarget*, *6*, 42749–42760.

Niefind, K., Guerra, B., Pinna, L. A., Issinger, O. G., & Schomburg, D. (1998). Crystal structure of the catalytic subunit of protein kinase CK2 from Zea mays at 2.1 A resolution. *The EMBO Journal*, *17*, 2451–2462.

Oshima, T., Niwa, Y., Kuwata, K., Srivastava, A., Hyoda, T., Tsuchiya, Y., et al. (2019). Cell-based screen identifies a new potent and highly selective CK2 inhibitor for modulation of circadian rhythms and cancer cell growth. *Science Advances*, *5*, eaau9060.

Padgett, C. S., Lim, J. K. C., Marschke, R. F., Northfelt, D. W., Andreopoulou, E., Hoff, D. D. V., et al. (2010). 414 Clinical pharmacokinetics and pharmacodynamics of CX-4945, a novel inhibitor of protein kinase CK2: Interim report from the phase 1 clinical trial. *European Journal of Cancer Supplements*, *8*, 131–132.

Pagano, M. A., Andrzejewska, M., Ruzzene, M., Sarno, S., Cesaro, L., Bain, J., et al. (2004). Optimization of protein kinase CK2 inhibitors derived from 4,5,6,7-tetrabromobenzimidazole. *Journal of Medicinal Chemistry*, *47*, 6239–6247.

Pagano, M. A., Bain, J., Kazimierczuk, Z., Sarno, S., Ruzzene, M., Di Maira, G., et al. (2008). The selectivity of inhibitors of protein kinase CK2: An update. *The Biochemical Journal*, *415*, 353–365.

Pagano, M. A., Poletto, G., Di Maira, G., Cozza, G., Ruzzene, M., Sarno, S., et al. (2007). Tetrabromocinnamic acid (TBCA) and related compounds represent a new class of specific protein kinase CK2 inhibitors. *Chembiochem: A European Journal of Chemical Biology*, *8*, 129–139.

Pencheva, N., de Gooijer, M. C., Vis, D. J., Wessels, L. F. A., Würdinger, T., van Tellingen, O., et al. (2017). Identification of a druggable pathway controlling glioblastoma invasiveness. *Cell Reports*, *20*, 48–60.

Perea, S. E., Baladrón, I., Valenzuela, C., & Perera, Y. (2018). CIGB-300: A peptide-based drug that impairs the protein kinase CK2-mediated phosphorylation. *Seminars in Oncology*, *45*, 58–67.

Perea, S. E., Reyes, O., Puchades, Y., Mendoza, O., Vispo, N. S., Torrens, I., et al. (2004). Antitumor effect of a novel proapoptotic peptide that impairs the phosphorylation by the protein kinase 2 (casein kinase 2). *Cancer Research*, *64*, 7127–7129.

Perera, Y., Ramos, Y., Padrón, G., Caballero, E., Guirola, O., Caligiuri, L. G., et al. (2020). CIGB-300 anticancer peptide regulates the protein kinase CK2-dependent phosphoproteome. *Molecular and Cellular Biochemistry*, *470*, 63–75.

Piazza, F., Manni, S., Ruzzene, M., Pinna, L. A., Gurrieri, C., & Semenzato, G. (2012). Protein kinase CK2 in hematologic malignancies: Reliance on a pivotal cell survival regulator by oncogenic signaling pathways. *Leukemia*, *26*, 1174–1179.

Pierre, F., Chua, P. C., O'Brien, S. E., Siddiqui-Jain, A., Bourbon, P., Haddach, M., et al. (2011). Discovery and SAR of 5-(3-chlorophenylamino)benzo[c][2,6]naphthyridine-8-carboxylic acid (CX-4945), the first clinical stage inhibitor of protein kinase CK2 for the treatment of cancer. *Journal of Medicinal Chemistry*, *54*, 635–654.

Pietsch, M., Viht, K., Schnitzler, A., Ekambaram, R., Steinkrüger, M., Enkvist, E., et al. (2020). Unexpected CK2β-antagonistic functionality of bisubstrate inhibitors targeting protein kinase CK2. *Bioorganic Chemistry*, *96*, 103608.

Pinna, L. A. (2002). Protein kinase CK2: A challenge to canons. *Journal of Cell Science, 115*, 3873–3878.
Prudent, R., Moucadel, V., Laudet, B., Barette, C., Lafanechere, L., Hasenknopf, B., et al. (2008). Identification of polyoxometalates as nanomolar noncompetitive inhibitors of protein kinase CK2. *Chemistry & Biology, 15*, 683–692.
Prudent, R., Sautel, C. F., & Cochet, C. (2010). Structure-based discovery of small molecules targeting different surfaces of protein-kinase CK2. *Biochimica et Biophysica Acta (BBA) - Proteins and Proteomics, 1804*, 493–498.
Purzner, T., Purzner, J., Buckstaff, T., Cozza, G., Gholamin, S., Rusert, J. M., et al. (2018). Developmental phosphoproteomics identifies the kinase CK2 as a driver of Hedgehog signaling and a therapeutic target in medulloblastoma. *Science Signaling, 11* eaau5147. https://doi.org/10.1126/scisignal.aau5147.
Qiao, Y., Chen, T., Yang, H., Chen, Y., Lin, H., Qu, W., et al. (2019). Small molecule modulators targeting protein kinase CK1 and CK2. *European Journal of Medicinal Chemistry, 181*, 111581.
Raaf, J., Brunstein, E., Issinger, O.-G., & Niefind, K. (2008). The CK2 alpha/CK2 beta interface of human protein kinase CK2 harbors a binding pocket for small molecules. *Chemistry & Biology, 15*, 111–117.
Rahnel, H., Viht, K., Lavogina, D., Mazina, O., Haljasorg, T., Enkvist, E., et al. (2017). A selective biligand inhibitor of CK2 increases caspase-3 activity in cancer cells and inhibits platelet aggregation. *ChemMedChem, 12*, 1723–1736.
Ruzzene, M., Bertacchini, J., Toker, A., & Marmiroli, S. (2017). Cross-talk between the CK2 and AKT signaling pathways in cancer. *Advances in Biological Regulation, 64*, 1–8.
Ruzzene, M., Penzo, D., & Pinna, L. A. (2002). Protein kinase CK2 inhibitor 4,5,6,7-tetrabromobenzotriazole (TBB) induces apoptosis and caspase-dependent degradation of haematopoietic lineage cell-specific protein 1 (HS1) in Jurkat cells. *The Biochemical Journal, 364*, 41–47.
Salvi, M., Sarno, S., Marin, O., Meggio, F., Itarte, E., & Pinna, L. A. (2006). Discrimination between the activity of protein kinase CK2 holoenzyme and its catalytic subunits. *FEBS Letters, 580*, 3948–3952.
Sarno, S., de Moliner, E., Ruzzene, M., Pagano, M. A., Battistutta, R., Bain, J., et al. (2003). Biochemical and three-dimensional-structural study of the specific inhibition of protein kinase CK2 by [5-oxo-5,6-dihydroindolo-(1,2-a)quinazolin-7-yl]acetic acid (IQA). *The Biochemical Journal, 374*, 639–646.
Scaglioni, P. P., Yung, T. M., Choi, S., Choi, S. C., Baldini, C., Konstantinidou, G., et al. (2008). CK2 mediates phosphorylation and ubiquitin-mediated degradation of the PML tumor suppressor. *Molecular and Cellular Biochemistry, 316*, 149–154.
Siddiqui, Y. H., Kershaw, R. M., Humphreys, E. H., Assis Junior, E. M., Chaudhri, S., Jayaraman, P.-S., et al. (2017). CK2 abrogates the inhibitory effects of PRH/HHEX on prostate cancer cell migration and invasion and acts through PRH to control cell proliferation. *Oncogenesis, 6*, e293.
Silva-Pavez, E., & Tapia, J. C. (2020). Protein kinase CK2 in cancer energetics. *Frontiers in Oncology, 10*, 893.
Singh, N. N., & Ramji, D. P. (2008). Protein kinase CK2, an important regulator of the inflammatory response? *Journal of Molecular Medicine (Berlin, Germany), 86*, 887–897.
Slaton, J. W., Unger, G. M., Sloper, D. T., Davis, A. T., & Ahmed, K. (2004). Induction of apoptosis by antisense CK2 in human prostate cancer xenograft model. *Molecular Cancer Research, 2*, 712–721.
Solares, A. M., Santana, A., Baladrón, I., Valenzuela, C., González, C. A., Díaz, A., et al. (2009). Safety and preliminary efficacy data of a novel casein kinase 2 (CK2) peptide inhibitor administered intralesionally at four dose levels in patients with cervical malignancies. *BMC Cancer, 9*, 146.

St-Denis, N. A., & Litchfield, D. W. (2009). Protein kinase CK2 in health and disease: From birth to death: The role of protein kinase CK2 in the regulation of cell proliferation and survival. *Cellular and Molecular Life Sciences: CMLS, 66*, 1817–1829.

Stefano, D. D., Villella, V. R., Esposito, S., Tosco, A., Sepe, A., Gregorio, F. D., et al. (2014). Restoration of CFTR function in patients with cystic fibrosis carrying the F508del-CFTR mutation. *Autophagy, 10*, 2053–2074.

Trembley, J. H., Chen, Z., Unger, G., Slaton, J., Kren, B. T., Van Waes, C., et al. (2010). Emergence of protein kinase CK2 as a key target in cancer therapy. *BioFactors (Oxford, England), 36*, 187–195.

Trembley, J. H., Kren, B. T., Abedin, M. J., Shaughnessy, D. P., Li, Y., Dehm, S. M., et al. (2019). CK2 pro-survival role in prostate cancer is mediated via maintenance and promotion of androgen receptor and NFκB p65 expression. *Pharmaceuticals (Basel, Switzerland), 12*, 89. https://doi.org/10.3390/ph12020089.

Trembley, J. H., Kren, B. T., Abedin, M. J., Vogel, R. I., Cannon, C. M., Unger, G. M., et al. (2017). CK2 molecular targeting-tumor cell-specific delivery of RNAi in various models of cancer. *Pharmaceuticals (Basel, Switzerland), 10*, 25. https://doi.org/10.3390/ph10010025.

Trembley, J. H., Wang, G., Unger, G., Slaton, J., & Ahmed, K. (2009). Protein kinase CK2 in health and disease: CK2: A key player in cancer biology. *Cellular and Molecular Life Sciences: CMLS, 66*, 1858–1867.

Unger, G. M., Davis, A. T., Slaton, J. W., & Ahmed, K. (2004). Protein kinase CK2 as regulator of cell survival: Implications for cancer therapy. *Current Cancer Drug Targets, 4*, 77–84.

Vaglio, P., Sarno, S., Marin, O., Meggio, F., Issinger, O. G., & Pinna, L. A. (1996). Mapping the residues of protein kinase CK2 alpha subunit responsible for responsiveness to polyanionic inhibitors. *FEBS Letters, 380*, 25–28.

Vazquez, F., Grossman, S. R., Takahashi, Y., Rokas, M. V., Nakamura, N., & Sellers, W. R. (2001). Phosphorylation of the PTEN tail acts as an inhibitory switch by preventing its recruitment into a protein complex. *The Journal of Biological Chemistry, 276*, 48627–48630.

Viht, K., Saaver, S., Vahter, J., Enkvist, E., Lavogina, D., Sinijärv, H., et al. (2015). Acetoxymethyl ester of tetrabromobenzimidazole-peptoid conjugate for inhibition of protein kinase CK2 in living cells. *Bioconjugate Chemistry, 26*, 2324–2335.

Villamañan, L., Alcaraz, E., Pinna, L. A., Ruzzene, M., Itarte, E., Arús, C., et al. (2019). Up-regulation of the alpha prime subunit of protein kinase CK2 as a marker of fast proliferation in GL261 cultured cells. *Pathology Oncology Research, 25*, 1659–1663.

Wadey, K. S., Brown, B. A., Sala-Newby, G. B., Jayaraman, P.-S., Gaston, K., & George, S. J. (2017). Protein kinase CK2 inhibition suppresses neointima formation via a proline-rich homeodomain-dependent mechanism. *Vascular Pharmacology, 99*, 34–44.

Yang, W., Gibson, S. A., Yan, Z., Wei, H., Tao, J., Sha, B., et al. (2020). Protein kinase 2 (CK2) controls CD4+ T cell effector function in the pathogenesis of colitis. *Mucosal Immunology, 13*, 788–798. https://doi.org/10.1038/s41385-020-0258-x.

Zandomeni, R. O. (1989). Kinetics of inhibition by 5,6-dichloro-1-beta-D-ribofuranosylbenzimidazole on calf thymus casein kinase II. *Biochemical Journal, 262*, 469–473.

Zanin, S., Sandre, M., Cozza, G., Ottaviani, D., Marin, O., Pinna, L. A., et al. (2015). Chimeric peptides as modulators of CK2-dependent signaling: Mechanism of action and off-target effects. *Biochimica et Biophysica Acta, 1854*, 1694–1707.

Zhang, S., Long, H., Yang, Y.-L., Wang, Y., Hsieh, D., Li, W., et al. (2013). Inhibition of CK2α down-regulates Notch1 signalling in lung cancer cells. *Journal of Cellular and Molecular Medicine, 17*, 854–862.

Zheng, Y., McFarland, B. C., Drygin, D., Yu, H., Bellis, S. L., Kim, H., et al. (2013). Targeting protein kinase CK2 suppresses prosurvival signaling pathways and growth of glioblastoma. *Clinical Cancer Research: An Official Journal of the American Association for Cancer Research, 19*, 6484–6494.

Zhong, B., Campagne, O., Salloum, R., Purzner, T., & Stewart, C. F. (2020). LC-MS/MS method for quantitation of the CK2 inhibitor silmitasertib (CX-4945) in human plasma, CSF, and brain tissue, and application to a clinical pharmacokinetic study in children with brain tumors. *Journal of Chromatography. B, Analytical Technologies in the Biomedical and Life Sciences*, *1152*, 122254.

Zhou, H., Zhu, P., Wang, J., Zhu, H., Ren, J., & Chen, Y. (2018). Pathogenesis of cardiac ischemia reperfusion injury is associated with CK2α-disturbed mitochondrial homeostasis via suppression of FUNDC1-related mitophagy. *Cell Death and Differentiation*, *25*, 1080–1093.

Zonta, F., Quezada Meza, C., P., Borgo, C., Masgras, I., Rasola, A., Salvi, M., … Ruzzene, M (n.d.). Protein kinase CK2 implication in the metabolic rewiring of tumor cells. Manuscript. In preparation.

Zou, J., Luo, H., Zeng, Q., Dong, Z., Wu, D., & Liu, L. (2011). Protein kinase CK2α is overexpressed in colorectal cancer and modulates cell proliferation and invasion via regulating EMT-related genes. *Journal of Translational Medicine*, *9*, 97.

CHAPTER THREE

Emerging role of protein kinases in diabetes mellitus: From mechanism to therapy

Preeti Gupta, Aaliya Taiyab, and Md Imtaiyaz Hassan*
Centre for Interdisciplinary Research in Basic Sciences, Jamia Millia Islamia, New Delhi, India
*Corresponding author: e-mail address: mihassan@jmi.ac.in

Contents

Abstract

Diabetes mellitus has emerged as a severe burden on the medical health system across the globe. Presently, around 422 million people are suffering from diabetes which is speculated to be expanded to about 600 million by 2035. Patients with type 2 diabetes are at increased risk of developing detrimental metabolic and cardiovascular complications. The scientific understanding of this chronic disease and its underlying root cause is not yet fully unraveled. Protein kinases are well known to regulate almost every cellular process through phosphorylation of target protein in diverse signaling pathways. The important role of several protein kinases including AMP-activated protein kinase, IκB kinase and protein kinase C have been well demonstrated in various animal models.

Advances in Protein Chemistry and Structural Biology, Volume 124
ISSN 1876-1623
https://doi.org/10.1016/bs.apcsb.2020.11.001

They modulate glucose tolerance, inflammation and insulin resistance in the cells via acting on diverse downstream targets and signaling pathways. Thus, modulating the activity of potential human kinases which are significantly involved in diabetes by targeting with small molecule inhibitors could be an attractive therapeutic strategy to tackle diabetes. In this chapter, we have discussed the potential role of protein kinases in glucose metabolism and insulin sensitivity, and in the pathogenesis of diabetes mellitus. Furthermore, the small molecules reported in the literature that can be potentially used for the treatment of diabetes have been discussed in detail.

Abbreviations

AMPK AMP-activated protein kinase
DN diabetic nephropathy
DR diabetic retinopathy
IRS insulin receptor substrate
MARK4 MAP/microtubule affinity-regulating kinase 4
PKC protein kinase C
PKR RNA dependent protein kinase
ROCK1 Rho kinase 1
PI3K phosphatidylinositol-3-kinase
SphK sphingosine kinase
T2D type 2 diabetes

1. Introduction

Protein and lipid kinases regulate the phosphorylation events in various signaling cascades and thus acts as key upstream regulators of diverse cellular pathways (Bononi et al., 2011). Till now, around 538 protein kinases have been identified that represent the largest family of genes encoded by the human genome (Bhullar et al., 2018). Aberrant expression/activity of kinases has been identified in inflammatory and metabolic diseases including cancer, diabetes, obesity, arthritis and asthma. More than 400 human diseases are linked directly or indirectly to kinases (Melnikova & Golden, 2004; Patterson, Nibbs, Mcinnes, & Siebert, 2014). Hence, a global research effort has been aimed in recent years to target protein kinases to combat an array of human ailments.

In recent years, a global consensus has become evident that dictates the use of protein kinase modulators for the treatment of various human diseases. Since the binding of ATP to the active site of kinases is a prerequisite for its function, and the structural features of ATP-binding pocket are well characterized, small molecules targeting this binding pocket were conceived as the premier choice for the therapeutic development. However, the high structural similarity of key functional domains including ATP-binding pocket among various protein kinases demands the generation of highly selective small-molecule inhibitors with optimum potency (Kuhn, Weskamp, Hüllermeier, & Klebe, 2007; Spitzer, Cleves, & Jain, 2011). Moreover, it is observed in some cases that the beneficial effects of modulating protein kinase in one system are paralleled with the deleterious effects in another. For instance, targeted inhibition of a protein kinase regulating the apoptotic process prevents ischemia-induced programmed cell death in cardiomyocytes; however, it also promotes tumorigenesis in other cell types. Another example is the inhibition of overexpressed HER2 tyrosine kinase receptors in breast cancer patients with trastuzumab (monoclonal antibody) which is associated with the severe dysfunction of cardiomyocytes in the patients (Mohan, Jiang, Dokmanovic, & Wu, 2018; Zeglinski, Ludke, Jassal, & Singal, 2011). All such cases indicate that a wise usage of kinase inhibitors is imperative so that they may not pose serious threats to the organs in which protein kinase is not dysregulated rather it performs vital functions.

Diabetes is a metabolic disease marked by elevated blood glucose or sugar levels. Uncontrolled blood glucose levels lead to hyperglycemia that increases the risk of various serious complications gradually over a period of time including hypertension, cardiovascular disease, kidney failure, diabetic neuropathy, and retinopathy. The inability of the pancreas to secrete sufficient amounts of the hormone, insulin, results in type I diabetes also known as insulin-dependent diabetes. The most common is type II diabetes or non-insulin-dependent diabetes wherein body cells lose their ability to sense insulin which is termed as insulin resistance (Association, 2009). Over the past three decades, the number of cases and prevalence of type II diabetes (T2D) has dramatically increased globally. Around 422 million people are inflicted with diabetes, most of which belong to low- and middle-income countries. Moreover, a study from the International Diabetes Federation indicates that these figures will expand to about 600 million by 2035. According to WHO figures, around 1.6 million mortalities each year are directly linked to diabetes and its associated long-term complications (https://www.who.int/health-topics/diabetes#tab=tab_1). The alarmingly high death rate and prevalence of diabetes make it a huge burden on the health care system and society.

2. Protein kinases, inflammation and insulin resistance

Diseases like diabetes (especially T2D) and obesity are characterized by low-grade chronic inflammation in the body that results in impaired metabolic stability and insulin receptor signaling called metaflammation (Ashcroft & Ashcroft, 1992; Bell et al., 1990). In response to inflammation, various intracellular signaling pathways are activated causing enhanced secretion of pro-inflammatory cytokines (IL-1β, IL-6 and TNF-α) and chemokines (MCP-1) which in turn mediates a chain of events/processes causing insulin resistance (13, 14). The cascade of events initiates with the stimulation of cell surface receptors including IL-6, IL-1, TNF-α, and toll-like receptor-4 by cytokines which in turn lead to the activation of specific downstream protein kinases. Further, activated intracellular kinases facilitate the enhanced secretion of additional mediators that enhances inflammation, and consequently potentiates insulin resistance (McArdle, Finucane, Connaughton, McMorrow, & Roche, 2013). Hence, the types of protein kinase activated and their downstream signaling targets determine the overall status of insulin resistance in the body (Itani, Ruderman, Schmieder, & Boden, 2002).

The activation of insulin receptor substrates (IRSs) crucial for insulin signaling and sensitivity is mediated by a variety of protein kinases as listed in Table 1. Some of them include AMP-activated protein kinase (AMPK), IκB kinase (IKK), mitogen-activated protein kinases (MAPKs), microtubule affinity-regulating kinase 4 (MARK4), protein kinase-c (PKC) and double-stranded RNA dependent protein kinase (PKR) (McArdle et al., 2013). Insulin binds to the insulin receptor thereby activating its intrinsic tyrosine kinase activity which further phosphorylates the multifunctional protein IRS1. This in turn stimulates its downstream target proteins viz., phosphatidylinositol 3-kinase (PI3K) and Akt leading to glucose transporter 4 (GLUT4)-mediated increase in glucose transport and glycogen synthesis (Li et al., 2004).

AMPK is an evolutionary conserved Ser/Thr kinase that functions as a positive regulator of insulin sensitivity by promoting the enhanced GLUT4 translocation with concomitant glucose uptake (Fazakerley et al., 2010). Substantial evidence exists that shows the dysregulation of AMPK as the key phenomenon underlying insulin resistance mediated T2D in animals and humans, and that the pharmacological activation of AMPK can ameliorate insulin sensitivity (Ruderman, Carling, Prentki, & Cacicedo, 2013;

Table 1 Human kinases involved in the pathogenesis of diabetes mellitus and its associated complications.

S. No	Kinase	Gene	Pathway/downstream targets	Implications in diseases	Structure	References
1.	AMP-activated protein kinase (AMPK)	PRKAA1	mTOR and JNK/NF-κB signaling	T2D	Heterotrimer (α-catalytic subunit, β and γ regulatory) (C69H)	Nandipati, Subramanian, and Agrawal (2017)
2.	IκB kinase (IKK)	CHUK	NF-κB signaling pathway	Diabetes-associated nephropathy and atherosclerosis (type I), Cocoon syndrome (mutation)	5EBZ	Hayden and Ghosh (2012)
3.	Microtubule affinity-regulating kinase 4 (MARK4)	MARK4	Induces inflammation and oxidative stress via IKKα/NF-κB signaling Activates JNK1 and inhibits the p38MAPK pathway	Obesity, T2D	5ES1	Liu et al. (2016)
4.	Protein kinase C (PKC)	PRKCH	Activates JNK or IKK subsequently causing inhibitory phosphorylation of the IRS.	Diabetes associated complications such as cardiomyopathy, retinopathy and nephropathy	2FK9	Nandipati et al. (2017)

Continued

Table 1 Human kinases involved in the pathogenesis of diabetes mellitus and its associated complications.—cont'd

S. No	Kinase	Gene	Pathway/downstream targets	Implications in diseases	Structure	References
5.	Double-stranded RNA dependent protein kinase (PKR)	EIF2AK2	Activation of JNK and disruption of IRS1/PI3K/Akt signaling pathway	T2D	Domain structure (33 – 103) (2DIX)	Chen et al. (2014a)
6.	Rho-kinase 1 (ROCK1)	ROCK1	Regulates PI3K associated IRS-1 signaling and phosphorylation of Akt	T2D associated diabetic nephropathy and retinopathy	Kinase domain and AGC-kinase C-terminal domain (6–415) (2ETK) RhoBD domain (947–1015) (1S1C)	Sung et al. (2018) and Lee et al. (2009)
7.	Phosphoinositide 3-kinases (PI3Ks)	PIK3R1	PI3K/AKT pathway	Diabetic nephropathy, neuropathy, encephalopathy, vasculopathy and cardiomyopathy	C-Terminal SH2 Dimain (617–224) (1PIC)	Huang, Liu, Guo, and Su (2018)
8	Sphingosine kinase (SphK)	SPHK	Akt signaling pathway	Diabetic nephropathy and cardiomyopathy	3VZB	Ng, Wadham, and Sukocheva (2017)

Saha, Coughlan, Valentine, & Ruderman, 2014). MAPK is known for regulating various cellular processes including inflammation through the activation of specific Ser/Thr kinases (Cargnello & Roux, 2011). Reduced insulin sensitivity is observed in the mice model deficient in MAPK signifying its important role in the pathogenesis of T2D (De Boer et al., 2014). Various protein kinases including AMPK and MAPK that plays important role in insulin sensitivity and thus the etiology of diabetes are discussed below in detail. The identification of specific protein kinases regulating the phosphorylation of target proteins responsible for inflammatory responses and insulin resistance will help in the development of therapeutic strategies to combat insulin-resistant diabetes and its associated complications.

2.1 AMP-activated protein kinase (AMPK)

AMPK is a heterotrimeric protein composed of α subunit possessing catalytic activity, β and γ subunit which are regulatory, and are important in providing stability to the complex. AMPK belongs to the serine/threonine kinase family and plays crucial role in glucose homeostasis (Towler & Hardie, 2007). AMPK gets activated due to depleting cellular energy levels and its activation is followed by alteration in several metabolic pathways to restore ATP (Carling, 2004). The mechanism of AMPK activation involved both phosphorylations at Thr172 residue of α subunit by upstream AMPK kinases (AMPKKs) and allosteric modification by AMP (Woods et al., 2003). The upstream kinases of AMPKs include liver kinase B1 (LKB1), the calcium/calmodulin-dependent protein kinase kinase β (CaMKKβ), and the transforming growth factor-beta-activated kinase 1 (TAK1), etc. (Willows et al., 2017). With the rise in intracellular Ca^{2+} and cAMP levels, these upstream kinases phosphorylate Thr172 of AMPK thereby activating it and further resulting in improved glucose uptake (Fig. 1).

The role of AMPK in glucose metabolism has been evidenced by several studies. AMPK activation enhances glucose transport in mice skeletal muscles involving the translocation of glucose transporter, GLUT4, whereas in $H\text{-}2K^b$ skeletal muscles GLUT1 and GLUT4 were translocated to the plasma membrane (Treebak et al., 2006). Along with stimulating glucose transport, AMPK is involved in elevating muscle fatty acid oxidation. AMPK activation causes phosphorylation of acetyl-CoA carboxylase (ACC) thereby declining its activity. The inactivation of ACC leads to reduce malonyl-CoA levels in muscles nullifying its suppressive effects on

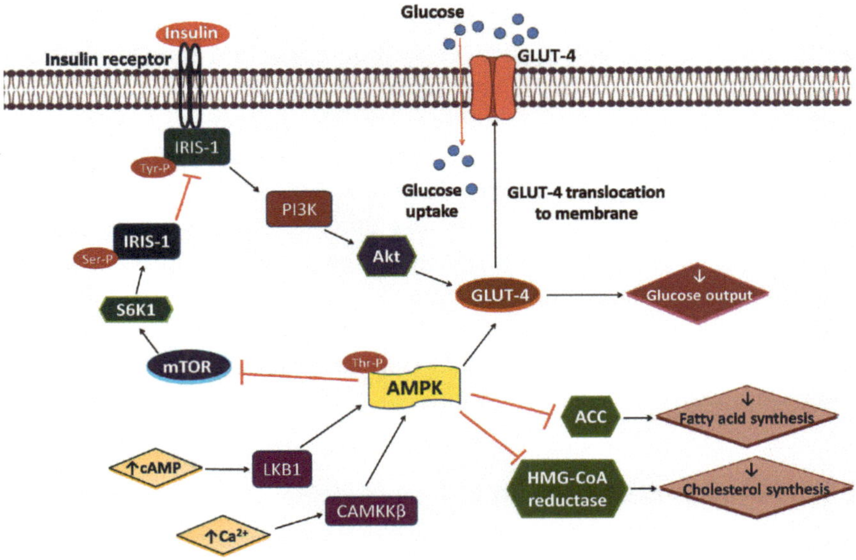

Fig. 1 Role of AMPK in glucose uptake and insulin sensitivity. The upstream kinases of AMPK include LKB1, CaMKKβ and TAK1. With the rise in intracellular Ca^{2+} and cAMP levels, these upstream kinases phosphorylate Thr172 of AMPK thereby activating it. Activated AMPK improves the glucose uptake either by phosphorylation of Akt or by increasing the GLUT-4 expression and translocation to the plasma membrane. AMPK activation also improves insulin sensitivity by impeding the negative feedback loop of IRS1 through inhibition of mTOR and its substrate S6K1. AMPK regulates fatty acid and cholesterol synthesis by modulating the activity of metabolic enzymes including HMG-CoA reductase and ACC. This plays a crucial role in fatty liver disease.

carnitine palmitoyltransferase 1 (CPT-1). The resultant enhanced CPT-1 activity increases the fatty acid uptake by mitochondria for oxidation. The AMPK stimulated clearance of myocellular fat improves insulin resistance (Kjøbsted et al., 2018; Musi, 2006). A similar mechanism is followed in the liver where AMPK activation enhances fatty acid oxidation and reduces cholesterol and triglycerides synthesis. Patients with diabetes, in general, display abnormally elevated cholesterol or fats (lipids) level in the blood which makes them more prone to cardiovascular diseases (Foretz, Even, & Viollet, 2018).

AMPK via adipocytokine signaling plays a crucial role in the pathogenesis of insulin resistance and T2D. The adipocyte-derived peripheral hormone leptin, on one side restricts the hypothalamic mediated AMPK to cut down food intake and regulate body weight, whereas on the other side it enhances the AMPK activation in peripheral tissues to advance lipid oxidation and drain triglyceride reserves (Foretz et al., 2018).

Hence, glucose and energy homeostasis necessitates the suitable balance between AMPK activation in central and peripheral tissues (Tom et al., 2014). Another adipokine namely Adiponectin (AD) marks the relevance of AMPK in cardioprotective activities. Activation of AMPK by adiponectin enhances cell viability in heart ischemia via instigating autophagy in cardiac muscle cells and impairing ER stress. AD phosphorylation carried out by AMPK activation further phosphorylate and activate endothelial nitric oxide synthase (eNOS) resulting in enhanced NO production, which accounts for several AMPK mediated cardioprotective effects (Potenza et al., 2019).

AMPK have been reported to be involved in the synthesis and release of insulin by pancreatic β cells, and release of glucagon from pancreatic α cell, regarding glucose concentration (Da Silva Xavier, 2012a). AMPK suppresses the phosphorylation of mTOR, which otherwise is linked with obesity pathogenesis and insulin resistance development in abnormal metabolic conditions. By modulating mTOR signaling, AMPK suppresses insulin resistance and T2D (Fig. 1). The AMPK mediated mTORC1 disruption shows a significant reduction in inflammation and insulin resistance via suppressing JNK/NFκB signaling in obese individuals (Nandipati et al., 2017). AMPK has been a target of glucose-reducing drugs such as metformin and thiazolidinediones that have been shown to prevent the progression to T2D by improving insulin sensitivity (Musi, 2006). These studies suggest the possible role of AMPK in obesity, insulin resistance and diabetes.

2.2 IκB kinase (IKK)

IKK is involved in the propagation of cellular responses to inflammation. The IKK enzyme complex belonging to the family of transferases is composed of three component subunits, IKK-α, IKK-β and IKK-ε, each of which are encoded by a distinct gene. The IKK-α and IKK-β subunits are both catalytically active, each bearing a kinase domain whereas IKK-ε component serves a regulatory function (Karin, 1999). The transcription factor NF-κB plays a key role in mediating inflammatory responses. It has been widely reported that the progression of NF-κB signal transduction pathway is through the IKK complex. NF-κB resides in the cytoplasm in an inactive form due to the masking of nuclear localization sequence/signal (NLS) via IκBα. For NF-κB activation, IKK phosphorylates IκBβ leading to its proteasomal degradation via ubiquitination, rendering NF-κB free for translocation into the nucleus and activating the expression of several genes (Hayden & Ghosh, 2012).

A study shows the activation of pathways with IκB kinase β (IKKβ) and NF-κB in insulin-sensitive tissues of obese and HFD animals. Their activation corresponds to the development of IR, implying the role of IKKβ/NF-kB pathway as a molecular mediator of insulin resistance in obese and T2D patients (Shoelson, Lee, & Yuan, 2003). IKK plays a crucial role in mediating insulin resistance by phosphorylating insulin receptor substrate-1 (IRS-1). The serine phosphorylation of IRS-1 further inhibits tyrosine phosphorylation of IRS-1 causing impairment in the activation of downstream PI3K cascade. Thus, IKK has a direct contributing role in IR, which marks the onset of T2D (Gao et al., 2002). Supporting the same, another study demonstrated the integration of IKKβ and JNK1 signals in the activation of Ribosomal protein S6 kinase beta-1 (S6K1), which in turn phosphorylate IRS-1. The serine-mediated S6K1 phosphorylation regulates the activity of histone deacetylases (HDACs) in the nucleus. IKKβ specifically enhances HDAC3 activation via NF-κB and JNK1 inhibits SIRT1. The mechanism suggested the role of IKKβ/JNK1 in progression of T2D from many risk factors including inflammation, free fatty acids or fatty acid derivatives (DAG or ceramide), ectopic fat (lipotoxicity), ER stress and oxidative stress (Ye, 2013).

Another study demonstrated that the acute activation of pro-inflammatory protein NF-κB, along with its upstream activator IKKβ, in the mediobasal hypothalamus, specifically in the hypothalamic pro-opiomelanocortin neurons, swiftly increases the blood pressure developing obesity-related hypertension as a result of hypothalamic inflammation. IKK-β/NF-κB/TNF-α –mediated hypothalamic inflammation can potentially lead to obesity-induced CVD, insulin resistance and T2D (Nieto-Vazquez et al., 2008; Purkayastha, Zhang, & Cai, 2011). Also, IKKβ/NF-κB pathway was reported to disrupt leptin signaling along with insulin signaling (Zhang et al., 2008). NF-κB and TNF-α mediated IKK-β activation is linked with endothelial dysfunction, inflammation, and insulin resistance in T2D murine hearts. The restricted activity of IKK-β preserves coronary arteriolar vasodilation along with the prevention of insulin resistance in diabetic murine hearts (Yang et al., 2009). IKK stimulated TNF-α activation in a p38 MAPK-dependent manner impair insulin signaling cascade by phosphorylating IRS-1 and subsequently inhibiting the activation of PI3K and Akt, leading to insulin resistance in skeletal muscles by restricting GLUT4 translocation (De Alvaro, Teruel, Hernandez, & Lorenzo, 2004; Nieto-Vazquez et al., 2008).

Several studies with IκBβ deficient mice showed a significant escalation in adiponectin content and enhanced hepatic insulin sensitivity

(Nandipati et al., 2017). IκBβ knockdown enhances local insulin sensitivity in hepatic cells, whereas, in myeloid cells of mice, it provides overall protection against insulin resistance (Arkan et al., 2005). The role of IKKβ in modulating IRF3 activity to preserve insulin sensitivity and prevent hepatic steatosis was observed. In normal conditioned hepatocytes, IRF3 interacts with the kinase domain of IKKβ thus blocking further activation of downstream signaling; whereas, in hepatocytes of an HFD induced fatty liver, IRF3 levels are significantly reduced and in the absence of IRF3, IKKβ gets activated, relaying signals to downstream mediators of inflammation. Thus, inhibition of IKKβ/NF-κB signaling via IRF3 resulted in mitigating hepatic inflammation, insulin resistance and hepatic steatosis (Wang et al., 2014). The obesity mediated overexpression of IRF3 resulted in the induction of insulin resistance whereas IRF3-deficient mice showed improved insulin sensitivity (Kumari et al., 2016). Recent studies report that inhibition of IKK by using SC-514 or an NBD (NEMO-binding domain) peptide, suppresses activation of NF-κB, thereby protecting against long terms diabetic complications like nephropathy, atherosclerosis and peripheral nerve dysfunction (Negi & Sharma, 2015; Oguiza et al., 2015).

In addition to the catalytic subunit, several recent studies report the role of the regulatory subunit, IKK-ε in insulin resistance. IKK-ε is the key mediator of insulin resistance in the hypothalamus of obese mice and its hypothalamic inhibition reverses obesity, leading to improved energy and glucose metabolism (Weissmann et al., 2014). IKK-ε deficient mice were secured from HFD-induced obesity, steatosis and chronic liver inflammation and had enhanced insulin sensitivity (Kumari et al., 2016). These results and literature suggest that IKK plays a crucial role in the development of insulin resistance leading to T2D and associated metabolic disorders.

2.3 Microtubule affinity-regulating kinase 4 (MARK4)

MARK4 is a human gene belonging to the family of serine/threonine kinases (Anwar et al., 2020; Drewes, Ebneth, Preuss, Mandelkow, & Mandelkow, 1997) This protein phosphorylates microtubule-associated proteins (MAP) and governs the transition between stable and dynamic microtubules causing a reorganization of microtubules into bundles (Naz et al., 2018; Peerzada et al., 2020). MARK4 is affiliated with the centrosome throughout mitosis and facilitates several dynamic biological functions like cell division, cell cycle progression precisely at G1/S transition, cell polarity fate determination, and cell shape alterations (Drewes et al., 1997; Rovina et al., 2014). The encoded protein plays an important role in cilium assembly by initiating the axoneme extension (Kuhns et al., 2013). Apart from reducing neuronal cell survival, MARK4 is required for maintaining energy homeostasis by regulating satiety and the metabolic rate (Sun et al., 2012). MARK4 is a negative regulator of the mammalian target of rapamycin complex 1 (mTORC1) where it selectively inhibits mTORC1 activation by Rag GTPases (Li & Guan, 2013).

More recently, MARK4 role in diabetes has surfaced. MARK4 is a member of the AMPK-related kinases involved in the regulation of glucose homeostasis and energy metabolism (Khan et al., 2017). MARK4 suppression enhances insulin sensitivity as observed by lower insulin serum levels after overnight fasting and augments insulin tolerance evidenced by an insulin tolerance test. Accordingly, MARK4 deficiency alleviates diet-induced insulin-resistance by significantly elevating insulin-stimulated AKT phosphorylation in all the major metabolic tissues. MARK4 deficient mice display a highly improved trait in glucose excursion and hence were secured from insulin resistance linked to the obesity onset (Sun et al., 2012). MARK4 knockdown efficiently enhances glucose homeostasis by increasing AMPK expression and activity. Also, low levels of AMPK phosphorylation in skeletal muscles has been reported in T2D patients (Sun et al., 2012).

In another study, it was observed that MARK4 induces oxidative stress and adipose inflammation via IKKα/NF-κB signal pathway. In individuals with obesity, oxidative stress hinders uptake of glucose and reduces insulin sensitivity. Obesity, insulin resistance and T2D have a close association with chronic inflammation induced by MARK4 (Liu et al., 2016). MARK4 is well supported for its role of fat deposition in testosterone-deficient human males. Low levels of testosterone are linked to obesity, metabolic syndrome and insulin sensitivity in men (Wang et al., 2017).

Other findings along the same line revealed that MARK4 enhances phosphorylation of JNK1 and increases Bax/Bcl-2 ratio leading to intensified caspase3 activation resulting in mitochondrial-mediated apoptotic signaling. Adipocyte apoptosis caused by MARK4 via stimulating the JNK1 pathway is crucial for IR, hepatic steatosis and infiltration of macrophages into adipose tissue in humans as well as in mice. MARK4 deficient mice showed reduced fat content and triglycerides in blood as well as decreased cholesterol levels, all of which are indicative markers of obesity and associated metabolic complexities. MARK4 aggravates adipogenesis via activating JNK1 and inhibiting the p38 mitogen-activated protein kinase (p38MAPK) pathways. MARK4 enhances expression of SREBP-1c, PPARγ and FAS required for adipocytes lipogenesis. The resultant increase in lipid accumulation leads to inflammation, impaired mitochondrial function and pathogenesis of insulin resistance in insulin-responsive tissues in humans with obesity. Additionally, a JNK inhibitor SP600125 approved the involvement of MARK4 in adipogenesis (Feng, Tian, Gan, Liu, & Sun, 2014).

Most recent data demonstrates MARK4 playing a crucial role in the regulation of NLRP3 inflammasome activation resulting in endothelial inflammation and cardiovascular complications which in turn are linked to vascular disorders in diabetic patients (Wang et al., 2020). Thus, MARK4 is involved in the regulation of both glucose homeostasis and energy metabolism. Drugs that specifically inhibit MARK4 activity are likely to be useful in the treatment of diabetes and obesity (Gupta et al., 2019; Khan et al., 2017; Parveen, Khan, Ali, Hassan, & Ahmed, 2018; Roy et al., 2020).

2.4 Protein kinase C (PKC)

PKC is a serine/threonine-related protein kinase, with 12 isoforms in humans which are broadly divided into three major groups: a conventional, novel, and atypical, depending on the requirement of their secondary messengers. The conventional PKCs (α, β_I, β_{II}, and γ) require Ca^{2+} and phospholipid for its activation. Novel PKCs (the δ, ε, η, θ and μ) are Ca^{2+} independent and phospholipid dependent. Atypical PKCs (ζ and ι/λ) require neither Ca^{2+} nor phospholipid (Singh, Cummings, Pantos, & Singh, 2017). The activation of specific PKC in a specific tissue is associated with insulin resistance mediated by fatty acid metabolites (Geraldes & King, 2010).

It has been widely reported that diabetic or hyperglycemic conditions lead to the accumulation of vascular diacylglycerol which subsequently activates PKC causing cardio- and microvascular defects (Das Evcimen & King, 2007). Activation of specific PKC isoforms is possibly involved in the execution of specific vascular pathologies, for example, PKC-α, -β1, -β2, PKC-δ, and –ε have been involved in ECM accumulation in glomerular mesangium and tubule-interstitium which is supposed to be a hallmark in a diabetic kidney patient (Graham et al., 2011). PKC-α, -β, PKC-δ, are involved in the induction of insulin resistance via mediating inhibitory Ser IRS-1 phosphorylation (Fig. 2). PKCs enhance growth factors like CTGF, TGF-*β*, VEGF leading to mesangium expansion and nephromegaly which eventually results in glomerular sclerosis due to matrix deposition and DN (Nandipati et al., 2017).

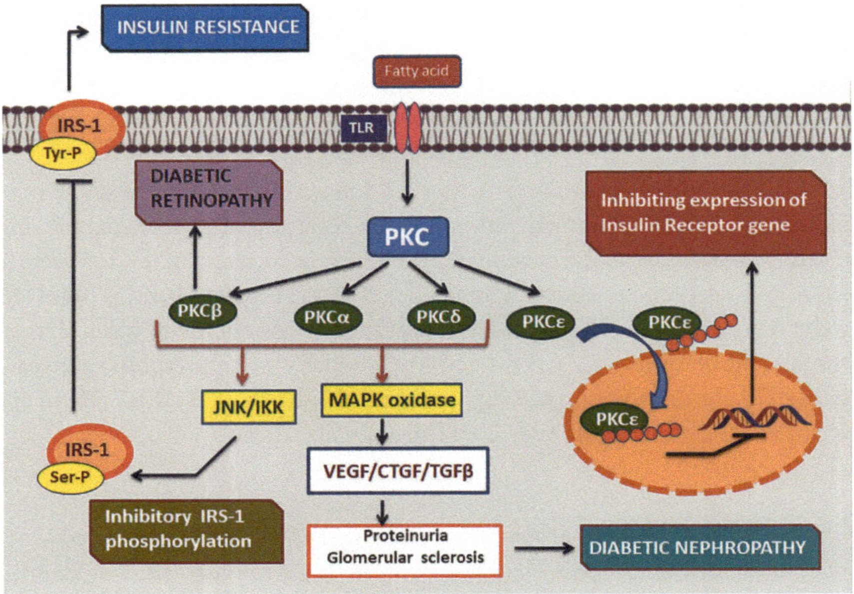

Fig. 2 PKC mediated insulin resistance. Fatty acids cause the activation of PKC α, β, δ that further activates JNK or IKK leading to inhibitory Ser phosphorylation of IRS-1. It also increases connective tissue growth factor and TGF-*β* resulting in mesangium expression and nephromegaly which eventually leads to glomerular sclerosis due to matrix deposition and DN, whereas the FA-activated PKCε via interacting with F-actin gets translocated to the nucleus where it inhibits the transcription of insulin receptors.

Similarly, PKC-β2 is responsible for monocyte activation followed by transformation into macrophages which are involved in atherosclerosis and inflammatory process (Geraldes & King, 2010). Also, activation of PKC-α, -β1, -β2, -δ and cPKC-β (conventional PKC-β), nPKC-δ (novel PKC-δ) are key players of endothelial cells and smooth muscle cells dysfunction, respectively (Das Evcimen & King, 2007; Idris, Gray, & Donnelly, 2001). These perturbations in vascular cell homeostasis notably lead to the development of macrovascular complications (cardiomyopathy and atherosclerosis) and microvascular complications (retinopathy and nephropathy) (Das Evcimen & King, 2007; Geraldes & King, 2010).

Diabetic patients have higher risks of developing congestive heart failure with active PKC (Idris et al., 2001). In a hyperglycemic environment, PKC-α, -β1, and -β2 expression were higher in the heart and aorta (Simran, Grewal, Arora, & Singh, 2019). PKC-β2 overexpression resulted in cardiomyopathy, assisting its role in myocardial hypertrophy and fibrosis

(Geraldes & King, 2010). PKC-α–knockdown mice displayed enhanced cardiac contractility and were less prone to cardiac failure (Min et al., 2017). Another study with transgenic PKC-β2 mice revealed the significant enhancement in connective tissue growth factor (CTGF) expression, which is likely to be a key contributor to cardiac fibrosis. PKC-δ inhibition aggravates CTGF expression leading to cardiac fibrosis in angiotensin II presence (Geraldes & King, 2010). Contrastingly, PKC-δ activation during myocardial ischemia/reperfusion injury is associated with pro-apoptotic signaling cascade in myocardiocytes (Jiang et al., 2017). In addition, cardiac-specific PKC-ε activation was found to prevent pathogenetic changes in the heart that have been caused due to diabetes which includes a ventricular function as well (Malhotra et al., 2005). Furthermore, PKC activation in diabetic patients showed impaired insulin-stimulated glucose uptake by cardiomyocytes by phosphorylating serine/threonine residues on the insulin receptors or their substrates (Idris et al., 2001). Loss of insulin action is linked with the reduced hypoxia-inducible factor-1α expression that further affects VEGF expression in the myocardium. Impaired VEGF signaling decreases collateral circulation formation which justifies large heart size and congestive cardiac failure after myocardial infarction in diabetic patients (Geraldes & King, 2010).

In diabetic retinopathy, activation of PKC-α, -β -δ and –ε are reported, where PKC-δ have been found to play a key role in inducing retinal pericyte apoptosis and further forming acellular capillaries (Aiello, 2002; Geraldes et al., 2009; Kim, Kim, Jun, Yu, & Kim, 2010). The potential contribution of PKC-β in atherosclerotic plaque formation is evident where PKC-β deficient mice showed reduced atherosclerosis via restricting early growth response-1 (Erg-1) protein that regulates expression of vascular cell adhesion molecule and activity of matrix metalloproteinase-2 favorably in endothelial cells (Geraldes & King, 2010; Rask-Madsen & King, 2005).

PKC activation is linked with alterations in the activity of certain enzymes like Na^+–K^+–ATPase, cPLA2, PI3Kinase and MAP kinase. Reports suggested that inhibition of Na^+–K^+–ATPase in diabetic patients was due to PKC activation and enhancement of phospholipase A2 (cPLA2) levels in the cytoplasmic matrix (Das Evcimen & King, 2007).

To conclude, PKC isoforms cause vascular dysfunctions and pathologies in diabetes. The DAG-PKC signaling cascade is one of the most considered pathways in cellular signaling induced by diabetes. The above literature shows how PKC actions and related signaling changes in the development of diabetic vascular complications. The hyperglycemic conditions disturb arteries and vascular cell functions. Clinical trials have been carried out for PKC β inhibitor as therapeutic management against diabetic vascular complications. More clinical trials targeting multiple PKC isoforms are required to test the effectiveness in delaying, halting and even reversing diabetic vascular complications.

2.5 Double-stranded RNA-dependent protein kinase

Protein kinase R (PKR) is a 551 amino acid long serine/threonine kinase, a gene in humans is located on chromosome 2 which is activated during various stress conditions. PKR's significance can be seen in central cellular processes like mRNA translation, transcriptional control, apoptosis regulation and proliferation (Gal-Ben-Ari, Barrera, Ehrlich, & Rosenblum, 2019). The first clone of PKR was achieved at the Pasteur Institute in 1990, and was also known as Protein kinase RNA-activated, interferon-induced, double-stranded RNA-domain kinase or eukaryotic translation initiation factor 2-alpha kinase 2 (EIF2AK2) (Watanabe, Imamura, & Hiasa, 2018).

The structural composition of PKR consists of two functional domains: an N-terminal dsRNA binding domain (dsRBD) and a C-terminal kinase domain. The N-terminal dsRBD is composed of two tandem repeats of a conserved dsRNA binding motif (dsRBM1 and dsRBM2) which are separated by a 23 amino acid linker. Both dsRBMs are essential for high-affinity interaction with dsRNA. There lies a flexible linker between N-terminal and C-terminal kinase domains. The C-terminal kinase domain contains a small and a large N- and C-terminal lobes which undergo a typical fold between the β-sheet of the N-terminal lobe and α-helical of the C-terminal lobe forming a catalytic domain where dimerization takes place. The C-terminal kinase domain provides major phosphorylation sites (Nallagatla, Toroney, & Bevilacqua, 2011).

PKR exhibits an important role in the development of insulin resistance in peripheral tissues in T2D. PKR is activated via glucolipotoxicity and pro-inflammatory cytokines leading to inhibition of pancreatic β-cells proliferation. PKR activation is followed by downregulation of cyclin D1 and D2 which are positive regulators of cell cycle and meanwhile upregulation of p27 and p53 which are negative controllers of cell cycle resulting in PKR-induced cell cycle arrest at G1 phase. The anti-proliferative effects of PKR were supported by proteasome-dependent degradation of cyclin D1 and D2 which is responsible for its low levels. Also, PKR activation is responsible for completely abolishing the pro-proliferative effects of

Insulin-like growth factor I (IGF—I) on pancreatic β-cells along with disrupting the IRS1/PI3K/Akt signaling pathway due to JNK activation. These conclude the participation of PKR in the pathogenesis of T2D (Chen et al., 2014a). PKR is involved in inflammation and immune dysfunction via regulating its mitogen-activated protein kinases, IRF 3, NF-κB and apoptosis (Kang & Tang, 2012).

PKR exhibits a major role in integrating nutrient and stress signaling, metaflammasome development, modulating insulin responsiveness and glucose homeostasis during obesity in mouse models and obese humans (Nakamura, Arduini, Baccaro, Furuhashi, & Hotamisligil, 2014). PKR is responsive towards both nutrient signals and ER stress and modulates the activity of kinases with inflammatory actions like the JNK to regulate insulin activity and other chronic metabolic pathologies. To control insulin signaling in a stressed condition, JNK cause serine phosphorylation of IRS1 and not the normal tyrosine phosphorylation. PKR stimulated activation of JNK also induces serine phosphorylation of IRS1 (Dhar, Reddy, Bhat, & Dhar, 2015).

PKR specifically targets and alters IRS1 and plays a critical role in the nutrient/pathogen-sensing interface (Marsollier, Ferré, & Foufelle, 2011). PKR causes inhibitory phosphorylation of IRS at site Ser312 resulting in activation of Foxo1 followed by the upregulation of protein IRS2 expression level. PKR knockdown provides shelter against insulin resistance and diabetes (Dhar et al., 2015). Obesity-induced via diet or by genetic means featured PKR activation in hepatic and adipose tissues and PKR deficiency mitigates metabolic deterioration because of the surplus amount of nutrient or energy excess in mice (Marsollier et al., 2011). PKR-knockout mice exhibited lower levels of fasting plasma glucose while a significant improvement was seen in insulin activity and insulin-stimulated Akt phosphorylation contrary to wild type control mice (Dhar et al., 2015).

Since all the above studies suggest the role of PKR activation in the pathogenesis of T2D, inhibition of PKR activation can be an effective therapeutic target for the treatment of insulin resistance and T2D. The same has been reported in a study where two structurally distinct small-molecule inhibitors of PKR were revealed to curtail inflammation of adipocyte tissues, enhance insulin sensitivity and improve glucose intolerance in mice model of obesity and insulin resistance, whereas in lean mice PKR inhibition did not affect either glucose tolerance or insulin sensitivity (Nakamura et al., 2014).

One of the prime issues in this regard is that the findings concerned with the role of PKR in T2D are, as yet, not complete, and new developments are emerging gradually. It appears as if it's still going to take some refined researches before we obtain a consolidated story of the role of PKR in metabolic aberrations and cardiovascular abnormality.

2.6 Rho-kinase 1 (ROCK1)

ROCK1 is a protein serine/threonine kinase with a molecular mass of 158 kDa. Its homodimeric structure consists of a catalytic kinase domain at its N terminus, Rho-binding domain and a pleckstrin-homology domain at its C terminus. The protein gets activated when the GTP-bound form of RhoA binds to its Rho-binding region disrupting the auto-inhibition interaction between the N-terminus and the C-terminus (Hartmann, Ridley, & Lutz, 2015). The ROCK1 signaling plays important roles in diabetes, neurodegenerative diseases and pulmonary hypertension.

ROCK1 is essential for maintaining normal insulin secretion in pancreatic β-cell. ROCK1 deletion led to a notable reduction in insulin secretion resulting in hyperglycemia in mice given a regular chow diet. Similarly, *in vitro* studies revealed that ROCK1 inhibition resulted in low levels of insulin secretion with no change in either β-cell death or proliferation. Additionally, tolerance to glucose was impaired in mice which were deficient in ROCK1. The role of ROCK1 becomes indispensable in glucose homeostasis as it was found that upon insulin stimulation and consequent inhibition of ROCK1 expression in INS-1 cells leads to impairment of factors involved in proximal insulin signaling like p-IRS2, p-PI3K and p-Akt (Sung et al., 2018).

It has been illustrated that peripheral ROCK1 governs insulin sensitivity and glucose metabolism whereas hypothalamic ROCK1 is prevalent in controlling body-weight balance and feeding habits. Hepatic ROCK1 deletion restrains obesity, insulin resistance, and fatty liver in HFD mice. Intriguingly ROCK1 deficient liver exhibit significant enhancement in FGF21 levels where faulty hepatic autophagic signaling exists. Moreover, the interaction between ROCK1 and LXR leading to enhancement in LXR serine phosphorylation showed the ROCK1 importance in LXR action. These findings suspect the essentiality of ROCK1 activation in regulating lipid and glucose homeostasis associated with LXR and FGF21 activation by autophagy (Bellomo, 2018).

ROCK1 is a prime modulator in functionally associating S1PR2 and mitochondrial fission via activating Drp1. S1PR2 antagonists enhances high glucose-induced fission and dysfunction of mitochondria via

RhoA/ROCK1/Drp1 signaling pathway in human renal glomerular endothelial cells, implying the importance of the S1PR2/ROCK1 pathway in high glucose milieu (Chen et al., 2019). Mitochondrial dysfunction is critically involved in the progression of microvascular complications of diabetes; including DN. ROCK1 triggers hyperglycemia-induced mitochondrial fragmentation by phosphorylating Drp1 at Serine 600 residue (Wang et al., 2012). High glucose-mediated activation of ROCK1 stimulates endothelial-to-mesenchymal transition leading to increased endothelial permeability. Hence targeting ROCK1 as a therapeutic approach can prevent glomerular endothelial malfunction and albuminuria in DN progression (Peng et al., 2016).

The involvement of the ROCK pathway has been notably observed in the progression of diabetic retinopathy (proliferative vitreoretinopathy and retinal neovascularization). ROCK activation in endothelial cells or hyalocytes is carried out by enhanced levels of glucose, TNF-α and TGF-β in diabetic serum or vitreous. ROCK activation causes eNOS disruption leading to endothelial damage. Furthermore, ROCK via enhancing ICAM-1 expression and stimulating ezrin, radixin, and moesin (ERM) activation in endothelial cells leads to stable leukocyte adhesion. ROCK is also involved in major functions like leukocyte adhesion, endothelial migration, and contraction of the proliferative membrane by mediating phosphorylation of myosin light chain in DR (Arita, Hata, & Ishibashi, 2010).

The specific role of the adipose ROCK1 isoform was demonstrated where ROCK1 deficient mice were observed to be protected from diet-induced insulin resistance by enhancing IRS-1/PI3K/Akt signaling. The causative reason for these effects was the inhibitory action of ROCK1 on insulin receptor which was notably observed when tyrosine phosphorylation of the insulin receptor was abrogated on the re-exposure of ROCK1 in ROCK1 deficient MEF cells. Hence adipose-specific ROCK1 disruption escalates tyrosine phosphorylation of insulin receptors in adipose tissue and enhances insulin sensitivity in HFD obese mice. However, adipose ROCK1 mediated increased insulin action had minimal effect on whole-body glucose metabolism (Lee et al., 2014).

ROCK1 is a key regulator in the pathogenesis of insulin resistance on glucose disposal in skeletal muscle of obese humans with T2D. ROCK1 activation via insulin administration in skeletal muscles of obese T2D patients was substantially impaired when compared to skeletal muscle of lean individuals since the interrelation between insulin-stimulated ROCK1 activity and rate at which glucose is disposed in lean individuals is absent in obese or T2D patients (Chun et al., 2011). The ROCK1 deficient mice exhibited a reduction in IRS-1 signaling linked with insulin-induced PI3K and also leading to a subsequent reduction in phosphorylation of Akt, AS160, GSK3, p70S6K and S6 in skeletal muscle (Lee et al., 2009). These studies suggest the involvement of ROCK1 in insulin resistance, glucose homeostasis, T2D and associated metabolic disorders. Pharmacological inhibition of ROCK1 activity is likely to be useful in the treatment of diabetes and its associated metabolic disorders.

2.7 Phosphoinositide 3-kinases (PI3Ks)

PI3Ks constitute a protein family typified by the activity of phosphorylation of lipids and proteins. Mammalian PI3Ks are classified into three groups depending on their structural characteristics and substrate preferences. Members of class I and class II are the most crucial in the insulin signaling pathway. The class I PI3Ks are heterodimers with a regulatory subunit and catalytic subunit (Jean & Kiger, 2014).

2.7.1 PI3Ks and insulin gene expression

The increase in insulin gene expression in response to elevated glucose concentration levels takes place in a PI3K-1a-dependent manner. Insulin gene expression is under the control of many transcription factors and out of them the transcription factors pancreatic duodenal homeobox-1 (PDX-1) and FoxO-1 are activated by insulin in a PI3K-1a-dependent manner. PDX-1 is translocated from the cytoplasm to the nucleus to bind to the upstream region of the insulin gene, A3 box. This binding is crucial in activating insulin gene expression which is dependent on glucose concentration and insulin response. The insulin-stimulated PDX-1 gene expression further implies that insulin gene expression is regulated in a feed-forward-mechanism with PI3K-1a and insulin (Da Silva Xavier, 2012b).

2.7.2 PI3K/AKT pathway in skeletal muscles

Under normal conditions in skeletal muscles, insulin stimulates PI3K/AKT pathway leading to glucose transport via AS160, glycogen synthesis by inactivating GSK3 and protein synthesis via S6K1 action and 4E-BP1 inhibition. Free fatty acid accumulation in skeletal muscles causes the formation of long-chain fatty acid CoAs (LCACoAs) that leads to either TAG synthesis or mitochondrial fatty acid oxidation which ultimately results in inhibition of PI3K/AKT signaling leading to skeletal muscle insulin resistance (Huang et al., 2018).

2.7.3 PI3K/AKT pathway in the brain

Under normal conditions in the brain, ligand binding activates PI3K/AKT signaling to restrain food intake in the hypothalamus along with inhibiting hepatic glucose production and lipolysis by regulating liver and adipose tissue. Whereas when the ligand is unable to cross the blood-brain barrier, and impairment of hypothalamic AKT activation owing to ER stress takes place, both contribute to the development of insulin resistance in the brain. Insulin resistance in the brain subsequently leads to aberrations in glucose metabolism, like hepatic glucose production and glucose translocation in organs which are insulin sensitive terminating in systemic insulin resistance (Huang et al., 2018).

2.7.4 PI3K/AKT pathway in the pancreas

The pancreatic β cell function and insulin secretion are regulated by insulin via the PI3K/AKT pathway. PI3K/AKT pathway activation stimulates β cells to secrete insulin. The constitutive or upregulated expression of pancreatic β cells AKT was found to enhance β cell mass, cell number and cell size. These effects were regulated by downstream targets of AKT like FoxO1, GSK3 and mTOR1 further proving the role of AKT in pancreatic cells. In conditions of insulin resistance, β cell activity is enhanced to supplement more insulin release for maintaining normal glucose levels which result in hyperinsulinemia development whereas β cell dysfunction impairs glucose tolerance leading to T2D development (Kahn et al., 1993). In obese individuals and T2D patients, the insulin-stimulated PI3K/AKT pathway is halted reducing insulin secretion and β cell activity (Huang et al., 2018).

2.7.5 PI3K and diabetic cardiomyopathy

PI3K γ isoform expression is significantly increased in diabetic heart mediating diabetic cardiomyopathy. Pharmacological inhibition of PI3K γ kinase activity can retain cardiac function after pressure overload. PI3K γ inhibition reduced inflammatory infiltration to the heart. PI3K γ was also responsible for diabetic-induced cardiac alterations like fibrosis and inflammation, along with diastolic and systolic dysfunction (Maffei, Lembo, & Carnevale, 2018).

2.7.6 PI3K and diabetic vasculopathy

PI3K is strongly implicated in the antiatherogenic and vasodilatory effects of insulin by mediating the phosphorylation of eNOS resulting in nitric oxide (NO) release. PI3K stimulates the release of vascular tissue compounds like HO-1 and VCAM-1 which further favors the antiatherogenic activity of insulin. Moreover, in PI3K γ mice, effects like enhanced ROS vascular production via increasing the activity of pro-oxidant enzymes like Rac and NADPH oxidase leads to quenching of the vasodilator action of NO, eventually terminating in vascular inflammation and atherosclerosis. Atherosclerosis majorly contributes to vascular compromises in DM (Maffei et al., 2018). PI3K and Diabetic Neuropathy and Encephalopathy: PI3K activity was decreased in the peripheral nerves of a diabetic patient causing a reduction in the transport of neurotrophins and nerve growth factors. Deficiency of PI3K γ signaling in locus coeruleus leads to a cerebral disease namely Attention Deficit Hyperactivity Disorder (ADHD) which is closely linked with diabetes mellitus (Maffei et al., 2018).

2.7.7 *PI3K and diabetic nephropathy*

PI3K supports the translocation of GLUT1 and GLUT4 to the plasma membrane of podocytes causing glucose uptake. Disruption of the PI3K-dependent pathway leads to diabetic nephropathy even without enhanced glycemic levels. PI3K and IRS2 association is enhanced by Bone Morphogenetic Protein-7 to secure kidney cells from the deleterious effects of DN (Maffei et al., 2018). Thus, PI3K has a fundamental role in many aspects of diabetes and its associated metabolic disorders.

2.8 Sphingosine kinase (SphK)

SphK is a lipid kinase that catalyzes the ATP-dependent phosphorylation of sphingosine to S1P (Cannavo et al., 2017; Khan et al., 2020). The activity of SphK is regulated by various agonists and stimuli that facilitate its phosphorylation at Ser225 residue followed by its translocation from cytoplasm to the plasma membrane where it interacts with calcium-myristoyl switch protein 1 that further facilitates the phosphorylation of sphingosine to S1P on the plasma membrane (Gupta et al., 2020; Hatoum, Haddadi, Lin, Nassif, & McGowan, 2017; Jarman, Moretti, Zebol, & Pitson, 2010). A range of external stimuli that activate SphK1 includes transforming growth factor β (TGF-β), tumor necrosis factor α (TNF-α), a pro-inflammatory cytokine, and various growth factors with both intra- and extracellular functions (Nishino et al., 2019; Roy et al., 2020).

2.8.1 *SphK signaling and hepatic insulin resistance*

One of the key factors for the onset of T2D is hepatic insulin resistance. In obese and diabetic individuals, SphK1 activation enhances hepatic insulin signaling (Lee et al., 2015; Tao, Sifuentes, & Holland, 2014). A study shows the minimal influence of SphK1 in glucose metabolism wherein liver of HFD mice low SphK1 activity was followed by elevated liver triacylglycerol (TAG) and diacylglycerol (DAG) levels and exhibited glucose intolerance. However SphK1 overexpression decreases hepatic TAG content but demonstrated no substantial change in glucose metabolism (Kowalski et al., 2015).

SphK2, an isoform of SphK, is dominant in the liver and its upregulation has been associated with enhanced hepatic S1P expression leading to the improved metabolism of glucose/lipid in KK/Ay diabetic mice. Hepatic SphK2 overexpression activates the Akt pathway that plays a crucial role in the insulin-induced regulation of glucose metabolism (Ng et al., 2017). Endoplasmic Reticulum stress also mediates the activation of SphK2 via ATF4 (Activating Transcription Factor 4) under inflammatory conditions, leading to improve hepatic glucose intolerance and lipid abnormalities (Lee et al., 2015).

2.8.2 SphK signaling and pancreatic β-cell death

Pancreatic β-cell death induced by lipotoxicity is a major factor in reducing insulin secretion and T2D. Increased levels of sphingolipid metabolites such as ceramide within the cell induce β-cell dysfunction leading to apoptosis in diabetic fatty rodent models of T2D *in vivo* and human β-cells *in vitro* (Boslem, Meikle, & Biden, 2012; Zhu et al., 2009). Sphingolipid rheostat defines the fact that the suppression of SphK leads to enhanced ceramide levels (Newton, Lima, Maceyka, & Spiegel, 2015). In contrast to the detrimental effects of ceramide, SphK/S1P has been reported to stimulate insulin release, enhance glucose sensing and inhibit islet-cell-mediated inflammatory responses (Jessup, Bonder, Pitson, Toby, & Coates, 2012).

2.8.3 SphK1 signaling and the development of peripheral insulin resistance

Skeletal muscle insulin resistance is a general marker of T2D onset. The pharmacological inhibition of SphK reduces insulin-stimulated glucose disposal as SphK1 overexpression promotes basal and insulin-mediated glucose absorption in **C2C12** cells, and significantly reduce blood glucose in diabetic mice. SphK1 deficiency augmented the gene expression of anti-inflammatory molecules, adiponectin and IL-10. The overall insulin sensitivity was increased in adipose and muscle tissues of SphK1 deficient mice. A study on the Zucker diabetic fatty rats showed that selective inhibition of SphK1 protects adipocyte cells from lipopolysaccharide (LPS) -mediated inflammation (Ng et al., 2017).

2.8.4 SphKs in diabetes-related complications

SphK/S1P signaling has been linked to several diabetic microvascular complications, such as neuropathy, retinopathy and nephropathy. Nearly 30% of diabetic patients are reported with DN. SphK/S1P signaling is crucially associated with glomerular sclerosis and/or tubulointerstitial (renal) fibrosis. SphK1 activation and S1P production stimulate glomerular mesangial cell proliferation and increases the level of the pro-fibrotic growth factors, collagen and fibronectin synthesis (Lan et al., 2011). The progression of DN is marked by a chain of pathological actions, such as early glomerular proliferation and hypertrophy, aggregation of extracellular matrix components and renal fibrosis (Liu, 2006). The pertinence of SphK was observed in SphK-deficient mice *in vivo*, wherein SphK2-deficient mice concentrations of plasma creatinine were significantly reduced, implying that SphK2 sheltered cells from renal ischaemia whereas in SphK1-deficient mice the nephropathy conditions were further deteriorated (Yaghobian et al., 2016). SphK1 knockdown in a mouse model of polycystic kidney disease and renal injury demonstrated worsening of cytogenesis whereas SphK1 overexpression in mice model of ischaemia-perfusion injury exhibited low tubular necrosis and lesser inflammation (Ng et al., 2017; Park et al., 2012).

Overall, recent literature suggests that both SphK1 and SphK2 are involved in moderating microvascular complications, like DN, albeit to varying degrees and affecting particular targets. The chances of cardiovascular disease, atherosclerosis and heart failure are significantly higher in diabetic patients. Supporting the same, SphK1 overexpression was correlated with improved heart function in a diabetic mouse subjected to increased myocardial glycogen accumulation leading to cardiomyopathy (Karliner, 2013). The role of SphK signaling in relation to atherosclerotic complications was seen in the non-obese diabetic mouse model NOD/LtJ where S1P was found to reduce glucose-stimulated monocyte-endothelial interaction (Whetzel, Bolick, & Hedrick, 2009).

Diabetic patients generally possess myocardial ischaemia in a silent form. SphK1 activation provides a protective strategy against ischaemia-associated injury by employing protein kinase Cε-dependent mechanisms (Ng et al., 2017). Also, SphK1 plays a crucial role in sustaining blood vessel integrity and improves the cutaneous wound healing process in diabetic rats. On the other side, SphK1 signifiacnnce in vascular health has been reported where SphK1 suppression enhances angiotensin II-induced acute hypertension (Furuya et al., 2013; Yaghobian et al., 2016).

3. Therapeutic molecules targeting protein kinases to manage DM

Diabetes is emerging as a major global threat to human health. Hence, understanding the cellular mechanisms and signaling pathways that regulate the glucose homeostasis and governs normal pancreatic islet function is of utmost importance to develop effective therapeutics for diabetes. Islet transplantation with functional β-cells could be a nice strategy to overcome pancreatic insufficiency in patients; however, the lack of suitable transplantable material makes it an infeasible treatment option for a large diabetics' population. Since various protein kinases have a well-established role in glucose uptake and homeostasis, targeted inhibition of specific intracellular kinases may prove helpful in improving pancreatic β-cells function in diabetes patients. An array of small therapeutic molecules targeting protein kinases are known to fight against T2D that are listed in Table 2.

3.1 AMPK modulators

Glucagon-like peptide-1 (GLP-1) and its agonist (exenatide) has been approved to treat diabetic hyperglycemia, and also improves hepatic insulin resistance and fatty acid β-oxidation in patients with nonalcoholic fatty liver disease and nonalcoholic steatohepatitis. Exenatide activates specific kinases including PI3K and AMPK causing (i) enhanced PPARα activity and reduced levels of free fatty acids in hepatic cells; (ii) increased expression of PPARg that further sensitize hepatocytes to insulin by decreasing Ser307 phosphorylation of JNK (Svegliati-Baroni et al., 2011). Exenatide treatment improves inflammation and glucose metabolism in hepatic tissues via increased expression of Sirt1 that further activates AMPK and its downstream targets (Lee et al., 2012).

A wide range of hormones and therapeutic agents are reported that activate AMPK when administered to cells or tissues. Some act directly by binding to AMPK holoenzyme whereas other acts in an indirect manner. A-769662, belonging to the thienopyridine family, has been identified as a direct AMPK activator. Upon short-term administration in obese mice model, A-769662 causes a decrease in body weight gain, lowered G6Pase, PEPCK and FAS expression, and markedly reduced triglyceride levels in both plasma and hepatic tissues (Zhang et al., 2009). Inspite of being beneficial effects, A-769662 suffers from certain disadvantages to being effectively used in antidiabetic therapy. It exerts AMPK-independent effects

Table 2 Potential small molecules to be used as therapeutics against diabetes mellitus with their respective target protein kinases.

S. No.	Molecule	Target kinase	Therapeutic effect	References
1.	GLP-1 and Exenatide	AMPK	Reduce the levels of FFA, and improve inflammation and glucose metabolism in hepatocytes	Lee et al. (2012) and Svegliati-Baroni et al. (2011)
2.	A-769662	AMPK	Decrease body-weight gain and markedly reduce triglyceride levels in both plasma and hepatic tissues	Zhang, Zhou, and Li (2009)
3.	PT1 and C24	AMPK	Reduce lipid and glucose levels in the blood, improve glucose intolerance	Li et al. (2013) and Nandipati, Subramanian, and Agrawal (2017)
4.	Bitter melon isolated terpenoids	AMPK	Ameliorate glucose disposal	Iseli et al. (2013)
5.	Adiponectin	AMPK	Improve glucose tolerance and insulin sensitivity	Kamon et al. (2004) and Tao et al. (2014)
6.	Metformin	AMPK	Inhibits PKA dependent inhibitory phosphorylation of AMPK	Aw, Sinha, Xie and Yen (2014)
7.	Atrovastatin	IKKβ	Increase GLUT4 translocation and insulin sensitivity in adipocytes	Furuya et al. (2010)
8.	IMD-0354	IKKβ	Ameliorate insulin resistance and increase plasma levels of an anti-diabetic hormone adiponectin	Kamon et al. (2004)
9.	Salicylate	IKKβ	Prevent FFA-mediated insulin signaling impairment in hepatic tissues	Park, Wong, Guan, Oprescu, and Giacca (2007)

10.	Hypocrellin	PKC	Improve the lipid and glucose metabolism abnormalities caused by insulin resistance	Qu, Dang, and Scale (2003)
12.	1H-imidazole-4-carboxamide, 5-amino-1-[2,3-dihydroxy-4-[(phosphonooxy)methyl] cyclopentyl-[1R-(1a,2b,3b,4a)] and aurothiomalate	PKC	Improve insulin signaling with the concomitant reversal in the metabolic abnormalities including hyperglycemia, glucose intolerance, hepatosteatosis and hypercholesterolemia	Sajan et al. (2012)
13.	5Z-7-oxozeaenol	TAK1	preserve pancreatic function and delay the onset of autoimmune diabetes	Cao et al. (2015)
14.	PHA666859	p38 MAPK	Prevent the onset of diabetic retinopathy and other associated complications of diabetes in rats	Du et al. (2010)
15.	Fish oil	AMPK	Prevent inflammation and insulin resistance in white adipose tissue of mice	Bargut, Mandarim-de-Lacerda, and Aguila (2015)

with off-target activity and also shows poor oral absorption (Moreno, Knecht, Viollet, & Sanz, 2008). PT1 is another small molecule allosteric activator of AMPK identified via screening of chemical databases. It directly interacts with the autoinhibitory domain of AMPKα1 thereby relieving the autoinhibition (Zhang et al., 2009; Pang et al., 2008). The lead molecule PT1 is structurally optimized further to improve its oral absorption to identify another AMPK activator, C24 (Li et al., 2013). Both PT1 and C24 facilitate increased AMPK functional activity and ACC phosphorylation (Nandipati et al., 2017).

There is a long list of AMPK activators that acts indirectly and modulates glucose metabolism and fatty acid oxidation. This includes metformin, thiazolidinediones (TZDs), adiponectin, leptin, ciliary neurotrophic factor (CNTF), ghrelin/cannabinoids, IL-6, α-lipoic acid, SIRT1 and natural products such as epigallocatechin-3-gallate and berberine (Zhang et al., 2009). Terpenoids isolated from bitter melon elicits anti-diabetic effects by modulating the activity of CaMKKβ, an upstream kinase regulating the activation of AMPK (Iseli et al., 2013). Metformin, a well established FDA approved drug for diabetes management, activates AMPK via an increase in LKB1 function (Aw et al., 2014) and inhibition of PKA dependent inhibitory phosphorylation of AMPK (Aw et al., 2014).

3.2 IKKβ modulators

Atorvastatin treatment is shown to reduce inflammatory responses and improves insulin sensitivity by increasing the GLUT4 expression. Because IKKβ phosphorylates IRS-1 and inhibits insulin signaling, it is quite reasonable that reduced phosphorylation of IKKβ by atorvastatin administration would lead to increased GLUT4 translocation and insulin sensitivity ultimately ameliorating insulin resistance in adipocytes. Atrovastatin exerts anti-inflammatory effects with a marked reduction in macrophage infiltration, IKKβ and plasma levels of IL-6 and TNF-α (Furuya et al., 2010). Administration of atrovastatin demonstrates improved pancreatic function in obese mice that is attributed to decreased ER stress and increased β-cell proliferation in the pancreas (Chen et al., 2014b). Likewise, IMD-0354, a novel IKKβ inhibitor, is shown to ameliorate insulin resistance, increases the plasma levels of anti-diabetic hormone adiponectin and up-regulate the Akt phosphorylation in adipocytes (Kamon et al., 2004). Salicylate, another IKKβ inhibitor, facilitates the ser phosphorylation of IRS-1 and IRS-2 and decreases phosphorylated Akt in hepatocytes. This, in turn, prevents free fatty acid-mediated insulin signaling impairment in the hepatic tissues (Park et al., 2007).

3.3 PKC modulators

Qu et al. (2003) showed the inhibitory action of hypocretin, derived from Chinese fungus *Hypocrella bambuase,* on PKC activity in insulin-targeted tissues of obese rats. Hypocrellin selectively inhibits PKC isoforms other than PKC-θ, and also improves the aberrations in lipid and glucose metabolism caused by insulin resistance. In another study, Sajan et al. studied the effects of two small PKC-ι/λ inhibitor molecules viz., 1H-imidazole-4-carboxamide, 5-amino-1-[2,3-dihydroxy-4-[(phosphonooxy)methyl] cyclopentyl-[1R-(1a,2b,3b,4a)] and aurothiomalate in a rodent model of T2D and obesity. They showed that the administration of either inhibitor causes a marked improvement in insulin signaling with the concomitant reversal in the metabolic abnormalities including hyperglycemia, glucose intolerance, hepatosteatosis and hypercholesterolemia (Sajan et al., 2012).

3.4 Molecules targeting other protein kinases

Transforming growth factor-β activated kinase-1 (TAK1), a member of the MAP3K family, plays a vital role in various immune responses. Administration of a TAK1 inhibitor, 5Z-7-oxozeaenol (OZ), has been shown to inhibit autoimmune diabetes in NOD mice. OZ was observed to preserve pancreatic function, increase α1- antitrypsin expression, and significantly inhibit the JNK/AP-1 and NF-κB signaling in the pancreas and immune organs (Cao et al., 2015).

The functional activity of p38 mitogen-activated protein kinase (p38 MAPK) is found to be raised in renal cells imparting diabetic nephropathy in diabetic mice model (Komers, Lindsley, Oyama, Cohen, & Anderson, 2007). Thus, the molecules specifically targeting p38 MAPK would envisage preventing diabetes-associated tissue damage. One such small molecule inhibitor of p38 MAPK is PHA666859 whose inhibitory action is studied by Du et al. (2010). They showed that the treatment of PHA666859 prevents the onset of diabetic retinopathy and other associated complications of diabetes in rats. Recently, Ozaki et al. (2016) showed that the oral administration of mitogen-activated protein kinase (MAPK) kinase (MEK) inhibitor improved glucose tolerance and insulin sensitivity thereby normalizing hyperglycemia in diabetic mice model. In another study, Bargut et al. demonstrated that fish oil activates AMPK and the three isoforms of PPAR in the obese mice model. It also prevents inflammation and insulin resistance and halts RAS and MAPK activation in the white adipose tissues of mice (Bargut et al., 2015). It is envisaged that these small molecules could prove useful in developing improved therapeutics against the protein kinases involved in the pathogenesis of diabetes and insulin resistance.

4. Concluding remarks

Protein kinases play a crucial role in regulating glucose metabolism, fatty acid oxidation and insulin sensitivity. The important role of several protein kinases including AMPK, IKK and PKC has been well demonstrated in various animal models. The protein kinases modulate glucose tolerance, inflammation and insulin resistance via acting on diverse downstream targets and signaling pathways which include mTOR, JNK, NF-κB and mTOR signaling. A large pool of studies has also established that kinase signaling is important for regulating various metabolic events, and modulation in the activity of specific intracellular kinases is linked to the deleterious effects in obesity that contribute to T2D. Despite having a direct correlation between kinases with glucose metabolism, the development of therapies targeting protein kinases for the treatment of diabetes has not been much successful yet. Hence, more intensive research needs to be carried out on kinases and their link to the insulin resistance to design novel therapeutic molecules to combat diabetes and its associated complications.

Acknowledgments

This work is supported by grants from the Indian Council of Medical Research, Government of India. PG thanks DST for the award of National-Post doctoral fellowship (File no. PDF/2017/001084).

References

Aiello, L. P. (2002). The potential role of PKC in diabetic retinopathy and macular edema. *Surv. Ophthalmol.* *47*(2), S263. https://www.sciencedirect.com/science/article/pii/S0039625702003910.

Anwar, S., Shamsi, A., Shahbaaz, M., Queen, A., Khan, P., Hasan, G. M., et al. (2020). Rosmarinic acid exhibits anticancer effects via MARK4 inhibition. *Scientific Reports*, *10*(1), 10300. https://doi.org/10.1038/s41598-020-65648-z.

Arita, R., Hata, Y., & Ishibashi, T. (2010). ROCK as a therapeutic target of diabetic retinopathy. *Journal of Ophthalmology*, *2010*, 1–9. https://doi.org/10.1155/2010/175163.

Arkan, M. C., Hevener, A. L., Greten, F. R., Maeda, S., Li, Z. W., Long, J. M., et al. (2005). IKK-β links inflammation to obesity-induced insulin resistance. *Nature Medicine*, *11*(2), 191–198. https://doi.org/10.1038/nm1185.

Ashcroft, F., & Ashcroft, S. (1992). *Insulin: Molecular biology to pathology*. IRL Press at Oxford University Press. 421 pp.

Association, A. D. (2009). Diagnosis and classification of diabetes mellitus. *Diabetes Care*, *32*(Suppl. 1), S62–S67. American Diabetes Association. https://doi.org/10.2337/dc09-S062.

Aw, D. K. L., Sinha, R. A., Xie, S. Y., & Yen, P. M. (2014). Differential AMPK phosphorylation by glucagon and metformin regulates insulin signaling in human hepatic cells. *Biochemical and Biophysical Research Communications*, *447*(4), 569–573. https://doi.org/10.1016/j.bbrc.2014.04.031.

Bargut, T. C. L., Mandarim-de-Lacerda, C. A., & Aguila, M. B. (2015). A high-fish-oil diet prevents adiposity and modulates white adipose tissue inflammation pathways in mice. *Journal of Nutritional Biochemistry*, *26*(9), 960–969. https://doi.org/10.1016/j.jnutbio.2015.04.002.

Bell, G. I., Kayano, T., Buse, J. B., Burant, C. F., Takeda, J., Lin, D., et al. (1990). Molecular biology of mammalian glucose transporters. *Diabetes Care*, *13*(3), 198–208. https://doi.org/10.2337/diacare.13.3.198.

Bellomo, C. (2018). *TGFβ and LXR signaling in hepatocellular carcinoma*. http://urn.kb.se/resolve?urn=urn:nbn:se:uu:diva-334408.

Bhullar, K. S., Lagarón, N. O., McGowan, E. M., Parmar, I., Jha, A., Hubbard, B. P., et al. (2018). Kinase-targeted cancer therapies: Progress, challenges and future directions. *Molecular Cancer*, *17*(1), 1–20. BioMed Central Ltd. https://doi.org/10.1186/s12943-018-0804-2.

Bononi, A., Agnoletto, C., De Marchi, E., Marchi, S., Patergnani, S., Bonora, M., et al. (2011). Protein kinases and phosphatases in the control of cell fate. *Enzyme Research*, *2011*(2), 329098. https://doi.org/10.4061/2011/329098.

Boslem, E., Meikle, P. J., & Biden, T. J. (2012). Roles of ceramide and sphingolipids in pancreatic β-cell function and dysfunction. *Islets*, *4*(3), 177–187. https://doi.org/10.4161/isl.20102.

Cannavo, A., Liccardo, D., Komici, K., Corbi, G., de Lucia, C., Femminella, G. D., et al. (2017). Sphingosine kinases and sphingosine 1-phosphate receptors: Signaling and actions in the cardiovascular system. *Frontiers in Pharmacology*, *8*(Aug). Frontiers Media S.A. https://doi.org/10.3389/fphar.2017.00556.

Cao, H., Lu, J., Du, J., Xia, F., Wei, S., Liu, X., et al. (2015). TAK1 inhibition prevents the development of autoimmune diabetes in NOD mice. *Scientific Reports*, *5*(1), 14593. https://doi.org/10.1038/srep14593.

Cargnello, M., & Roux, P. P. (2011). Activation and function of the MAPKs and their substrates, the MAPK-activated protein kinases. *Microbiology and Molecular Biology Reviews*, *75*(1), 50–83. https://doi.org/10.1128/mmbr.00031-10.

Carling, D. (2004). The AMP-activated protein kinase cascade—A unifying system for energy control. *Trends in Biochemical Sciences*, *29*(1), 18–24. Elsevier Ltd. https://doi.org/10.1016/j.tibs.2003.11.005.

Chen, S.-S., Jiang, T., Wang, Y., Gu, L.-Z., Wu, H.-W., Tan, L., et al. (2014). Activation of double-stranded RNA-dependent protein kinase inhibits proliferation of pancreatic β-cells. *Biochemical and Biophysical Research Communications*, *443*(3), 814–820. https://doi.org/10.1016/j.bbrc.2013.12.051.

Chen, W., Xiang, H., Chen, R., Yang, J., Yang, X., Zhou, J., et al. (2019). S1PR2 antagonist ameliorate high glucose-induced fission and dysfunction of mitochondria in HRGECs via regulating ROCK1. *BMC Nephrology*, *20*(1), 135. https://doi.org/10.1186/s12882-019-1323-0.

Chen, Z. Y., Liu, S. N., Li, C. N., Sun, S. J., Liu, Q., Lei, L., et al. (2014b). Atorvastatin helps preserve pancreatic β cell function in obese C57BL/6 J mice and the effect is related to increased pancreas proliferation and amelioration of endoplasmic-reticulum stress. *Lipids in Health and Disease*, *13*(1), 98. https://doi.org/10.1186/1476-511X-13-98.

Chun, K. H., Choi, K. D., Lee, D. H., Jung, Y. S., Henry, R. R., Ciaraldi, T. P., et al. (2011). In vivo activation of ROCK1 by insulin is impaired in skeletal muscle of humans with type 2 diabetes. *American Journal of Physiology—Endocrinology and Metabolism*, *300*(3), E536–E542. https://doi.org/10.1152/ajpendo.00538.2010.

Das Evcimen, N., & King, G. L. (2007). The role of protein kinase C activation and the vascular complications of diabetes. *Pharmacological Research*, *55*(6), 498–510. https://doi.org/10.1016/j.phrs.2007.04.016.

De Alvaro, C., Teruel, T., Hernandez, R., & Lorenzo, M. (2004). Tumor necrosis factor α produces insulin resistance in skeletal muscle by activation of inhibitor κB kinase in a p38 MAPK-dependent manner. *Journal of Biological Chemistry*, *279*(17), 17070–17078. https://doi.org/10.1074/jbc.M312021200.

De Boer, J. F., Dikkers, A., Jurdzinski, A., Felden, J. V., Gaestel, M., Bavendiek, U., et al. (2014). Mitogen-activated protein kinase-activated protein kinase 2 deficiency reduces insulin sensitivity in high-fat diet-fed mice. *PLoS One*, *9*(9), e106300. https://doi.org/10.1371/journal.pone.0106300.

Dhar, A., Reddy, P., Bhat, A., & Dhar, I. (2015). Double stranded RNA dependent protein kinase (PKR) and type 2 diabetes. *Pharmacy & Pharmacology International Journal*, *2*(2), 41–43. https://doi.org/10.15406/ppij.2015.02.00015.

Drewes, G., Ebneth, A., Preuss, U., Mandelkow, E. M., & Mandelkow, E. (1997). MARK, a novel family of protein kinases that phosphorylate microtubule- associated proteins and trigger microtubule disruption. *Cell*, *89*(2), 297–308. https://doi.org/10.1016/S0092-8674(00)80208-1.

Du, Y., Tang, J., Li, G., Berti-Mattera, L., Lee, C. A., Bartkowski, D., et al. (2010). Effects of p38 MAPK inhibition on early stages of diabetic retinopathy and sensory nerve function. *Investigative Ophthalmology and Visual Science*, *51*(4), 2158–2164. https://doi.org/10.1167/iovs.09-3674.

Jessup, C. F., Bonder, C. S., Pitson, S. M., Toby, H., & Coates, P. (2012). The sphingolipid rheostat: A potential target for improving pancreatic islet survival and function. *Endocrine, Metabolic & Immune Disorders Drug Targets*, *11*(4), 262–272. https://doi.org/10.2174/187153011797881201.

Fazakerley, D. J., Holman, G. D., Marley, A., James, D. E., Stöckli, J., & Coster, A. C. F. (2010). Kinetic evidence for unique regulation of GLUT4 trafficking by insulin and AMP-activated protein kinase activators in L6 myotubes. *Journal of Biological Chemistry*, *285*(3), 1653–1660. https://doi.org/10.1074/jbc.M109.051185.

Feng, M., Tian, L., Gan, L., Liu, Z., & Sun, C. (2014). Mark4 promotes adipogenesis and triggers apoptosis in 3T3-L1 adipocytes by activating JNK1 and inhibiting p38MAPK pathways. *Biology of the Cell*, *106*(9), 294–307. https://doi.org/10.1111/boc.201400004.

Foretz, M., Even, P. C., & Viollet, B. (2018). AMPK activation reduces hepatic lipid content by increasing fat oxidation in vivo. *International Journal of Molecular Sciences*, *19*(9), 2826. https://doi.org/10.3390/ijms19092826.

Furuya, D. T., Poletto, A. C., Favaro, R. R., Martins, J. O., Zorn, T. M. T., & Machado, U. F. (2010). Anti-inflammatory effect of atorvastatin ameliorates insulin resistance in monosodium glutamate-treated obese mice. *Metabolism: Clinical and Experimental*, *59*(3), 395–399. https://doi.org/10.1016/j.metabol.2009.08.011.

Furuya, H., Wada, M., Shimizu, Y., Yamada, P. M., Hannun, Y. A., Obeid, L. M., et al. (2013). Effect of sphingosine kinase 1 inhibition on blood pressure. *The FASEB Journal*, *27*(2), 656–664. https://doi.org/10.1096/fj.12-219014.

Gal-Ben-Ari, S., Barrera, I., Ehrlich, M., & Rosenblum, K. (2019). PKR: A kinase to remember. *Frontiers in Molecular Neuroscience*, *11*, 480. Frontiers Media S.A. https://doi.org/10.3389/fnmol.2018.00480.

Gao, Z., Hwang, D., Bataille, F., Lefevre, M., York, D., Quon, M. J., et al. (2002). Serine phosphorylation of insulin receptor substrate 1 by inhibitor κB kinase complex. *Journal of Biological Chemistry*, *277*(50), 48115–48121. https://doi.org/10.1074/jbc.M209459200.

Geraldes, P., Hiraoka-Yamamoto, J., Matsumoto, M., Clermont, A., Leitges, M., Marette, A., et al. (2009). Activation of PKC-and SHP-1 by hyperglycemia causes vascular cell apoptosis and diabetic retinopathy. *Nature Medicine, 15*(11), 1298–1306. https://doi.org/10.1038/nm.2052.

Geraldes, P., & King, G. L. (2010). Activation of protein kinase C isoforms and its impact on diabetic complications. *Circulation Research, 106*(8), 1319–1331. Lippincott Williams & Wilkins https://doi.org/10.1161/CIRCRESAHA.110.217117.

Graham, S., Gorin, Y., Abboud, H. E., Ding, M., Lee, D. Y., Shi, H., et al. (2011). Abundance of TRPC6 protein in glomerular mesangial cells is decreased by ROS and PKC in diabetes. *American Journal of Physiology-Cell Physiology, 301*(2), C304–C315. https://doi.org/10.1152/ajpcell.00014.2011.

Gupta, P., Khan, F. I., Roy, S., Anwar, S., Dahiya, R., Alajmi, M. F., et al. (2020). Functional implications of pH-induced conformational changes in the Sphingosine kinase 1. *Spectrochimica Acta Part A: Molecular and Biomolecular Spectroscopy, 225*, 117453. https://doi.org/10.1016/j.saa.2019.117453.

Gupta, P., Mohammad, T., Khan, P., Alajmi, M. F., Hussain, A., Rehman, M. T., et al. (2019). Evaluation of ellagic acid as an inhibitor of sphingosine kinase 1: A targeted approach towards anticancer therapy. *Biomedicine and Pharmacotherapy, 118*, 109245. https://doi.org/10.1016/j.biopha.2019.109245.

Hartmann, S., Ridley, A. J., & Lutz, S. (2015). The function of rho-associated kinases ROCK1 and ROCK2 in the pathogenesis of cardiovascular disease. *Frontiers in Pharmacology, 6*(Nov), 1. Frontiers Research Foundation https://doi.org/10.3389/fphar.2015.00276.

Hatoum, D., Haddadi, N., Lin, Y., Nassif, N. T., & McGowan, E. M. (2017). Mammalian sphingosine kinase (SphK) isoenzymes and isoform expression: Challenges for SphK as an oncotarget. *Oncotarget, 8*(22), 36898–36929. https://doi.org/10.18632/oncotarget.16370.

Hayden, M. S., & Ghosh, S. (2012). NF-κB, the first quarter-century: Remarkable progress and outstanding questions. *Genes and Development, 26*(3), 203–234. https://doi.org/10.1101/gad.183434.111.

Huang, X., Liu, G., Guo, J., & Su, Z. Q. (2018). The PI3K/AKT pathway in obesity and type 2 diabetes. *International Journal of Biological Sciences, 14*(11), 1483–1496. Ivyspring International Publisher. https://doi.org/10.7150/ijbs.27173.

Idris, I., Gray, S., & Donnelly, R. (2001). Protein kinase C activation: Isozyme-specific effects on metabolism and cardiovascular complications in diabetes. *Diabetologia, 44*(6), 659–673. Springer https://doi.org/10.1007/s001250051675.

Iseli, T. J., Turner, N., Zeng, X.-Y., Cooney, G. J., Kraegen, E. W., Yao, S., et al. (2013). Activation of AMPK by bitter melon triterpenoids involves CaMKKβ. *PLoS One, 8*(4), e62309. https://doi.org/10.1371/journal.pone.0062309.

Itani, S. I., Ruderman, N. B., Schmieder, F., & Boden, G. (2002). Lipid-induced insulin resistance in human muscle is associated with changes in diacylglycerol, protein kinase C, and IκB-α. *Diabetes, 51*(7), 2005–2011. https://doi.org/10.2337/diabetes.51.7.2005.

Jarman, K. E., Moretti, P. A. B., Zebol, J. R., & Pitson, S. M. (2010). Translocation of sphingosine kinase 1 to the plasma membrane is mediated by calcium- and integrin-binding protein 1. *Journal of Biological Chemistry, 285*(1), 483–492. https://doi.org/10.1074/jbc.M109.068395.

Jean, S., & Kiger, A. A. (2014). Classes of phosphoinositide 3-kinases at a glance. *Journal of Cell Science, 127*(5), 923–928. https://doi.org/10.1242/jcs.093773.

Jiang, P., Zhang, D., Qiu, H., Yi, X., Zhang, Y., Cao, Y., et al. (2017). Tiron ameliorates high glucose-induced cardiac myocyte apoptosis by PKCδ-dependent inhibition of osteopontin. *Clinical and Experimental Pharmacology and Physiology, 44*(7), 760–770. https://doi.org/10.1111/1440-1681.12762.

Kahn, S. E., Prigeon, R. L., McCulloch, D. K., Boyko, E. J., Bergman, R. N., Schwartz, M. W., et al. (1993). Quantification of the relationship between insulin sensitivity and -cell function in human subjects: Evidence for a hyperbolic function. *Diabetes*, *42*(11), 1663–1672. https://doi.org/10.2337/diab.42.11.1663.

Kamon, J., Yamauchi, T., Muto, S., Takekawa, S., Ito, Y., Hada, Y., et al. (2004). A novel IKKβ inhibitor stimulates adiponectin levels and ameliorates obesity-linked insulin resistance. *Biochemical and Biophysical Research Communications*, *323*(1), 242–248. https://doi.org/10.1016/j.bbrc.2004.08.083.

Kang, R., & Tang, D. (2012). PKR-dependent inflammatory signals. *Science Signaling*, *5*(247), pe47. https://doi.org/10.1126/scisignal.2003511.

Karin, M. (1999). How NF-κB is activated: The role of the IκB kinase (IKK) complex. *Oncogene*, *18*(49), 6867–6874. Nature Publishing Group https://doi.org/10.1038/sj.onc.1203219.

Karliner, J. S. (2013). Sphingosine kinase and sphingosine 1-phosphate in the heart: A decade of progress. *Biochimica et Biophysica Acta—Molecular and Cell Biology of Lipids*, *1831*(1), 203–212. https://doi.org/10.1016/j.bbalip.2012.06.006.

Khan, F. I., Gupta, P., Roy, S., Azum, N., Alamry, K. A., Asiri, A. M., et al. (2020). Mechanistic insights into the urea-induced denaturation of human sphingosine kinase 1. *International Journal of Biological Macromolecules*, *161*, 1496–1505. https://doi.org/10.1016/j.ijbiomac.2020.07.280.

Khan, P., Rahman, S., Queen, A., Manzoor, S., Naz, F., Hasan, G. M., et al. (2017). Elucidation of dietary polyphenolics as potential inhibitor of microtubule affinity regulating kinase 4: In silico and in vitro studies. *Scientific Reports*, 7(1), 1–15. https://doi.org/10.1038/s41598-017-09941-4.

Kim, J. H., Kim, J. H., Jun, H. O., Yu, Y. S., & Kim, K. W. (2010). Inhibition of protein kinase C δ attenuates blood-retinal barrier breakdown in diabetic retinopathy. *American Journal of Pathology*, *176*(3), 1517–1524. https://doi.org/10.2353/ajpath.2010.090398.

Kjøbsted, R., Hingst, J. R., Fentz, J., Foretz, M., Sanz, M., Pehmøller, C., et al. (2018). AMPK in skeletal muscle function and metabolism. *The FASEB Journal*, *32*(4), 1741–1777. https://doi.org/10.1096/fj.201700442R.

Komers, R., Lindsley, J. N., Oyama, T. T., Cohen, D. M., & Anderson, S. (2007). Renal p38 MAP kinase activity in experimental diabetes. *Laboratory Investigation*, *87*(6), 548–558. https://doi.org/10.1038/labinvest.3700549.

Kowalski, G. M., Kloehn, J., Burch, M. L., Selathurai, A., Hamley, S., Bayol, S. A. M., et al. (2015). Overexpression of sphingosine kinase 1 in liver reduces triglyceride content in mice fed a low but not high-fat diet. *Biochimica et Biophysica Acta - Molecular and Cell Biology of Lipids*, *1851*(2), 210–219. https://doi.org/10.1016/j.bbalip.2014.12.002.

Kuhn, D., Weskamp, N., Hüllermeier, E., & Klebe, G. (2007). Functional classification of protein kinase binding sites using Cavbase. *ChemMedChem*, *2*(10), 1432–1447. https://doi.org/10.1002/cmdc.200700075.

Kuhns, S., Schmidt, K. N., Reymann, J., Gilbert, D. F., Neuner, A., Hub, B., et al. (2013). The microtubule affinity regulating kinase MARK4 promotes axoneme extension during early ciliogenesis. *Journal of Cell Biology*, *200*(4), 505–522. https://doi.org/10.1083/jcb.201206013.

Kumari, M., Wang, X., Lantier, L., Lyubetskaya, A., Eguchi, J., Kang, S., et al. (2016). IRF3 promotes adipose inflammation and insulin resistance and represses browning. *Journal of Clinical Investigation*, *126*(8), 2839–2854. https://doi.org/10.1172/JCI86080.

Lan, T., Liu, W., Xie, X., Xu, S., Huang, K., Peng, J., et al. (2011). Sphingosine kinase-1 pathway mediates high glucose-induced fibronectin expression in glomerular mesangial cells. *Molecular Endocrinology*, *25*(12), 2094–2105. https://doi.org/10.1210/me.2011-0095.

Lee, D. H., Shi, J., Jeoung, N. H., Kim, M. S., Zabolotny, J. M., Lee, S. W., et al. (2009). Targeted disruption of ROCK1 causes insulin resistance in vivo. *Journal of Biological Chemistry*, *284*(18), 11776–11780. https://doi.org/10.1074/jbc.C900014200.

Lee, J., Hong, S.-W., Chae, S. W., Kim, D. H., Choi, J. H., Bae, J. C., et al. (2012). Exendin-4 improves steatohepatitis by increasing Sirt1 expression in high-fat diet-induced obese C57BL/6J mice. *PLoS One*, 7(2), e31394. https://doi.org/10.1371/journal.pone.0031394.

Lee, S. H., Huang, H., Choi, K., Lee, D. H., Shi, J., Liu, T., et al. (2014). ROCK1 isoform-specific deletion reveals a role for diet-induced insulin resistance. *American Journal of Physiology—Endocrinology and Metabolism*, *306*(3), E332–E343. https://doi.org/10.1152/ajpendo.00619.2013.

Lee, S. Y., Hong, I. K., Kim, B. R., Shim, S. M., Sung Lee, J., Lee, H. Y., et al. (2015). Activation of sphingosine kinase 2 by endoplasmic reticulum stress ameliorates hepatic steatosis and insulin resistance in mice. *Hepatology*, *62*(1), 135–146. https://doi.org/10.1002/hep.27804.

Li, L., & Guan, K. L. (2013). Microtubule-associated protein/microtubule affinity-regulating kinase 4 (MARK4) is a negative regulator of the mammalian target of rapamycin complex 1 (mTORC1). *Journal of Biological Chemistry*, *288*(1), 703–708. https://doi.org/10.1074/jbc.C112.396903.

Li, Y., Soos, T. J., Li, X., Wu, J., DeGennaro, M., Sun, X., et al. (2004). Protein kinase C θ inhibits insulin signaling by phosphorylating IRS1 at Ser1101. *Journal of Biological Chemistry*, *279*(44), 45304–45307. https://doi.org/10.1074/jbc.C400186200.

Li, Y. Y., Yu, L. F., Zhang, L. N., Qiu, B. Y., Su, M. B., Wu, F., et al. (2013). Novel small-molecule AMPK activator orally exerts beneficial effects on diabetic db/db mice. *Toxicology and Applied Pharmacology*, *273*(2), 325–334. https://doi.org/10.1016/j.taap.2013.09.006.

Liu, Y. (2006). Renal fibrosis: New insights into the pathogenesis and therapeutics. *Kidney International*, *69*(2), 213–217. Elsevier https://doi.org/10.1038/sj.ki.5000054.

Liu, Z., Gan, L., Chen, Y., Luo, D., Zhang, Z., Cao, W., et al. (2016). Mark4 promotes oxidative stress and inflammation via binding to PPARγ and activating NF-κB pathway in mice adipocytes. *Scientific Reports*, *6*(1), 1–12. https://doi.org/10.1038/srep21382.

Maffei, A., Lembo, G., & Carnevale, D. (2018). PI3kinases in diabetes mellitus and its related complications. *International Journal of Molecular Sciences*, *19*(12), 4098. MDPI AG https://doi.org/10.3390/ijms19124098.

Malhotra, A., Begley, R., Kang, B. P. S., Rana, I., Liu, J., Yang, G., et al. (2005). PKC-ε-dependent survival signals in diabetic hearts. *American Journal of Physiology. Heart and Circulatory Physiology*, *289*(4), H1343–H1350. https://doi.org/10.1152/ajpheart.01200.2004.

Marsollier, N., Ferré, P., & Foufelle, F. (2011). Novel insights in the interplay between inflammation and metabolic diseases: A role for the pathogen sensing kinase PKR. *Journal of Hepatology*, *54*(6), 1307–1309. Elsevier B.V. https://doi.org/10.1016/j.jhep.2010.11.037.

McArdle, M. A., Finucane, O. M., Connaughton, R. M., McMorrow, A. M., & Roche, H. M. (2013). Mechanisms of obesity-induced inflammation and insulin resistance: Insights into the emerging role of nutritional strategies. *Frontiers in Endocrinology*, *4*(May), 52. https://doi.org/10.3389/fendo.2013.00052.

Melnikova, I., & Golden, J. (2004). Targeting protein kinases. *Nature Reviews Drug Discovery*, *3*(12), 993–994. https://doi.org/10.1038/nrd1600.

Min, Q., Bai, Y., Zhang, Y., Yu, W., Zhang, M., Liu, D., et al. (2017). Hawthorn leaf flavonoids protect against diabetes-induced cardiomyopathy in rats via PKC- α signaling pathway. *Evidence-Based Complementary and Alternative Medicine*, *2017*, 2071952. https://doi.org/10.1155/2017/2071952.

Mohan, N., Jiang, J., Dokmanovic, M., & Wu, W. J. (2018). Trastuzumab-mediated cardiotoxicity: Current understanding, challenges, and frontiers. *Antibody Therapeutics*, *1*(1), 13–17. https://doi.org/10.1093/abt/tby003.

Moreno, D., Knecht, E., Viollet, B., & Sanz, P. (2008). A769662, a novel activator of AMP-activated protein kinase, inhibits non-proteolytic components of the 26S proteasome by an AMPK-independent mechanism. *FEBS Letters*, *582*(17), 2650–2654. https://doi.org/10.1016/j.febslet.2008.06.044.

Musi, N. (2006). AMP-activated protein kinase and type 2 diabetes. *Current Medicinal Chemistry*, *13*(5), 583–589. https://doi.org/10.2174/092986706776055724.

Nakamura, T., Arduini, A., Baccaro, B., Furuhashi, M., & Hotamisligil, G. S. (2014). Small-molecule inhibitors of PKR improve glucose homeostasis in obese diabetic mice. *Diabetes*, *63*(2), 526–534. https://doi.org/10.2337/db13-1019.

Nallagatla, S. R., Toroney, R., & Bevilacqua, P. C. (2011). Regulation of innate immunity through RNA structure and the protein kinase PKR. *Current Opinion in Structural Biology*, *21*(1), 119–127. NIH Public Access https://doi.org/10.1016/j.sbi.2010.11.003.

Nandipati, K. C., Subramanian, S., & Agrawal, D. K. (2017). Protein kinases: Mechanisms and downstream targets in inflammation-mediated obesity and insulin resistance. *Molecular and Cellular Biochemistry*, *426*(1–2), 27–45. Springer New York LLC. https://doi.org/10.1007/s11010-016-2878-8.

Naz, F., Khan, F. I., Mohammad, T., Khan, P., Manzoor, S., Hasan, G. M., et al. (2018). Investigation of molecular mechanism of recognition between citral and MARK4: A newer therapeutic approach to attenuate cancer cell progression. *International Journal of Biological Macromolecules*, *107*, 2580–2589. https://doi.org/10.1016/j.ijbiomac.2017.10.143.

Negi, G., & Sharma, S. S. (2015). Inhibition of IκB kinase (IKK) protects against peripheral nerve dysfunction of experimental diabetes. *Molecular Neurobiology*, *51*(2), 591–598. https://doi.org/10.1007/s12035-014-8784-8.

Newton, J., Lima, S., Maceyka, M., & Spiegel, S. (2015). Revisiting the sphingolipid rheostat: Evolving concepts in cancer therapy. *Experimental Cell Research*, *333*(2), 195–200. Academic Press Inc. https://doi.org/10.1016/j.yexcr.2015.02.025.

Ng, M. L., Wadham, C., & Sukocheva, O. A. (2017). The role of sphingolipid signalling in diabetes-associated pathologies (Review). *International Journal of Molecular Medicine*, *39*(2), 243–252. Spandidos Publications. https://doi.org/10.3892/ijmm.2017.2855.

Nieto-Vazquez, I., Fernández-Veledo, S., Krämer, D. K., Vila-Bedmar, R., Garcia-Guerra, L., & Lorenzo, M. (2008). Insulin resistance associated to obesity: The link TNF-alpha. *Archives of Physiology and Biochemistry*, *114*(3), 183–194. Taylor & Francis. https://doi.org/10.1080/13813450802181047.

Nishino, S., Yamashita, H., Tamori, M., Mashimo, M., Yamagata, K., Nakamura, H., et al. (2019). Translocation and activation of sphingosine kinase 1 by ceramide-1-phosphate. *Journal of Cellular Biochemistry*, *120*(4), 5396–5408. https://doi.org/10.1002/jcb.27818.

Oguiza, A., Recio, C., Lazaro, I., Mallavia, B., Blanco, J., Egido, J., et al. (2015). Peptide-based inhibition of IκB kinase/nuclear factor-κB pathway protects against diabetes-associated nephropathy and atherosclerosis in a mouse model of type 1 diabetes. *Diabetologia*, *58*(7), 1656–1667. https://doi.org/10.1007/s00125-015-3596-6.

Ozaki, K. I., Awazu, M., Tamiya, M., Iwasaki, Y., Harada, A., Kugisaki, S., et al. (2016). Targeting the ERK signaling pathway as a potential treatment for insulin resistance and type 2 diabetes. *American Journal of Physiology—Endocrinology and Metabolism*, *310*(8), E643–E651. https://doi.org/10.1152/ajpendo.00445.2015.

Pang, T., Zhang, Z. S., Gu, M., Qiu, B. Y., Yu, L. F., Cao, P. R., et al. (2008). Small molecule antagonizes autoinhibition and activates AMP-activated protein kinase in cells. *Journal of Biological Chemistry*, *283*(23), 16051–16060. https://doi.org/10.1074/jbc.M710114200.

Park, E., Wong, V., Guan, X., Oprescu, A. I., & Giacca, A. (2007). Salicylate prevents hepatic insulin resistance caused by short-term elevation of free fatty acids in vivo. *Journal of Endocrinology*, *195*(2), 323–331. https://doi.org/10.1677/JOE-07-0005.

Park, S. W., Kim, M., Kim, J. Y., Brown, K. M., Haase, V. H., D'Agati, V. D., et al. (2012). Proximal tubule sphingosine kinase-1 has a critical role in A 1 adenosine receptor-mediated renal protection from ischemia. *Kidney International*, *82*(8), 878–891. https://doi.org/10.1038/ki.2012.224.

Parveen, I., Khan, P., Ali, S., Hassan, M. I., & Ahmed, N. (2018). Synthesis, molecular docking and inhibition studies of novel 3-N-aryl substituted-2-heteroarylchromones targeting microtubule affinity regulating kinase 4 inhibitors. *European Journal of Medicinal Chemistry*, *159*, 166–177. https://doi.org/10.1016/j.ejmech.2018.09.030.

Patterson, H., Nibbs, R., Mcinnes, I., & Siebert, S. (2014). Protein kinase inhibitors in the treatment of inflammatory and autoimmune diseases. *Clinical and Experimental Immunology*, *176*(1), 1–10. John Wiley & Sons, Ltd. https://doi.org/10.1111/cei.12248.

Peerzada, M. N., Khan, P., Khan, N. S., Avecilla, F., Siddiqui, S. M., Hassan, M. I., et al. (2020). Design and development of small-molecule Arylaldoxime/5-Nitroimidazole hybrids as potent inhibitors of MARK4: A promising approach for target-based cancer therapy. *ACS Omega*, *5*(36), 22759–22771. https://doi.org/10.1021/acsomega.0c01703.

Peng, H., Li, Y., Wang, C., Zhang, J., Chen, Y., Chen, W., et al. (2016). ROCK1 induces endothelial-to-mesenchymal transition in glomeruli to aggravate albuminuria in diabetic nephropathy. *Scientific Reports*, *6*(1), 1–10. https://doi.org/10.1038/srep20304.

Potenza, M. A., Sgarra, L., Nacci, C., Leo, V., De Salvia, M. A., & Montagnani, M. (2019). Activation of AMPK/SIRT1 axis is required for adiponectin-mediated preconditioning on myocardial ischemia-reperfusion (I/R) injury in rats. *PLoS One*, *14*(1), e0210654. https://doi.org/10.1371/journal.pone.0210654.

Purkayastha, S., Zhang, G., & Cai, D. (2011). Uncoupling the mechanisms of obesity and hypertension by targeting hypothalamic IKK-Î 2 and NF-Î°B. *Nature Medicine*, *17*(7), 883–887. https://doi.org/10.1038/nm.2372.

Qu, X., Dang, L., & Scale, J. P. (2003). Inhibitory effect of hypocrellin A on protein kinase C in liver and skeletal muscle of obese Zucker rats. *American Journal of Chinese Medicine*, *31*(6), 871–878. https://doi.org/10.1142/S0192415X03001624.

Rask-Madsen, C., & King, G. L. (2005). Proatherosclerotic mechanisms involving protein kinase C in diabetes and insulin resistance. *Arteriosclerosis, Thrombosis, and Vascular Biology*, *25*(3), 487–496. Lippincott Williams & Wilkins. https://doi.org/10.1161/01.ATV.0000155325.41507.e0.

Rovina, D., Fontana, L., Monti, L., Novielli, C., Panini, N., Sirchia, S. M., et al. (2014). Microtubule-associated protein/microtubule affinity-regulating kinase 4 (MARK4) plays a role in cell cycle progression and cytoskeletal dynamics. *European Journal of Cell Biology*, *93*(8–9), 355–365. https://doi.org/10.1016/j.ejcb.2014.07.004.

Roy, S., Mohammad, T., Gupta, P., Dahiya, R., Parveen, S., Luqman, S., et al. (2020). Discovery of harmaline as a potent inhibitor of sphingosine Kinase-1: A chemopreventive role in lung cancer. *ACS Omega*, *5*(34), 21550–21560. https://doi.org/10.1021/acsomega.0c02165.

Ruderman, N. B., Carling, D., Prentki, M., & Cacicedo, J. M. (2013). AMPK, insulin resistance, and the metabolic syndrome. *Journal of Clinical Investigation*, *123*(7), 2764–2772. American Society for Clinical Investigation https://doi.org/10.1172/JCI67227.

Saha, A., Coughlan, K., Valentine, R., & Ruderman, N. (2014). AMPK activation: A therapeutic target for type 2 diabetes? *Diabetes, Metabolic Syndrome and Obesity: Targets and Therapy*, 7, 241. https://doi.org/10.2147/DMSO.S43731.

Sajan, M. P., Nimal, S., Mastorides, S., Acevedo-Duncan, M., Kahn, C. R., Fields, A. P., et al. (2012). Correction of metabolic abnormalities in a rodent model of obesity, metabolic syndrome, and type 2 diabetes mellitus by inhibitors of hepatic protein kinase C-ι. *Metabolism: Clinical and Experimental*, *61*(4), 459–469. https://doi.org/10.1016/j.metabol.2011.12.008.

Shoelson, S. E., Lee, J., & Yuan, M. (2003). Inflammation and the IKKβ/IκB/NF-κB axis in obesity- and diet-induced insulin resistance. *International Journal of Obesity*, *27*(S3), S49–S52. https://doi.org/10.1038/sj.ijo.0802501.

Silva Xavier, G. D. (2012). Protein kinases and pancreatic islet function. In *Protein kinases* InTech. https://doi.org/10.5772/37282.

Simran, Grewal, A. K., Arora, S., & Singh, T. G. (2019). Role of protein kinase C in diabetic complications. *Journal of Pharmaceutical Technology, Research and Management*, 7(2), 87–95. https://doi.org/10.15415/jptrm.2019.72011.

Singh, R. M., Cummings, E., Pantos, C., & Singh, J. (2017). Protein kinase C and cardiac dysfunction: A review. *Heart Failure Reviews*, *22*(6), 843–859. Springer New York LLC. https://doi.org/10.1007/s10741-017-9634-3.

Spitzer, R., Cleves, A. E., & Jain, A. N. (2011). Surface-based protein binding pocket similarity. *Proteins: Structure, Function, and Bioinformatics*, *79*(9), 2746–2763. https://doi.org/10.1002/prot.23103.

Sun, C., Tian, L., Nie, J., Zhang, H., Han, X., & Shi, Y. (2012). Inactivation of MARK4, an AMP-activated protein kinase (AMPK)-related kinase, leads to insulin hypersensitivity and resistance to diet-induced obesity. *Journal of Biological Chemistry*, *287*(45), 38305–38315. https://doi.org/10.1074/jbc.M112.388934.

Sung, B.-J., Hur, K. Y., Kim, J. H., Kim, Y.-B., & Lee, M.-K. (2018). Role of ROCK1 in pancreatic ß cells. *Diabetes*, *67*(Suppl. 1), 196-OR. https://doi.org/10.2337/db18-196-or.

Svegliati-Baroni, G., Saccomanno, S., Rychlicki, C., Agostinelli, L., de Minicis, S., Candelaresi, C., et al. (2011). Glucagon-like peptide-1 receptor activation stimulates hepatic lipid oxidation and restores hepatic signalling alteration induced by a high-fat diet in nonalcoholic steatohepatitis. *Liver International*, *31*(9), 1285–1297. https://doi.org/10.1111/j.1478-3231.2011.02462.x.

Tao, C., Sifuentes, A., & Holland, W. L. (2014). Regulation of glucose and lipid homeostasis by adiponectin: Effects on hepatocytes, pancreatic β cells and adipocytes. *Best Practice and Research: Clinical Endocrinology and Metabolism*, *28*(1), 43–58. https://doi.org/10.1016/j.beem.2013.11.003.

Tom, R. Z., Garcia-Roves, P. M., Sjögren, R. J. O., Jiang, L. Q., Holmström, M. H., Deshmukh, A. S., et al. (2014). Effects of AMPK activation on insulin sensitivity and metabolism in leptin-deficient ob/ob mice. *Diabetes*, *63*(5), 1560–1571. https://doi.org/10.2337/db13-0670.

Towler, M. C., & Hardie, D. G. (2007). AMP-activated protein kinase in metabolic control and insulin signaling. *Circulation Research*, *100*(3), 328–341. https://doi.org/10.1161/01.RES.0000256090.42690.05.

Treebak, J. T., Glund, S., Deshmukh, A., Klein, D. K., Long, Y. C., Jensen, T. E., et al. (2006). AMPK-mediated AS160 phosphorylation in skeletal muscle is dependent on AMPK catalytic and regulatory subunits. *Diabetes*, *55*(7), 2051–2058. https://doi.org/10.2337/db06-0175.

Wang, J., Shen, X., Liu, J., Chen, W., Wu, F., Wu, W., et al. (2020). High glucose mediates NLRP3 inflammasome activation via upregulation of ELF3 expression. *Cell Death and Disease*, *11*(5), 1–14. https://doi.org/10.1038/s41419-020-2598-6.

Wang, K., Li, W., Bai, Y., Yang, W., Ling, Y., & Fang, M. (2017). Ssc-miR-7134-3p regulates fat accumulation in castrated male pigs by targeting MARK4 gene. *International Journal of Biological Sciences*, *13*(2), 189–197. https://doi.org/10.7150/ijbs.17386.

Wang, W., Wang, Y., Long, J., Wang, J., Haudek, S. B., Overbeek, P., et al. (2012). Mitochondrial fission triggered by hyperglycemia is mediated by ROCK1 activation in podocytes and endothelial cells. *Cell Metabolism*, *15*(2), 186–200. https://doi.org/10.1016/j.cmet.2012.01.009.

Wang, X.-A., Zhang, R., She, Z.-G., Zhang, X.-F., Jiang, D.-S., Wang, T., et al. (2014). Interferon regulatory factor 3 constrains IKKβ/NF-κB signaling to alleviate hepatic steatosis and insulin resistance. *Hepatology*, *59*(3), 870–885. https://doi.org/10.1002/hep.26751.

Watanabe, T., Imamura, T., & Hiasa, Y. (2018). Roles of protein kinase R in cancer: Potential as a therapeutic target. *Cancer Science*, *109*(4), 919–925. Blackwell Publishing Ltd. https://doi.org/10.1111/cas.13551.

Weissmann, L., Quaresma, P. G. F., Santos, A. C., De Matos, A. H. B., Pascoal, V. D.Á. B., Zanotto, T. M., et al. (2014). IKKε is key to induction of insulin resistance in the hypothalamus, and its inhibition reverses obesity. *Diabetes*, *63*(10), 3334–3345. https://doi.org/10.2337/db13-1817.

Whetzel, A. M., Bolick, D. T., & Hedrick, C. C. (2009). Sphingosine-1-phosphate inhibits high glucose-mediated ERK1/2 action in endothelium through induction of MAP kinase phosphatase-3. *American Journal of Physiology-Cell Physiology*, *296*(2), C339–C345. https://doi.org/10.1152/ajpcell.00293.2008.

Willows, R., Sanders, M. J., Xiao, B., Patel, B. R., Martin, S. R., Read, J., et al. (2017). Phosphorylation of AMPK by upstream kinases is required for activity in mammalian cells. *Biochemical Journal*, *474*(17), 3059–3073. https://doi.org/10.1042/BCJ20170458.

Woods, A., Vertommen, D., Neumann, D., Türk, R., Bayliss, J., Schlattner, U., et al. (2003). Identification of phosphorylation sites in AMP-activated protein kinase (AMPK) for upstream AMPK kinases and study of their roles by site-directed mutagenesis. *Journal of Biological Chemistry*, *278*(31), 28434–28442. https://doi.org/10.1074/jbc.M303946200.

Yaghobian, D., Don, A. S., Yaghobian, S., Chen, X., Pollock, C. A., & Saad, S. (2016). Increased sphingosine 1-phosphate mediates inflammation and fibrosis in tubular injury in diabetic nephropathy. *Clinical and Experimental Pharmacology and Physiology*, *43*(1), 56–66. https://doi.org/10.1111/1440-1681.12494.

Yang, J., Park, Y., Zhang, H., Xu, X., Laine, G. A., Dellsperger, K. C., et al. (2009). Feed-forward signaling of TNF-α and NF-κB via IKK-β pathway contributes to insulin resistance and coronary arteriolar dysfunction in type 2 diabetic mice. *American Journal of Physiology - Heart and Circulatory Physiology*, *296*(6), 1850–1858. https://doi.org/10.1152/ajpheart.01199.2008.

Ye, J. (2013). Mechanisms of insulin resistance in obesity. *Frontiers of Medicine in China*, 7(1), 14–24. NIH Public Access. https://doi.org/10.1007/s11684-013-0262-6.

Zeglinski, M., Ludke, A., Jassal, D. S., & Singal, P. K. (2011). Trastuzumab-induced cardiac dysfunction: A "dual-hit.". *Experimental and Clinical Cardiology*, *16*(3), 70–74. Pulsus Group. /pmc/articles/PMC3209542/?report=abstract.

Zhang, B. B., Zhou, G., & Li, C. (2009). AMPK: An emerging drug target for diabetes and the metabolic syndrome. *Cell Metabolism*, *9*(5), 407–416. https://doi.org/10.1016/j.cmet.2009.03.012.

Zhang, X., Zhang, G., Zhang, H., Karin, M., Bai, H., & Cai, D. (2008). Hypothalamic IKKβ/NF-κB and ER stress link overnutrition to energy imbalance and obesity. *Cell*, *135*(1), 61–73. https://doi.org/10.1016/j.cell.2008.07.043.

Zhu, Q., Shan, X., Miao, H., Lu, Y., Xu, J., You, N., et al. (2009). Acute activation of acid ceramidase affects cytokine-induced cytotoxicity in rat islet β-cells. *FEBS Letters*, *583*(12), 2136–2141. https://doi.org/10.1016/j.febslet.2009.05.047.

CHAPTER FOUR

Dual roles of ATP-binding site in protein kinases: Orthosteric inhibition and allosteric regulation

Mingyu Li[a], Ashfaq Ur Rehman[a], Yaqin Liu[b], Kai Chen[c,*], and Shaoyong Lu[a,b,*]

[a]Department of Pathophysiology, Key Laboratory of Cell Differentiation and Apoptosis of Chinese Ministry of Education, Shanghai Jiao Tong University, School of Medicine, Shanghai, China
[b]Medicinal Chemistry and Bioinformatics Centre, Shanghai Jiao Tong University, School of Medicine, Shanghai, China
[c]Department of Orthopedics, Changhai Hospital, Naval Military Medical University, Shanghai, China
*Corresponding authors: e-mail address: ch_kai@163.com; lushaoyong@yeah.ent

Contents

Advances in Protein Chemistry and Structural Biology, Volume 124
ISSN 1876-1623
https://doi.org/10.1016/bs.apcsb.2020.09.005

Abstract

Protein kinases use ATP to phosphorylate other proteins. Phosphorylation (p) universally orchestrates a fine-tuned network modulating a multitude of biological processes. Moreover, the start of networks, ATP-binding site, has been recognized dual roles to impact protein kinases function: (i) orthosteric inhibition, via being blocked to interference ATP occupying and (ii) allosteric regulation, via being altered first to induce further conformational changes. The above two terminologies are widely used in drug design, which has acquired quite a significant progress in the protein kinases field over the past 2 decades. Most small molecular inhibitors directly compete with ATP to implement orthosteric inhibition, still exhibiting irreplaceable and promising therapeutic effects. Additionally, numerous inhibitors can paradoxically lead protein kinases to hyper-phosphorylation, even activation, indicative of the allosteric regulation role of the ATP-binding site. Here, we review the quintessential examples that apply for the dual roles in diverse ways. Our work provides an insight into the molecular mechanisms under the dual roles and will be promisingly instructive for future drug development.

1. Introduction

The family of human protein kinase genes, consisting of a total 518 along with 106 pseudogenes encoding the second largest enzyme family, is now being specially termed as ***Kinome*** (Manning, Whyte, Martinez, Hunter, & Sudarsanam, 2002). Protein kinases function as transferring the terminal phosphate group of adenosine triphosphate (ATP) to the hydroxyl (OH) group of Serine (Ser), Threonine (Thr) or Tyrosine (Tyr) residues in substrate proteins. Protein kinases with this residue specificity can be widely classified as Ser/Thr kinases, Tyr kinases, and dual specific kinases (Hanks, 2003). Simple but grand, those universal Kinome-mediated phosphotransfer cascades successively triggers signal molecules responsible for transducing critical biological information (Johnson & Lewis, 2001). Even subtle alterations in just one protein kinase would disable the signal pathway, leading to the pathogenesis of various diseases, including inflammatory disorder, endocrine dyscrasia, neurologic abnormality, and cancer. Due to the significant role of protein kinases, therefore, they have always being one of the most attractive targets for therapeutics in the pharmaceutical industry over a couple of decades (Cohen, 2002).

A vast majority of protein kinase inhibitors were designed, which had the potential to target a well-known studied "druggable" pocket, an ATP-binding site, due to its highly conservative characteristics (Vulpetti & Bosotti, 2004). Interestingly, it was first thought that targeting the ATP-binding site of protein kinases was unselective and ineffective on account of the likelihood of those sites in the Kinome. However, imatinib, the first protein kinase inhibitor approved by the Food and Drug Administration (FDA), is achieving enormous success in showing 80% clinical validity in chronic CML patients (Hehlmann et al., 2017). Also, the structural level studies have identified several critical residues within the ATP-binding site useful for small molecular inhibitors design. It demolishes the regrets of promiscuity inhibitors targeting the ATP-binding site, and mobilizes more zeal in researchers to investigate potential inhibitors.

The rapid growth of small molecular inhibitors gradually reveals yet another role of ATP-binding site, namely allosteric regulation, except the natural orthosteric inhibitory role. The fundamental difference between these two roles was whether or not structural changes would emerge at another site, usually on the regulatory site, topologically away from the ATP-binding site. To understand the notion of dual roles would give us a detailed understanding of the ATP-binding site and navigate us in developing optimal use of small molecular inhibitors. With the aims of this notion, here, we first depict the predominant catalytic domain of protein kinases and a close-up of the ATP-binding site to gain a preliminary understanding of their structural configurations. Thereupon this review will summarize with a few quintessential examples on the dual functions of the ATP-binding site, highlighting multiple pathophysiological mechanisms. Also, it will address future drug design expectations associated with this ATP-binding site. This review further emphasizes the significance of the ATP-binding site and, therefore, will ideally be a foundation stone aid for the development of a new generation potential drugs.

2. Protein kinases catalytic domain

Protein kinases generally consist of two domains: the regulatory and the catalytic domain. The regulatory domain, usually recognized as unique evolutionary additions to the catalytic domain, may yield essential interactions with the catalytic domain for regulation of kinases or with other proteins for higher-order assembly (Gógl, Kornev, Reményi, & Taylor, 2019). The first protein kinase get crystallized was protein kinase A (PKA), a number of AGC kinase groups (named after protein kinase A, G, C), which remains today as a model for interpreting the general intrinsic properties of the catalytic domain among the Kinome as shown in Fig. 1 (Knighton, Zheng, Ten Eyck, Ashford, et al., 1991; Knighton, Zheng, Ten Eyck, Xuong, et al., 1991).

The catalytic domain is a highly conserved, uneven bi-lobal structure zconsisting of a smaller N-terminal lobe with the minimal five stranded

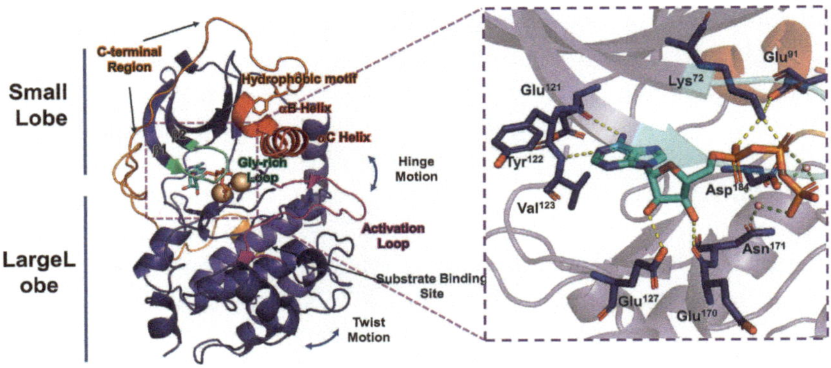

Fig. 1 Overall catalytic domain of the protein kinase. Cartoon representation of protein kinase catalytic domain (PDB: 1ATP) with interactions between ATP and crucial residues nearby in close-up. Pivotal structures are highlighted in different colors and labeled by the sites. Moreover, unique to the AGC kinases and in PKA, the shorter αB helix separates αC helix from the upper half of the β-sheets, generating a deep hydrophobic groove. This groove, named as PDK1-interacting fragment pocket (PIF-pocket), is a regulatory hotspot. The hydrophobic motif (HM), consisting of the general motif Phe/Tyr-X-X-Phe/Tyr-Ser/Thr/Glu/Asp Phe/Tyr, here is shortened in the C-terminal region, with the Phe residues (shown in stick and colored in orange) docking to the PIF-pocket. Furthermore, divalent cation such as Mn^{2+} (the balls colored in wheat) is required for the catalysis reaction, due to the activity of some protein kinases in vitro is greater with Mn^{2+} than Mg^{2+}. In the close-up, it clearly shows that Glu^{91} of the αC helix forms a salt-bridge with Lys^{72} of the β3-strand serving to hold the α- and β-phosphates in position. The 6-amino group of the adenine base of ATP forms a hydrogen bond with the carbonyl oxygen of Glu^{121}, the first hinge residue of PKA. Moreover, the N-1 of the adenine ring also forms a hydrogen bond with –NH group of the hinge residue Val^{123}. Respectively, 2-OH of the ATP-ribose interact with Glu^{127} (the first residue following the hinge), and the 3-OH binds to Glu^{170} of the catalytic loop. Most small-molecule protein kinase inhibitors make contacts with residues of the ATP-binding pocket that are homologous to the residues mentioned in this legend.

anti-parallel β-sheets, αB and αC helices, and a larger α-helical C-terminal lobe with at least eight α-helices. Notably, one should always keep in mind that protein kinases are not static, but with significant flexibility, associating with diverse function (Taylor & Kornev, 2011). The typical example for this notion is the activation loop, one of the crucial structural elements regulating the protein kinases enzymatic activity. This loop is marked by conserved DFG (usually Asp Phe Gly) and APE motifs (usually Asp Phe Glu) at the start and end of the loop, respectively. It can undertake multiple conformations, especially the use of extended active conformer that is permissive for catalysis, while the inactive conformer turned into collapsed through inter-lobe motions to block the binding of the substrate. Moreover, the signature motif and two spines of protein kinase discussed next are also flexible and important.

2.1 The signature motif of protein kinase

Among the highlighted structures assembled into the functional core of protein kinases, the **K/E/D/D** motif ($Lys^{72}/Glu^{91}/Asp^{166}/Asp^{184}$ # in PKA), is the skeleton aiding in the catalytic action of virtually all the protein kinases (Hanks & Hunter, 1995). In the protein kinase signature motif, the conserved **K residue** locates in β3-strand. The **E residue** in the middle of the αC helix can form a salt-bridge with the K residue. Then the side chain of K residue will form hydrogen bonds with the α- and β-phosphate of ATP molecule and stabilize it during catalysis. Hence, this salt-bridge serves as a signature of active protein kinase by defining a "**αC-in**" conformation. In contrast, the salt-bridge trim-off corresponds to the "**αC-out**" conformation (Fig. 2A). The conformational conversion from the αC-in to the αC-out is acquired for a reduction of kinase catalytic activity (Taylor, Shaw, Kannan, & Kornev, 2015).

Moreover, the first **D residue** in the motif is positioned in the catalytic loop (HRD(x)$_4$N, x refers to any residues), enabling to hold the –OH group of the protein substrate in place to facilitate the process of nucleophilic attack by γ-phosphorous atom of the ATP molecule. In the active state, protein kinases usually adopt an "**DFG-in**" configuration in which the second **D residue** is inserted into the ATP-binding site, coordinating with the β-phosphate and chelating Mg^{2+} ions, however, the aromatic ring of phenylalanine (Phe) residue removes from the pocket; primarily in contact with

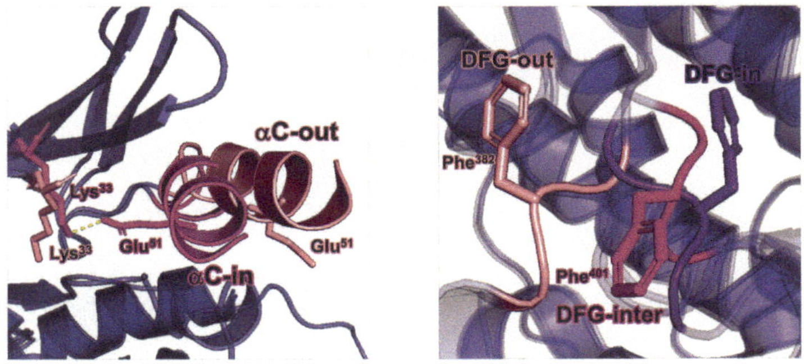

Fig. 2 DFG motif in/out and αC helix in/out in the active/inactive protein kinase. Different orientations of DFG motif and αC helix in the active or inactive conformation of ABL1(bottom) and CDK2(top). In the active state, protein kinases usually adopt a "DFG-in" and "αC-in" configuration and vice versa. Notably, DFG-inter is an intermediate conformation, previously named DFG-up. PDB: DFG-in (2F4J); DFG-inter (2G2H); DFG-out (2E2B); αC-in (1JST); and αC-out (1HCL).

the hydrophobic interface of αC-helix. Conversely, it changes to a dormant "**DFG-out**" configuration through swapping the positions of the Phe and the aspartic acid residue (Asp). In addition, the DFG-in configuration actually contains a set of active and inactive conformations since the Phe orientation may be diverse in different protein kinases. There is also a small group called "**DFG-inter,**" of which the Phe is just in the intermediate position (Fig. 2B). Thus, the DFG motif and αC helix endorse both the **-in** and **-out** conformation with highly dynamic pattern, precisely aligned sets of switches that regulate protein kinase active and inactive (Modi & Dunbrack, 2019).

2.2 Regulatory and catalytic spines of protein kinase

Besides, Kornev et al. discovered two stacks of hydrophobic spinal residues (Fig. 3) that define the internal architecture of almost the whole Kinome,

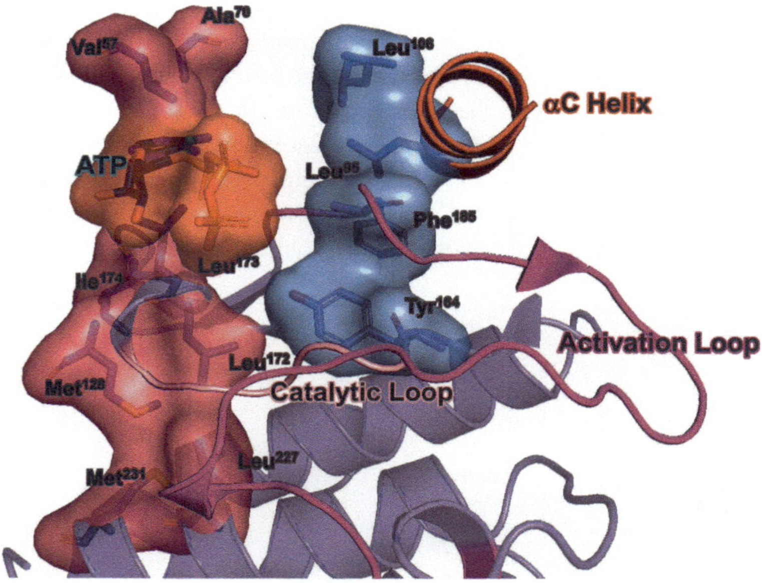

Fig. 3 Surface representation of two spines of the protein kinase in Cartoon mode. The R-spine (blue) consists typically of two residues in the N-lobe, one near the beginning of the β4-strand (Leu^{106} in PKA), the other near the end of the αC helix (Leu^{95} in PKA), and two residues in the C-lobe, involving DFG-F (Phe^{185} in PKA) in the activation loop, HRD-H (but Tyr^{164} in PKA) in the catalytic loop. The C-spine (warm pink) is generally composed of two residues from the N-lobe, one from β2-strand (Val^{57} in PKA), the other from conserved AxK motif of β3-strand (Ala^{70} in PKA), and six residues from C-lobe, three adjacent residues (Leu^{172}, Leu^{173}, Ile^{174} in PKA) from β7-strand, one (Met^{128} in PKA) from the αD helix, the others (Leu^{227}, Met^{231} in PKA) from αF helix.

span both N- and C-lobes, and enable both lobes to move in a fluid way as the cleft opens and closes (Kornev, Haste, Taylor, & Eyck, 2006; Kornev, Taylor, & Ten Eyck, 2008). These spines are stable but assembled dynamically. The regulatory spine (R-spine) not only connects the activation segment with the regulatory hotspot but is precisely aligned in the active kinases to locate the protein substrates. It can be broken in many ways. For instance, Phe in the DFG-out configuration can disrupt the integrity of R-spine by moving away from the hydrophobic interface with αC helix. Also, DFG-in conformation can also be inactive if any other spine residues are aligned by mistake.

The catalytic spine (C-spine) is assembled with an ATP molecule occupying the front cleft, and it then enables the kinase to catalyze.

3. ATP-binding site

There is a deep cleft between two lobes in the crystallographic coordinate of PKA. In the front of the cleft is the ATP-binding site, whose "ground" corresponds to the surface of the large C-terminal lobe, whereas the "roof" consists of two sheets (i.e., β1 & β2) with a Glycine-(G) rich loop (GxGxΦG, Φ refers to a hydrophobic residue) linking both the sheets (Fig. 1). The ATP-binding site is conventionally partitioned into five regions: adenine region, sugar region, phosphate region, hydrophobic I and II region (Fig. 4A) (Vulpetti & Bosotti, 2004).

1. *The adenine region* of the ATP-binding site is occupied by a heterocyclic planar ring system, mainly hydrophobic. This hydrophobic core can form non-polar interactions with hydrophobic residues around the pocket.

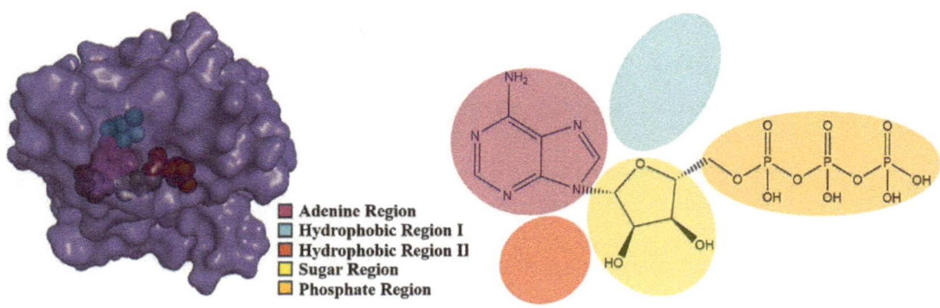

Fig. 4 The ATP-binding site. The overlook of the ATP-binding site of the kinase domain is in surface representation with ball filling indicating each subregion (left) and the particular colored subregion on ATP molecular structure (right). These subregions are adenine region, sugar region, phosphate region, hydrophobic region I and hydrophobic region II. Precisely, the hydrophobic region I and II are not occupied by ATP.

In addition, at least two hydrogen bonds could be formed in the hinge region between N^1, N^6 of adenine, and peptide backbone (Fig. 1). Such interactions make a significant contribution to the ATP molecule occupancy in the binding site; hence adenine moiety seems to be a standard feature for many protein kinase inhibitors.

2. *The sugar region* of the ATP-binding site accommodates the ATP-ribose part, which contains two OH-groups appropriate for hydrogen bonds formation (Fig. 1). Many researchers have indeed exploited this region in order to provide both the sensitive and selective inhibitors. And the covalent modification of type VI inhibitors usually takes place in the sugar region.

3. *The phosphate region*, the conserved residues, i.e., Glu^{91}, Lys^{72}, and Asp^{171} in the PKA, coordinate with the divalent cations to position the phosphate for catalysis (Fig. 1). Additionally, this region includes the highly dynamic DFG motif we discussed above and the Gly-rich loop, the most mobile part which always changes its orientation to suit the binding compound. As a result of highly flexible conformation, this region has poor drugability.

4 and 5. The hydrophobic region I and II of the ATP-binding site are also termed as "buried region" and "solvent-accessible region." In particular, the region I exhibits large residue diversity among the protein kinase superfamily and is therefore considered mostly to improve the selectivity of a specific potential inhibitor. The shape and volume of region I are inversely correlated with the size of the gatekeeper, first residue of hinge linker. The smaller gatekeeper would leave considerable space for further antagonists, not only to enhance the efficiency but also to exhibit higher selectivity (Zuccotto, Ardini, Casale, & Angiolini, 2010).

Mostly, the kinase inhibitors discovered to date yet are ATP competitive and show ~1–3 hydrogen interactions with the key residues in the adenine region of targeted kinase. Other regions are being exploited as a complement to improve the pharmacological properties of inhibitors.

4. Orthosteric inhibition

It has become a consensus that since ligands can co-evolve with their target for optimal docking pattern, molecules resembling their structural features ought to retain biological function. In light of this, if we artificially synthesize molecules lacking the essential functional groups without affecting the structural interaction-dependent features, these ligands would prevent the receptors from fulfilling their functions. Indeed, researchers have got a few inhibitors that can eject ATP from its binding site with higher affinity, making substrate protein phosphorylation unlikely without a donor of phosphates. To date, four of total six types of small molecular inhibitors have been developed underlining this thought. Type-I inhibitors bind directly to the ATP pocket of the active protein kinase, type-II unify a specific conformational segment nearby the pocket (reviewed in Dar & Shokat, 2011), while type-V conjunctly bound to another site through a long intermolecular linker (reviewed in Lamba & Ghosh, 2012), and type-VI adopts irreversibly covalent modification with the pocket (reviewed in Roskoski, 2016). Next, we will review the typical inhibitors in detail to better understand the orthosteric inhibition role of the ATP-binding site.

4.1 Orthosteric inhibition with inhibitors that reversibly, directly bind to the active state

In drug discovery schemes, a widespread approach is to find molecules from natural resources, particularly in the genomic era (Harvey, Edrada-Ebel, & Quinn, 2015). As a matter of fact, over 80% of recent drugs available in the market have indeed been derived explicitly or implicitly from the natural products (Sneader, 1996). The development of protein kinase inhibitors can trace back to the 1980s with a natural-product-based screen. One of the very first potent inhibitors; *Staurosporine*, which belongs to the indolocarbazole alkaloids, with a heteroaromatic ring displaying an ATP-competitive nature. Due to its large massive molecular size, the structural analysis has revealed that its benzene ring can stretch to the contiguous ATP unoccupied, hydrophobic region I (as well as ATP back pocket) and region II to gain higher binding affinity (Fig. 5A) (Lawrie et al., 1997). This promising inhibitor thus successfully captures the ATP-binding site in competition with high concentration ATP in vivo. Nevertheless, the Kinome-wide selectivity screening observation revealed that this potent inhibitor could potentially prevent (~87%) the kinases (Karaman et al., 2008). This promiscuity often induces toxic side effects (i.e., off-target effect) and makes it challenging as a candidate for drug design.

The inhibition of staurosporine, despite its wide range, is indeed a significant advantage in the discovery of selective kinase inhibitors. For example, for Chk1, a G2-checkpoint kinase in the cell cycle, the 7-*hydroxystaurosporine* derived from staurosporine becomes more selective (Fig. 5B). The combination of this derivative with the topotecan hydrochloride is a promising dosage for the treatment of patients with solid refractory tumors and is now this is in the Phase-I clinical trial (Hotte et al., 2006).

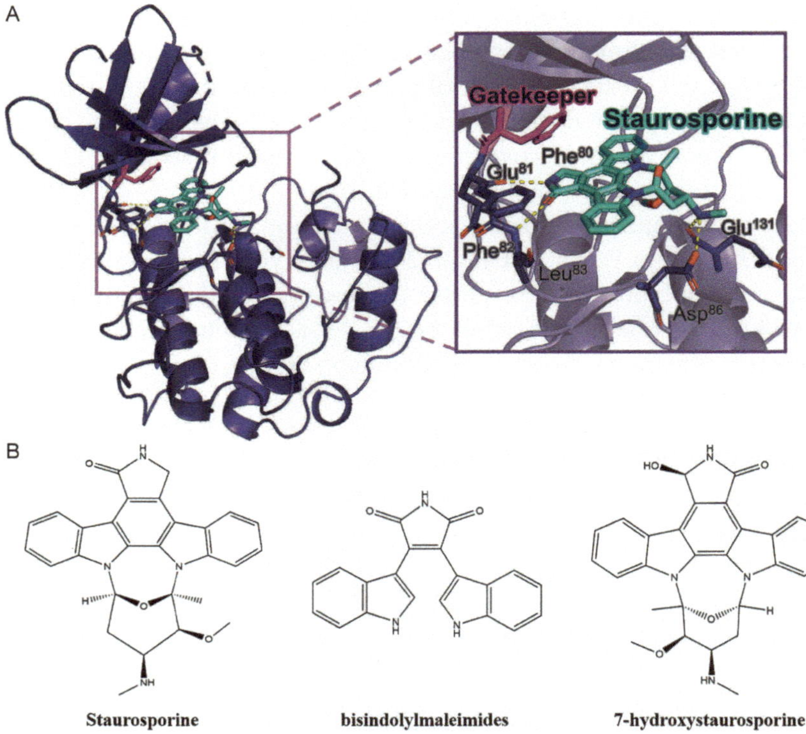

Fig. 5 Type I inhibitor Staurosporine in CDK2. (A) Cartoon representation of the CDK2 catalytic domain with Staurosporine and its interactions with residues nearby in close-up (PDB: 1AQ1). In the close-up, the Staurosporine makes interactions with critical residues nearby showed in the sticks, with yellow dashes indicating hydrogen bond. The heteroaromatic rings of Staurosporine function as a substitute of the adenine moiety, mimicking ATP in making similar hydrogen bonds with the Glu^{81} and Leu^{82} in the hinge region pharmacophore. (B) Molecular structure of staurosporine, bisindolylmaleimides, and 7-hydroxystaurosporine.

In addition, motivated by the lactam of staurosporine mimicking the adenine ring of the ATP molecule, several inhibitors, now referred to as type-I inhibitors, were designed using a heterocyclic ring system that occupies the adenine region. Some subtle but unique differences in the ATP back pocket are favorable to highly selective type-I kinase protein inhibitors, i.e., erlotinib (Liao, 2007).

Besides, we can obtain *bisindolylmaleimides* (Fig. 5B), the main ingredients to give a series of macrocyclic bisindolylmaleimides, by disrupting the planarity of the indolocarbazole. Such conformationally diverse macrocyclic bisindolylmaleimides occupy distinct regions within the ATP-binding site to generate inhibition profiling, allowing for the comparison of ATP-binding sites in different protein kinases without their structural features in science investigation (Bartlett et al., 2005).

4.2 Orthosteric inhibition with inhibitors that reversibly unify specific conformational portion

Later, unifying specific conformational portions has attracted considerable attention since some valid inhibitors have emphasized the potent pocket adjacent to the ATP-binding site. It figured out that the Asp of the DFG motif in the phosphate region adopts an "**out**" conformation, contributing to the creation of an additional hydrophobic cavity, which is termed as an **allosteric site**. This cavity has quite a superb complementarity with the ATP-binding site, aiding in the development of type-II inhibitor. The *Imatinib* was the first to apply this mechanism and achieved a substantial therapeutic efficacy, which demonstrating incredibly high binding preference toward the Abl in the biological system (Druker et al., 1996). The hinge-hydrogen binding type-I pharmacophore stabilizes the pyridine, pyrimidine, and the methylbenzene moieties. Additionally, in the allosteric site, peptide linkage moieties form a pair of hydrogen bonds with the Glu residue of αC helix and Asp of DFG motif. The following is a hydrophobic pyrimidine moiety, which is incorporated further into the hydrophobic cavity via van der Waals interaction and R-spine reformation participation (Fig. 6) (Schindler et al., 2000). However, type-II inhibitors cannot achieve full selectivity and are susceptible to resistance mutations within the binding site (i.e., gatekeeper T315I mutation in Abl for imatinib) (Shah et al., 2002). As well as the inactive DFG-D out conformational kinases only take up a small proportion of the Kinome (Modi & Dunbrack, 2019).

Furthermore, a combination of both type I and type II inhibitors, the type I $^{1}\!/_{2}$ inhibitors are developed (Zuccotto et al., 2010). The type I inhibitor binds to the DFG-D in/αC-in active conformation. Type II binds to the DFG-D out/αC-out inactive conformation. Most inhibitors of type I $^{1}\!/_{2}$ target protein kinases that have the αC-out and DFG-D in conformation. They occupy the adenine region by hinge-hydrogen binding type I pharmacophore and can extend to the ATP back cavity to form allosteric type II pharmacophore. The hydrophobic portion interacts with and stabilizes the R-spine, merely resulting in phenylalanine residue of the DFG motif moving out a little.

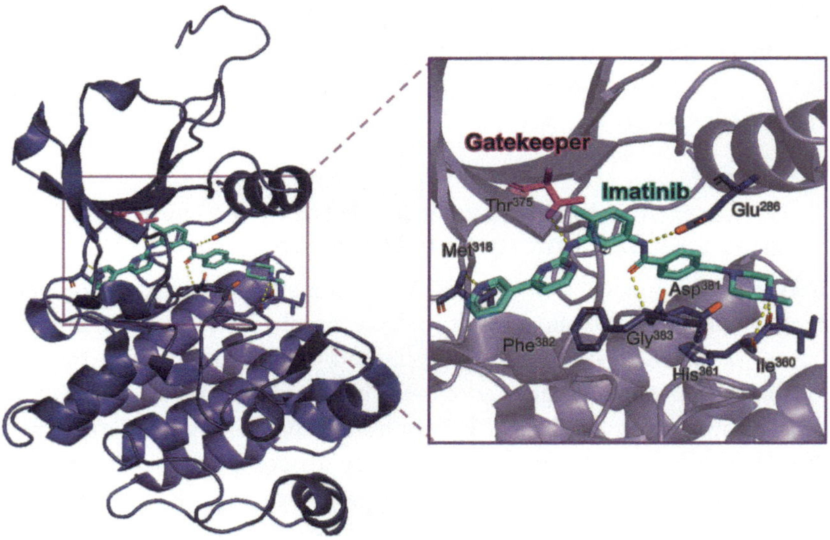

Fig. 6 Type II inhibitor Imatinib in c-Abl. Cartoon representation of the c-Abl catalytic domain with Imatinib and its interactions with nearby residues in close-up (PDB: 1IEP). The activation loop is collapsed onto the surface of the protein, blocking substrate binding and rendering the kinase catalytically inactive. In the close-up, Imatinib makes interactions with crucial residues nearby showed in the sticks. The aromatic rings of Imatinib function as a substitute for the adenine moiety, mimicking ATP in making hydrogen bonds with the Met^{318} in the hinge region pharmacophore and Glu^{286} in the αC helix. Imatinib can also form a hydrogen bond with Asp^{381} in the DFG-out motif.

4.3 Orthosteric inhibition with reversible, bivalent inhibitors

Bivalent inhibitors can target any two sites on the protein kinase surface and combining the advantage of both ligands can boost their potency and selectivity in theory. Bivalent inhibitors, now called type-V inhibitors. As is known to all, protein substrate-binding site exists in any protein kinase and show large diversity. Bi-substrate analog inhibitors, therefore, appear to be more attractive and feasible.

The classic example is the bi-substrate inhibitor (compound 2) for insulin receptor kinase (IRK) proposed by Parang et al. (2001). The inhibitor is synthesized by linking ATPγS (ATP analog) to IRS727 (peptide substrate analog) through a short two carbon spacer between the tyrosine nitrogen atom (used to be the phenolic oxygen atom) of the IRS727 analog and γ-phosphorous of the ATP analog. This design is inspired by observing the dissociative transition state catalyzed by protein kinases during phosphoryl transfer process. In order to be linked to the linker without abandoning its hydrogen acceptor role, they replace the phenolic oxygen atom with

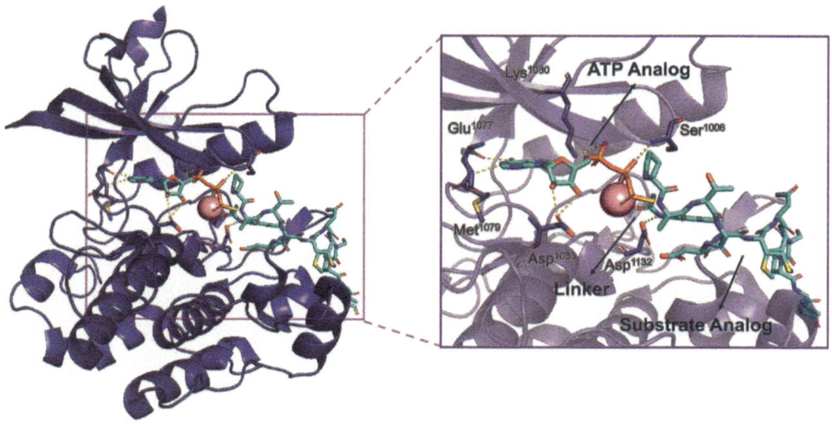

Fig. 7 Type-V inhibitors in IRK. Cartoon representation of the IRK catalytic domain with bi-substrate analog inhibitors and its interactions with residues nearby in close-up (PDB: 1GAG). In the close-up, the bi-substrate analogs make interactions with pivotal residues nearby showed in the sticks. The ATP analog functions as a substitute of the adenine moiety mimicking ATP in making hydrogen bonds with the Glu^{1077} and Met^{1079} in the hinge region pharmacophore and Ser^{1006} in the Gly-rich loop. Meanwhile, the substrate analog occupies the substrate-binding site to improve the efficiency of the inhibitor.

a nitrogen atom (Fig. 7). There is a linear competition relationship between the ATP analog as well as the peptide substrate analog of the bi-substrate inhibitor, and the Ki value is 370 nM, markedly lower than the respective substrate Km value (Km_{app}(ATP) = 71 μM; Km_{app} (IRS 727) = 280 μM). This exhibits a significant improvement in effectiveness. In a similar vein, a bi-substrate analog inhibitor for PKA by covalently tethering the ATPγS to the peptide substrate kemptide analog is 20 times more selective against closely associated kinase PKC with an unsatisfactory Ki value of 3.8 μM, though (Hines & Cole, 2004).

Regrettably, peptide moieties commonly exhibit unfavorable pharmacokinetic properties due to their conformational versatility. On certain occasions, peptide sequences parallel to the activation loop result in a secondary β-sheet restricting the regulatory capability of the loop. These dramatically limit the development of type-V inhibitors.

4.4 Orthosteric inhibition with irreversible inhibitors

Recently, it has drawn broad interest in developing irreversible inhibitors (i.e., afatinib) that form the covalent bonds with mainly cysteine (Cys) or other electronic-rich residues by Michael reaction around the ATP-binding site. Structural analysis reveals that hydrogen bonds and hydrophobic interactions with related residues in the ATP-binding site configure the afatinib in the proper hydrophilic attack orientation, leading to the α,β-unsaturated ketone moiety of afatinib covalently bond with Cys797 in the sugar region of epidermal growth factor receptor (EGFR) (Fig. 8) (Solca et al., 2012). Compared to previously clinically approved EGFR inhibitor drugs (erlotinib, gefitinib, and lapatinib) in EGFR/HER2 family, except for JHC7 ($IC_{50} = 1.346\,\mu mol$), afatinib has demonstrated persistent, effective pharmacodynamics for almost all cell lines (Magnaghi et al., 2018).

Type-VI inhibitors with a prolonged occupancy of the target are irreversible due to the covalent bond formation. However, the prolonged period of occupancy, which is defined as the reciprocal of the dissociation rate and concretely means the average time a ligand stayed at the binding site, could be toxic. For example, Clopidogrel irreversibly inhibits the aggregation of platelets to reduce thrombosis significantly; meanwhile, platelets mainly contribute toward hemostasis. Thus prolonged inhibition caused by more extended periods of occupancy will cause excessive bleeding (De Cesco, Kurian, Dufresne, Mittermaier, & Moitessier, 2017). In the healthcare facility, this restriction should always be a clear indication to consider whether or not covalent inhibitors have always been the most suitable choice.

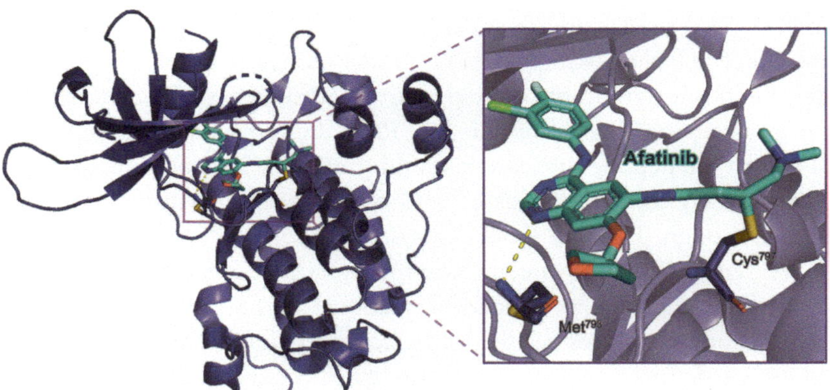

Fig. 8 Type-VI inhibitor Afatinib in EGFR. Cartoon representation of the EGFR catalytic domain with afatinib and its interactions with residues nearby in close-up. In the close-up, the 2-chloro-3-fluorobenze moiety binds to the ATP-binding site. The quinazoline ring forms the hydrogen bonds with Met^{793} in the hinge region pharmacophore. Above all, the afatinib can covalently bind with Cys^{797} in the sugar region.

5. Allosteric regulation

Protein kinases have evolved to be highly dynamic molecules but with strictly arranged relative positions. Because rotational, reciprocating motions of the flexible helixes, loops, and domains within protein kinases are precisely regulated by cofactors, autoinhibitory domains, or the coupling proteins interactions (Taylor & Kornev, 2011). Moreover, each scattered flexible fraction is strung together in an interactions-line. So ligands binding events at the direct site immediately perturb the biochemical environment of the site, the "first layer" of atoms have to readjust their interactions as a response to the binding events, then migrate layers by layers, leading to changes at the topologically distinct site, and macroscopically accompanying altered biological macromolecular activity (i.e., active and inactive state) (Lu, He, Ni, & Zhang, 2019). This is allosteric regulation, "the second secret of life," ranking only second to the genetic code (Fenton, 2008). The communication between the allosteric site and the active site is achieved through the allosteric signal pathways, including atomic fluctuations, residues networks, domain movements, or other macromolecules participations, and these are generally a superposition relationship (He, Ni, Lu, & Zhang, 2019). Currently, the reverse allosteric communication between ATP-binding site and regulatory site of the catalytic domain is closely concerned, which will provide innovations for disturbing the interactions between the target protein kinases and their regulatory partners (Leroux & Biondi, 2020). In addition, the ATP molecule itself can as an allosteric modulator, for it exists in two distinct conformations: compact or extended (Lu et al., 2014), however, not the point in this article. Herein, we summed up the allosteric regulatory role of the ATP-binding site, highlighted the structural and molecular mechanism, and show how compounds trigger this significant role in boosting or destabilizing the interactions at the active site.

5.1 Allosteric modulation through intramolecular communication

5.1.1 Through the residues networks

Akt (as well as protein kinase B) is a member of AGC protein kinase family, sharing numerous similarities with PKA. We have discussed in the earlier para, i.e., phosphorylation at the activation loop (i.e., Thr^{308}) for kinase activity, but differing from PKA, whereas, the phosphorylation is necessary at the HM-site (i.e., Ser^{473}) to reach the maximum kinase activity

(Manning & Toker, 2017; Sarbassov, Guertin, Ali, & Sabatini, 2005). Cytosolic Akt exists in a compact, autoinhibited "**PH-in**" configuration (PH refer to pleckstrin domain), whereby the Phosphatidylinositol (3,4,5)-trisphosphate (PIP_3) (or $PI(3,4)P_2$) binding site along with phosphorylation site either of activation loop or HM-site are buried within the interface between regulatory and kinase domain (Fig. 9A). PIP_3 binding sites would be uncovered in a preliminarily opening conformation, can sense and bind PIP_3 on the cellular membrane. The Akt is PIP_3-hypersensitive. Only a small portion of Akt in the preliminarily opening conformation can shift the equilibrium to the extended open conformation, which the substrate-binding pocket is relived and, together with activation loop and HM exposed to phosphoinositide-dependent kinase-1 (PDK1) and mammalian target of rapamycin (mTOR) for phosphorylation to promote substrate binding (Alessi & Cohen, 1998; Coffer, Jin, & Woodgett, 1998; Wu et al., 2010; Yang et al., 2002). In addition, the contents of phosphorylation are finely calibrated by the inverse abilities of upstream kinases and phosphatases. Co-localization of the Akt to an intracellular membrane will turn up the phosphorylation level by drastically reducing the Akt sensitivity to phosphatase (Fig. 9B). Furthermore, ATP-binding event would amplify this tendency (Fig. 9D) (Chan et al., 2011). The ATP-binding site is spatially far away from the phosphorylation sites, hence, it is suggested that allosteric regulation originated from the ATP-binding site takes place, and eventually leads to this hyperphosphorylation modulation.

Using the molecular dynamics simulation Lu et al. reveal the pivotally interconnected residues on the ATP → $pThr^{308}$ allosteric signal pathway (ATP → Thr^{160} → Glu^{191} → His^{194} → $pThr^{308}$) as shown in Fig. 9G (Lu et al., 2015). Notably, the carboxyl group of Glu^{191} residue forms a hydrogen bond with the imidazole ring of His^{194} residue, which is likely to weaken the interaction between the His^{194} residue and $pThr^{308}$ residue. These residue networks make it possible to transduce the ATP occupying effects allosterically through the Gly-rich loop and the αC-helix to the activation loop, and eventually promote the burial of the $pThr^{308}$ in the "locking cages" to avoid accessibility of phosphatase. This "locking cages" is constructed by two salt bridges with His^{194} and Arg^{273}, together with a hydrogen bond with Lys^{297} (Fig. 9E). Additionally, mutant analysis further shows that a positive electrostatic group and a structural match are essential for this construction. Meanwhile, the residue $pSer^{473}$ of the HM engages in a hydrophobic groove in the regulatory hotspot, hidden away from the phosphatase in the regulatory hotspot, as shown in Fig. 9F (Lin et al., 2012; Yang et al., 2002).

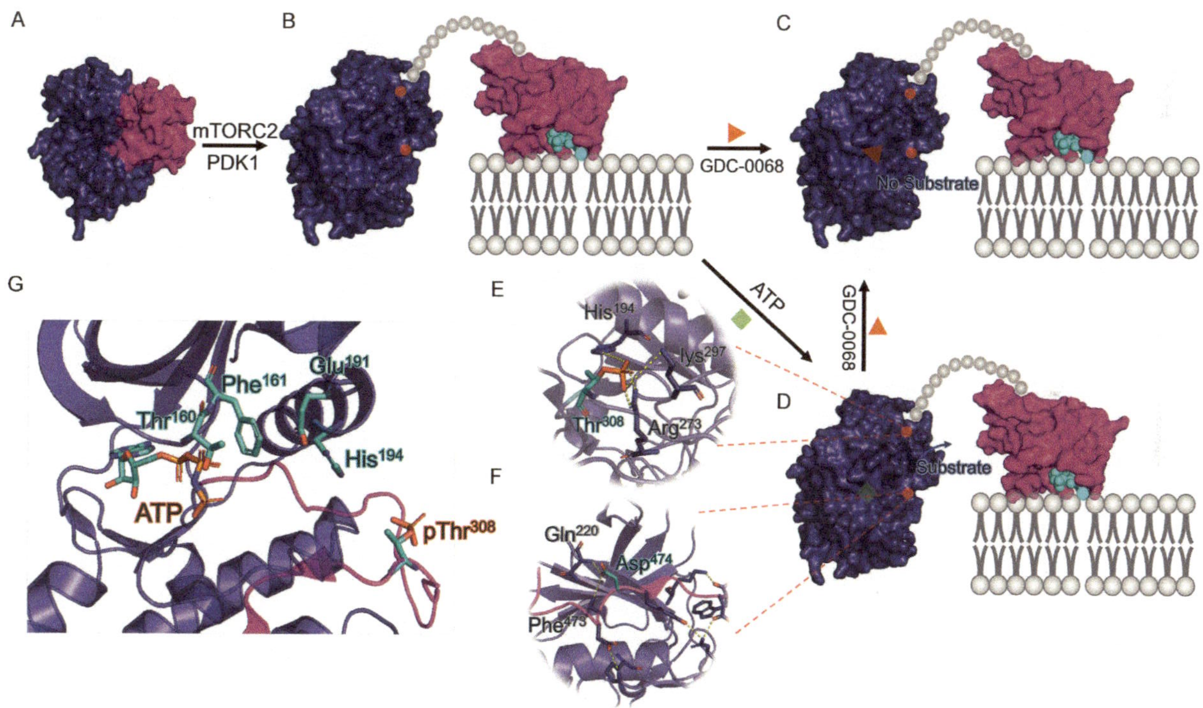

Fig. 9 See figure legend on next page.

Such new residue networks may provide a general model on the allosteric regulation role of ATP-binding site, and a valuable framework for residue networks of other AGC kinase group protein kinases.

Surprisingly, ATP-competitive inhibitors (i.e.GDC-0068) also exhibit the ability to hyper-phosphorylate the residue Thr^{308} and Ser^{473} of Akt (Fig. 9C) (Lin et al., 2012; Okuzumi et al., 2009). Experimental analysis shows that the GDC-0068 selectively favors the activated "**DFG-in**" Akt, on account of the steric hindrance between the GDC-0068 and the Phe residue of "**DFG-out**" conformation. Hence, the ATP-competitive inhibitors may interfere with Akt autoinhibition reformation, locking Akt in a phosphorylated but non-functional state. Along the same lines, $pThr^{308}$ forms multiple interactions with residues nearby, and $pSer^{473}$ hides into the hydrophobic cavity to shield from dephosphorylation by phosphatase.

5.1.2 Through domain movements

AMP-activated protein kinase (AMPK) is a trimeric complex consisting of two regulatory subunits (refer to β and γ) and a catalytic subunit (refer to α), which plays a crucial role in holding energy homeostasis (Ross, MacKintosh, & Hardie, 2016). The AMPK complex could sensitively detect the physiological energy stress that is characterized by a larger ratio between AMP/ADP and ATP, after which activated to facilitate the ATP-generating processes, while also suppress the ATP-consuming pathways (Oakhill, Scott, & Kemp, 2012). There are four cystathionine-B synthase (CBS)

Fig. 9 Akt allosteric regulation model. Surface representation of the stepwise hyper-phosphorylation process. (A) In the quiescent state, Akt preferentially adopts the auto-inhibited "PH-in" conformation, in which the PH domain (magenta) stabilized by interacting with the regulatory hotpot in the Akt (deep blue). (B) A small portion of conformation in which PIP_3 binding site exposed can be induced to bind PIP_3 (cyan) on the plasma membrane. As a consequence, $pThr^{308}$ (red circle) in the activation loop and $pSer^{473}$ (red circle) in the hydrophobic motif (mimicking by S474D substitution in Akt2) are exposed to be phosphorylated. Protected from dephosphorylating by phosphatase by the "shielding interactions." (C) ATP-competitive inhibitor (orange triangle) (D) ATP (green square) binding in the ATP-binding site contributes to protecting $pThr^{308}$ (E) and $pSer^{473}$ (F) from dephosphorylating by phosphatase via the "shielding interactions." (G) Cartoon representation of Akt allosteric residues networks from the ATP-binding site to the active site with residues shown in the sticks. Two residues, Thr^{160} and Phe^{161}, are in the Gly-rich loop, and the other two, Glu^{191} and His^{194}, are in the αC helix. PDB code used: 1O6L, 1UNQ, 3O96, 6BUU. Kinase domain colored in deep blue; PH domain colored in magenta.

adenine nucleotide binding sites in AMPK. However, only the CBS1 and CBS3 are interchangeable, CBS4 is permanently bound to AMP, while CBS2 is empty due to the point mutation, which lost the ribose-stabilizing interactions [51–53]. Excessive AMP binding in the γ-subunits at CBS1, CBS3, and CBS4 sites shift the kinase from inactive to the active state manifested by Thr of the activation loop phosphorylation in the α-catalytic subunit. Also, researchers find that CBS3 is the sensor site (Chen et al., 2013; Xiao et al., 2011; Xin, Wang, Zhao, Wang, & Wu, 2013), allosterically transducing AMP binding information from the regulatory domain to the kinase domain and subsequently eliminating the interactions between the kinase domain and an autoinhibited domain (AID). Moreover, AMP binding at CBS1 and CBS4 contributes to strengthening AMP binding at CBS3, albeit much higher ATP concentration in vivo (Chen et al., 2012; Gu et al., 2017; reviewed in Yan, Zhou, Xu, & Melcher, 2018).

Structurally, the resolved AMP-bound AMPK crystalized structures reveal a linker, now called α-linker, linking the AID and α-β subunit-interacting C-terminal domain (α-CTD) to the γ-subunit (Xiao et al., 2011). On the one side, in the α-linker, 10 amino acid sequences at the boundary of the AID can connect to the empty CBS2, now named as regulatory subunit-interacting motif 1 (αRIM1) (Chen et al., 2013; Xiao et al., 2013, 2011; Xin et al., 2013). On the other side, another 10 amino acid sequence of the α-linker, αRIM2 (Chen et al., 2013; Xin et al., 2013), is later validated to be the adenine nucleotide sensor segment by interacting with AMP at CBS3 directly. It would, therefore, be evident that the allosteric communication from αRIM2/CBS3 to αRIM1-AID/CBS2 takes place when the AMP binding at CBS3 (Fig. 10E).

Ja-Wei Wu et al. adopt a mutational analysis to determine the pivotal residues in the intermolecular residues networks (Xin et al., 2013). First, Lys^{170} (in γ1) directly interact with the AMP at CBS3 and αRIM2 Glu^{364} (in α1) via electrostatic bond, the following residue Arg^{171} connect with αRIM1 F342 (in α1); Next, Lys^{170}-interacting (through hydrogen bond) residues Lys^{174} and Phe^{175} form Van der Waals bonds with Phe^{342} and Phe^{179} (in γ1). The latter is the heart of the interface between the regulatory domain and AID, directly interacting with all four αRIM1 residues (Ile^{335}, Met^{336}, Tyr^{343}, and Phe^{342}) required for the relief of AMPK autoinhibition. Similarly, Glu^{364}, Arg^{171}, and Phe^{179} are also required for the relief of AMPK autoinhibition (Fig. 10E). This mechanism is responsible for shifting the AID equilibrium from the inactive, kinase domain-bound conformation to the active, γ/CBS2-bound conformation.

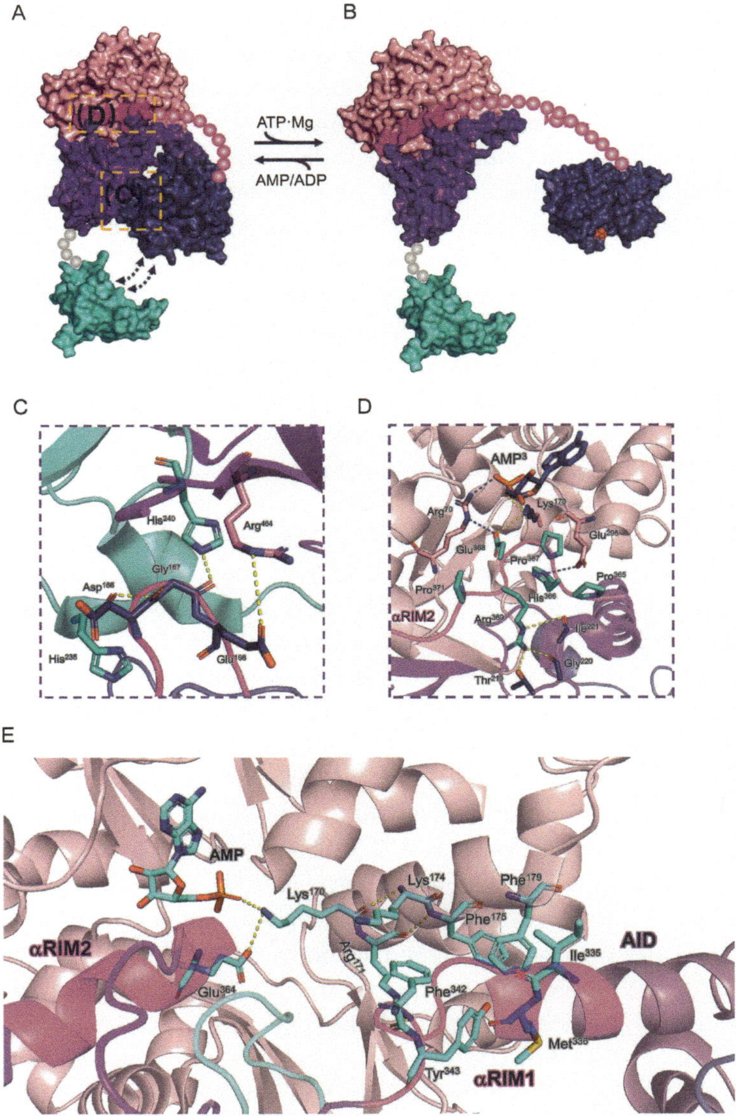

Fig. 10 AMPK allosteric regulation model. Surface representation of the hyperphosphorylation stepwise process. (A) Binding AMP/ADP can instead protect the phosphorylated residues. (B) Binding ATP at the CBS sites may displace the αRIM2, induce the kinase domain movement, and then expose the phosphorylated residues. The close-up (C) shows how the activation loop interacts with residues in the regulatory core. The close-up (D) shows how αRIM2 interacts with residues nearby, and each position containing a Pro will form a turn. (E) Cartoon representation of AMPK allosteric residues networks from the CBS 3 to the AID with residues shown in the sticks. PDB code used: 4CFF, 4CFH, 4GFG. The γ-subunit is colored in pink; α-regulatory domain is colored in blue-violet and the kinase domain colored in deep blue; β subunit is colored in teal; α-linker is represented by magenta balls; β-linker is represented by gray balls.

Intriguingly, in AMPK, compared to Akt, not only the ATP-binding event has a different consequence in phosphorylation/dephosphorylation circle, but the phosphorylated threonine of the activation loop is observed to be protected tighter with AMP/ADP binding (Xiao et al., 2013, 2011). The former is a result of the unique function of AMPK, and the latter relies on the different "docking pattern." In detail, AMP/ADP bound to CBS3 causes position changing of αRIM2, induces the displacement of α-linker (Fig. 10D), and results in kinase domain movement toward the regulatory domain. Thus, the activation loop of the kinase domain is buried into the core of AMPK, around which is the stable β subunit C-terminal domain (β-CTD) (Fig. 10C), stabilizing the activation loop. However, this also presents a conundrum that how can the activation loop buried in the AMPK core be mostly inaccessible to protein kinase without affecting access to the upstream kinase. Without activator, the enzyme exhibits less activity, because the interactions between the CBM and the kinase domain are weaker (Fig. 10A). Replacing AMP (or ADP) by Mg^{2+}/ATP reversely leads to the replacement of the α-linker and then driven the kinase domain movement, dissociated from the regulatory domain. In this conformation, the kinase is no longer allosterically active and susceptible to dephosphorylation (Fig. 10B).

5.2 Allosteric enhancement through protein-protein communication

5.2.1 Through other macromolecules participation—Upstream kinase phosphorylation

Prior to this, protein kinase D (PKD) was misclassified as a member of the PKC group, mainly owing to their kinase domains structural homology, and they both bind to the lipid diacylglycerol (DAG) on the intracellular membrane (Baron & Malhotra, 2002). The N-terminal regulatory domain of PKD comprises two cysteine-rich C1 domains (C1a and C1b or solely CRD domain) and pH-domain. Likewise, the regulatory domain can exert autoinhibition toward the C-terminal kinase domain, since disruption of any component of regulatory domains results in increased kinase activity (Van Lint et al., 2002). This cytosolic autoinhibition conformation can be released dependent on DAG-sensor C1, localizing the PKD to the membranes, whereby its upstream kinases novel PKCs (nPKCs) facilitate phosphorylations at the activation loop (Ser^{738} and Ser^{742} in human PKD1) (Rozengurt, Rey, & Waldron, 2005). Moreover, autophosphorylation can take place within a PDZ-binding motif at the C-terminal site (Ser^{910} in protein kinase).

Additionally, phosphorylations at other sites may also be required for PKD to perform other functions (reviewed in Storz, 2012).

Both ATP-competitive inhibitors (i.e., Go 6976) and noncompetitive inhibitors (i.e., CID 755673) targeting PKD have the ability to hyper-phosphorylate the two serine residues of the activation loop. The underlying mechanism is distinguished from Akt, as it is proved not to be an intrinsic property of PKD (Kunkel & Newton, 2015). Paradoxical inhibitors bound to the ATP-binding site (other sites for noncompetitive inhibitors but is not the point here) induce a conformational change at the regulatory hotspot, unmasking the DAG-sensor C1 domain so that PKD can be recruited to the DAG-rich membranes, subsequently activating its upstream kinase nPKC to promote phosphorylation at the activation loop of PKD (Fig. 11). Notably, because PKD gradually autophosphorylates at its C-terminal site, the substrates would still be tethered at this site, accumulating phosphorylation in a long way to act.

5.2.2 Through other macromolecules participation—Dimer formation

RAF kinases are composed of a typical protein kinase architecture with the C-terminal kinase domain and N-terminal regulatory domain, transmitting

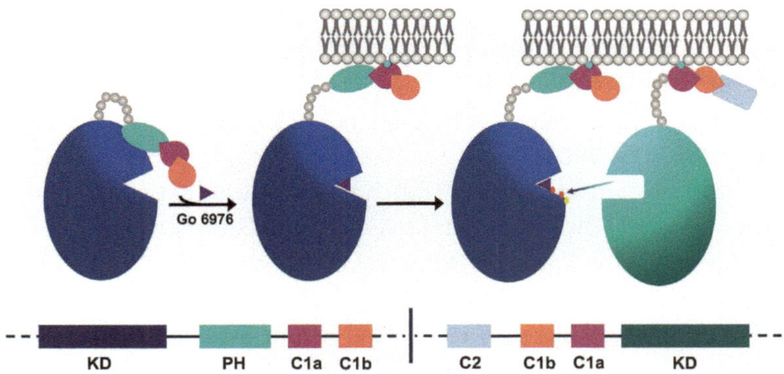

Fig. 11 PKD allosteric regulation model. Cartoon representation of the hyper-phosphorylation stepwise process by upstream kinases participants. Inhibitors occupy the ATP-binding site and allosterically disrupt the autoinhibited, inactive state via PH domain releasement. Following that, PKD is recruited to the membrane by exposed DAG-sensor C1 domain, downstream signal forces nPKC binding membrane nearby, which constantly phosphorylate the activation loop of protein kinase (red circle). Notably, because PKD can autophosphorylate at the C-terminal site, the substrate will still accumulate phosphates at the tail of the C-terminal domain (yellow circle). PKD colored in deep blue; nPKC colored in teal; functional domain in the cartoon corresponds to the below domain array.

signals that downstream of RAS-GTP generation on the intracellular membranes to activate the downstream MEK/ERK pathways (reviewed in Lavoie & Therrien, 2015). Besides the conserved kinase domain (as well as conserved region 3: CR3), RAF kinases contain two other conserved regions in their regulatory domain. The first conserved region (CR1) is made up of a Cys-rich domain (CRD) along with a RAS-binding domain (RBD), and the second (CR2) is characterized by Ser/Thr richness. Similarly, in the cytosolic quiescent state, RAF prefers to adopt an autoinhibited conformation (Fig. 12A). Current researches show that this autoinhibited interaction is further fixed by binding 14-3-3 cradle to the phosphorylated sites of the CR2 ($pSer^{365}$ in BRAF) and the carboxyl terminus ($pSer^{729}$ in BRAF) (Park et al., 2019). Moreover, when RAF turns to be active, the 14-3-3 protein reorganizes to link $pSer^{729}$ binding sites of two primers and enforce RAF dimerization. This mechanism gives a satisfactory answer to the paradox that ATP is essential for RAF activation, while the physiological concentration of ATP is prejudice to RAF dimerization (Liau et al., 2020; Park et al., 2019).

Canonical activation of RAF starts with GTP loading and RAS-GTP accumulating preparations. Then RAF is translocated to the membrane upon CRD and RBD concertedly binding to distinct sites in RAS. However, the CR1 reliving mechanism remains elusive (Tran, Wu, & Frost, 2005). Sequentially, RAF is generally regarded to stimulate activity by contacting with another RAF kinase domain via a side-to-side contact mode (reviewed in Lavoie, Li, Thevakumaran, Therrien, & Sicheri, 2014), the linchpin of which is exploited to be Arg^{509} (in BRAF) through extended interaction networks between two promoters (Fig. 12D) (Rajakulendran, Sahmi, Lefrançois, Sicheri, & Therrien, 2009). The productive αC helix holds the dimerization crucial residue Arg^{509} positioned within the dimer interface. With the crystal structure of monomeric BRAF (Thevakumaran et al., 2015) and BRAF-MEK complex (Park et al., 2019), we can gain an insight into the mechanism. In the monomeric state, the αC helix presents an expected "αC-out" inactive conformation. Furthermore, the structures of BRAF dimer docking with MEK shows that the activation loop of BRAF starches out, losing its stabilizing interaction with αC-out conformation, and αC helix turns in to mediate dimerization. Notably, the DFG motif adopts only DFG-in conformation in both active and inactive states.

Surprisingly, in clinical treatment, almost all the ATP-competitive inhibitors targeting the BRAF kinases cause an opposite effect, leading to paradoxically enhanced kinase activity. Some even accordingly activate

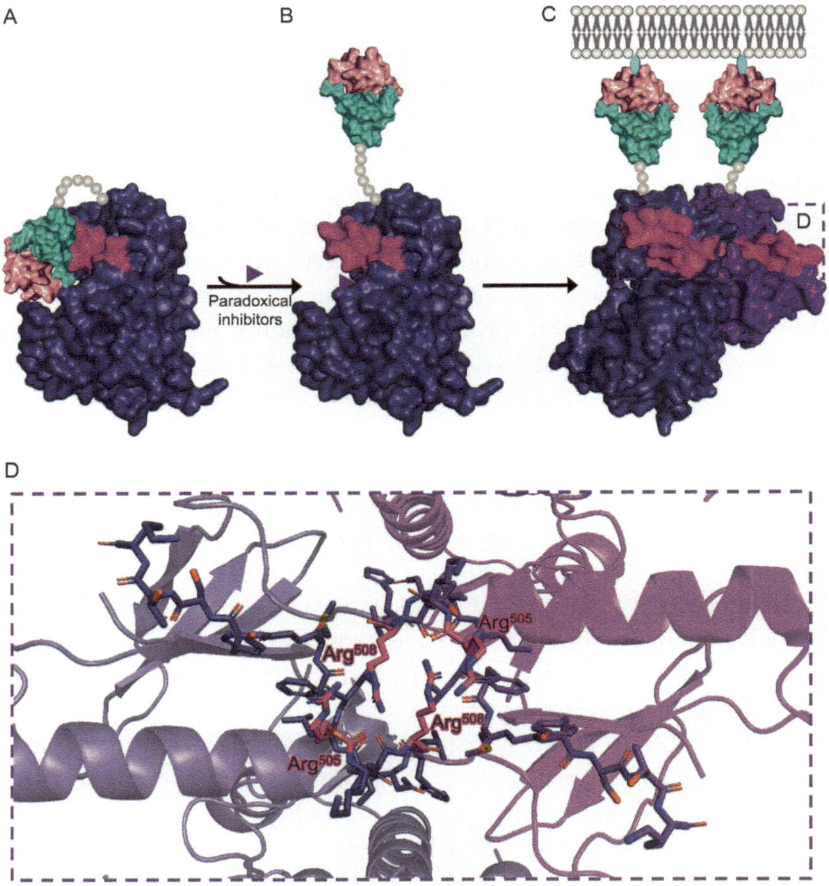

Fig. 12 RAF allosteric regulation model. Surface representation of hyperphosphorylation stepwise process by dimer formation. Inhibitors occupy the ATP-binding site and thus allosterically disrupt (A) the autoinhibited, inactive state via (B) negative regulatory domain release. The activated RAF kinases then form (C) a side-to-side dimer, mediated by dimer interface interactions. (D) The crucial dimer interface residues are shown in stick in the close-up below. Arg^{505} is crucial for RAF-RAS-GTP interaction, and Arg^{508} with adjacent residues are crucial for dimerization. Both are highlighted in magenta. PDB code used: 1FAR, 3Q4C, 5J17. RBD domain is colored in pink; the CRD domain is colored in teal; the kinase domain is colored in deep blue, and the other one is colored in purple violet. αC helix is colored in magenta.

the downstream effector ERK for further signal transduction (Hatzivassiliou et al., 2010; Poulikakos, Zhang, Bollag, Shokat, & Rosen, 2010). Nevertheless, this is understandable through the dimerization-depend allosteric regulation mechanism. Some inhibitors, termed as RAF αC-in inhibitors, contribute to stabilizing the αC-in conformation, promoting RAF to interact with RAS-GTP (Fig. 12B), and facilitating the subsequent phosphorylation and dimerization process (Fig. 12C). However, some drugs stabilizing αC-out conformation can also increase the Raf-RAS-GTP interactions. A reasonable hypothesis corresponds to the unexpected phenomenon that not the whole αC helix part, but a small region near Arg^{506} positions in an active orientation (Fig. 12D), which is adequate for strengthening the RAF-RAS-GTP interactions (Karoulia et al., 2016).

Moreover, RAF αC-out inhibitors induce stronger negative allostery for the second promoter binding. In other words, the other is more likely to participate in a αC-in conformation, owing to the unfavorable steric clash when two outward αC helix concurrence. Therefore, αC-out inhibitors show low affinity to the second promoter, ineffectively inhibiting the active dimer. On the contrary, αC-in inhibitors have a similar, high affinity toward both promoters at medium to high concentration, and they inhibit the downstream MEK activation more effectively (Karoulia et al., 2016).

5.3 Allosteric disruption through protein-protein communication

Aurora kinase A (AURKA) is essential for cell division, regulating mitosis initialization, centrosome maturation, and separation, bipolar spindle formation, evolutionarily closely related to AGC kinases (reviewed in de Souza & Kawano, 2020).

Similar to AGC kinases, threonine phosphorylation ($pThr^{288}$) in AURKA of the activation loop is crucial to AURKA activation. However, without Xenopuskinesin-like protein 2 (TPX2) binding, $pThr^{288}$ locates outwards and is susceptible to hydrolysis. AURKA-TPX2 interactions establishment stabilizes a water-mediated allosteric network with rare polar glutamine in the hydrophobic spine (Cyphers, Ruff, Behr, Chodera, & Levinson, 2017), causing an alteration of the activation loop, displacing $pThr^{288}$ inward to form a salt-bridge with Arg^{255} of the HRD motif, and keeping it away from phosphatase (Bayliss, Sardon, Vernos, & Conti, 2003). Moreover, binding with TPX2 will not only lock AURKA in the active state but target AURKA to the spindle (Kufer et al., 2002). Structurally, McIntyre et al. show that Tyr^{8}, Tyr^{10}, Phe^{16}, Trp^{34}, Tyr^{8} is responsible for cohering TPX2 with AURKA (Fig. 13C) (McIntyre et al., 2017).

Recently, it has been shown that ATP-competitive inhibitors CD532 could induce a conformational change that disrupts the interactions between the regulatory hotspot of AURKA and TPX2 (Fig. 13A and B)

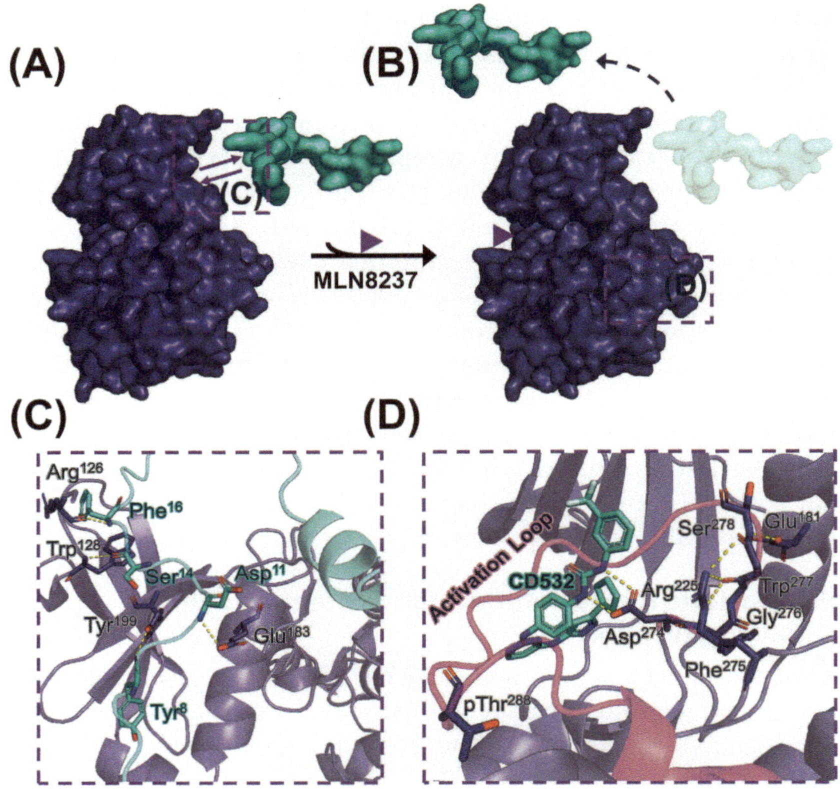

Fig. 13 AURKA allosteric regulation model. Surface representation of the AURKA-TPX2 dissociation stepwise process. (A) TPX2 and AURKA tightly interact. Inhibitors CD532 bound to the ATP-binding site induces N-lobe movement and regulatory hotspot disorders through residues networks of regulatory spines, (B) leading to the disrupted interaction between the TPX2 and the regulatory hotspot of the kinase domain. The close-up (C) shows how TPX2 interacts with residues in the hotspot, and (D) shows how CD532 stabilizes the dormant state of AURKA. PDB code used: 3HA6. TPX2 is colored in teal; the Kinase domain is colored in deep blue.

(Gustafson et al., 2014). The N-terminal domain moves 6.2 Å relative to the C-terminal domain, and regulatory hotspot disorders through residues networks of regulatory spine lead to AURKA-TPX2 interaction disruptions. In more detail, the polar interactions between the urea moiety of CD532 and Asp^{274} in the DFG motif fix CD532 in a configuration, whereby the β1, β2 strands have to avoid the steric clash caused by the trifluoromethylphenyl moiety of CD532. As a result, the displacement of β1, β2 strands, the major portion of the rigid core of the N-terminal domain, induces the N-terminal domain movements. Besides, the displaced αC helix and Arg^{255} together trap the most N-terminal portion of the activation loop in a network of the hydrogen bonds, holding the activation loop in its inactive state, also manifested by a typical DFG-out conformation (Fig. 13D). Consistent with such a model, inhibitors stabilizing the DFG-in conformation has positive cooperativity with TPX2, while stabilizing DFG-out conformation has negative cooperativity with TPX2 (Lake et al., 2018).

Furthermore, the disruption of protein-protein communication could add a second level of inhibition to a signal unit in AURKA signaling because the detachment of AURKA with TPX2 will further alter TPX2-multiprotein complex colocalization. Accordingly, one ought to realize that the longer the residues time of the drug, the lesser the structural dimension of integration into a functional signal unit, even though the AURKA is completely active (Leroux & Biondi, 2020).

6. Conclusions and perspectives

We have come a long way since the first FDA-approved protein kinase inhibitor launched on the market about 20 years ago. As we can see, ATP-binding site of protein kinases, was, is, and always will be the "secret weapon" to win the war between human beings and diseases, particularly cancer. To make the best of it, we ought to understand their language. Here, we decode the dual role of the ATP-binding site in today's perspective.

Researchers first designed orthosteric inhibitors targeting the ATP-binding site to restrict the action of uncontrolled protein kinase. Four recapitulative applications of orthosteric inhibition have been used extensively up to now. Afterward, the research on the reverse allosteric regulation commences at the ATP-binding site and picks up momentum. Compounds that bind to the ATP - binding site may allosterically enhance or disrupt intra- or outra-molecular interactions through a variety of mechanisms, including residue networks, domain movements, upstream kinase-dependent, dimerization, protein-protein interactions, and so forth.

Collectively, the words from the ATP-binding site are written in interactions. Electrostatic interactions, hydrogen bonds, van der Waals interactions, hydrophobic contacts, and covalent interaction exits in the ATP-binding site contributing to either the orthosteric core or the allosteric regulation pathway.

Notwithstanding significant progress has been made in the field of protein kinases, researchers now actually focus on a small portion of protein kinases in the face of the vast Kinome. Nevertheless, quintessential examples listed in this review might enlighten us to exploit resembling interactions in the protein kinases responsible for orthosteric inhibition or allosteric regulation of the ATP-binding site, so that researchers can develop rational design modulators to orthosterically inhibit protein kinase activity or produce desired allosteric effect that regulates the intramolecular or protein-protein interactions. Slowly but surely, we will gradually reveal the mystery of the Kinome as a whole.

Conspicuously, in the light of the "paradoxical activation" frequent occurrences, researchers should take care of the small molecular inhibitors when they try to explain the phosphorylation result in the conducting experiments (Kunkel & Newton, 2015).

Besides, in recent years, drug resistance has become the leading obstacle in this field. This obstacle is generally divided into two groups, target-dependent resistance causes ineffective binding modes triggered by alteration in the protein kinase targets and pathway-dependent resistance that invalidates the inhibitors through abnormally activating targets in compensation signal transduction ways. Here, summarizing the dual roles of the ATP-binding site may provide some suggestions on the underlying mechanisms of any orthosteric and allosteric drugs, so that we can better combine those two kinds of drugs to overcome the resistances (reviewed in Agnello, Brand, Chellat, Gazzola, & Riedl, 2019).

Last but not the end, since ATP is a universal, essential mediator of cellular homeostasis, the language we deciphering from ATP-binding sites may be conservative to other proteins, and we hopefully offer some references on regulatory mechanisms in related to ATP-binding site in other protein families.

Acknowledgment

This work was supported by the National Natural Science Foundation of China (21778037).

Declaration of competing interest

The authors declare that they do not have any conflict of interest related to this manuscript.

References

Agnello, S., Brand, M., Chellat, M. F., Gazzola, S., & Riedl, R. (2019). A structural view on medicinal chemistry strategies against drug resistance. *Angewandte Chemie (International Ed. in English)*, *58*(11), 3300–3345. https://doi.org/10.1002/anie.201802416.

Alessi, D. R., & Cohen, P. (1998). Mechanism of activation and function of protein kinase B. *Current Opinion in Genetics & Development*, *8*(1), 55–62. Retrieved from https://pubmed.ncbi.nlm.nih.gov/9529606.

Baron, C. L., & Malhotra, V. (2002). Role of diacylglycerol in PKD recruitment to the TGN and protein transport to the plasma membrane. *Science (New York, NY), 295*(5553), 325–328. Retrieved from https://pubmed.ncbi.nlm.nih.gov/11729268.

Bartlett, S., Beddard, G. S., Jackson, R. M., Kayser, V., Kilner, C., Leach, A., et al. (2005). Comparison of the ATP binding sites of protein kinases using conformationally diverse bisindolylmaleimides. *Journal of the American Chemical Society, 127*(33), 11699–11708. Retrieved from https://pubmed.ncbi.nlm.nih.gov/16104747.

Bayliss, R., Sardon, T., Vernos, I., & Conti, E. (2003). Structural basis of aurora-A activation by TPX2 at the mitotic spindle. *Molecular Cell, 12*(4), 851–862. Retrieved from https://pubmed.ncbi.nlm.nih.gov/14580337.

Chan, T. O., Zhang, J., Rodeck, U., Pascal, J. M., Armen, R. S., Spring, M., et al. (2011). Resistance of Akt kinases to dephosphorylation through ATP-dependent conformational plasticity. *Proceedings of the National Academy of Sciences of the United States of America, 108*(46), E1120–E1127. https://doi.org/10.1073/pnas.1109879108.

Chen, L., Wang, J., Zhang, Y.-Y., Yan, S. F., Neumann, D., Schlattner, U., et al. (2012). AMP-activated protein kinase undergoes nucleotide-dependent conformational changes. *Nature Structural & Molecular Biology, 19*(7), 716–718. https://doi.org/10.1038/nsmb.2319.

Chen, L., Xin, F.-J., Wang, J., Hu, J., Zhang, Y.-Y., Wan, S., et al. (2013). Conserved regulatory elements in AMPK. *Nature, 498*(7453), E8–10. https://doi.org/10.1038/nature12189.

Coffer, P. J., Jin, J., & Woodgett, J. R. (1998). Protein kinase B (c-Akt): a multifunctional mediator of phosphatidylinositol 3-kinase activation. *The Biochemical Journal, 335*(Pt. 1), 1–13. Retrieved from https://pubmed.ncbi.nlm.nih.gov/9742206.

Cohen, P. (2002). Protein kinases—The major drug targets of the twenty-first century? *Nature Reviews. Drug Discovery, 1*(4), 309–315. Retrieved from https://pubmed.ncbi.nlm.nih.gov/12120282.

Cyphers, S., Ruff, E. F., Behr, J. M., Chodera, J. D., & Levinson, N. M. (2017). A water-mediated allosteric network governs activation of Aurora kinase A. *Nature Chemical Biology, 13*(4), 402–408. https://doi.org/10.1038/nchembio.2296.

Dar, A. C., & Shokat, K. M. (2011). The evolution of protein kinase inhibitors from antagonists to agonists of cellular signaling. *Annual Review of Biochemistry, 80*, 769–795. https://doi.org/10.1146/annurev-biochem-090308-173656.

De Cesco, S., Kurian, J., Dufresne, C., Mittermaier, A. K., & Moitessier, N. (2017). Covalent inhibitors design and discovery. *European Journal of Medicinal Chemistry, 138*, 96–114. https://doi.org/10.1016/j.ejmech.2017.06.019.

de Souza, V. B., & Kawano, D. F. (2020). Structural basis for the design of allosteric inhibitors of the Aurora kinase A enzyme in the cancer chemotherapy. *Biochimica et Biophysica Acta, General Subjects, 1864*(1). https://doi.org/10.1016/j.bbagen.2019.129448, 129448.

Druker, B. J., Tamura, S., Buchdunger, E., Ohno, S., Segal, G. M., Fanning, S., et al. (1996). Effects of a selective inhibitor of the Abl tyrosine kinase on the growth of Bcr-Abl positive cells. *Nature Medicine, 2*(5), 561–566. Retrieved from https://pubmed.ncbi.nlm.nih.gov/8616716.

Fenton, A. W. (2008). Allostery: An illustrated definition for the 'second secret of life'. *Trends in Biochemical Sciences, 33*(9), 420–425. https://doi.org/10.1016/j.tibs.2008.05.009.

Gógl, G., Kornev, A. P., Reményi, A., & Taylor, S. S. (2019). Disordered protein kinase regions in regulation of kinase domain cores. *Trends in Biochemical Sciences, 44*(4), 300–311. https://doi.org/10.1016/j.tibs.2018.12.002.

Gu, X., Yan, Y., Novick, S. J., Kovach, A., Goswami, D., Ke, J., et al. (2017). Deconvoluting AMP-activated protein kinase (AMPK) adenine nucleotide binding and sensing. *The Journal of Biological Chemistry, 292*(30), 12653–12666. https://doi.org/10.1074/jbc.M117.793018.

Gustafson, W. C., Meyerowitz, J. G., Nekritz, E. A., Chen, J., Benes, C., Charron, E., et al. (2014). Drugging MYCN through an allosteric transition in Aurora kinase A. *Cancer Cell, 26*(3), 414–427. https://doi.org/10.1016/j.ccr.2014.07.015.

Hanks, S. K. (2003). Genomic analysis of the eukaryotic protein kinase superfamily: A perspective. *Genome Biology, 4*(5), 111. Retrieved from https://pubmed.ncbi.nlm.nih.gov/12734000.

Hanks, S. K., & Hunter, T. (1995). Protein kinases 6. The eukaryotic protein kinase superfamily: Kinase (catalytic) domain structure and classification. *FASEB Journal: Official Publication of the Federation of American Societies for Experimental Biology, 9*(8), 576–596. Retrieved from https://pubmed.ncbi.nlm.nih.gov/7768349.

Harvey, A. L., Edrada-Ebel, R., & Quinn, R. J. (2015). The re-emergence of natural products for drug discovery in the genomics era. *Nature Reviews. Drug Discovery, 14*(2), 111–129. https://doi.org/10.1038/nrd4510.

Hatzivassiliou, G., Song, K., Yen, I., Brandhuber, B. J., Anderson, D. J., Alvarado, R., et al. (2010). RAF inhibitors prime wild-type RAF to activate the MAPK pathway and enhance growth. *Nature, 464*(7287), 431–435. https://doi.org/10.1038/nature08833.

He, X., Ni, D., Lu, S., & Zhang, J. (2019). Characteristics of allosteric proteins, sites, and modulators. *Advances in Experimental Medicine and Biology, 1163*, 107–139. https://doi.org/10.1007/978-981-13-8719-7_6.

Hehlmann, R., Lauseker, M., Saußele, S., Pfirrmann, M., Krause, S., Kolb, H. J., et al. (2017). Assessment of imatinib as first-line treatment of chronic myeloid leukemia: 10-year survival results of the randomized CML study IV and impact of non-CML determinants. *Leukemia, 31*(11), 2398–2406. https://doi.org/10.1038/leu.2017.253.

Hines, A. C., & Cole, P. A. (2004). Design, synthesis, and characterization of an ATP-peptide conjugate inhibitor of protein kinase A. *Bioorganic & Medicinal Chemistry Letters, 14*(11), 2951–2954. Retrieved from https://pubmed.ncbi.nlm.nih.gov/15125966.

Hotte, S. J., Oza, A., Winquist, E. W., Moore, M., Chen, E. X., Brown, S., et al. (2006). Phase I trial of UCN-01 in combination with topotecan in patients with advanced solid cancers: A princess margaret hospital phase II consortium study. *Annals of Oncology: Official Journal of the European Society for Medical Oncology, 17*(2), 334–340. Retrieved from https://pubmed.ncbi.nlm.nih.gov/16284058.

Johnson, L. N., & Lewis, R. J. (2001). Structural basis for control by phosphorylation. *Chemical Reviews, 101*(8), 2209–2242. Retrieved from https://pubmed.ncbi.nlm.nih.gov/11749371.

Karaman, M. W., Herrgard, S., Treiber, D. K., Gallant, P., Atteridge, C. E., Campbell, B. T., et al. (2008). A quantitative analysis of kinase inhibitor selectivity. *Nature Biotechnology, 26*(1), 127–132. https://doi.org/10.1038/nbt1358.

Karoulia, Z., Wu, Y., Ahmed, T. A., Xin, Q., Bollard, J., Krepler, C., et al. (2016). An integrated model of RAF inhibitor action predicts inhibitor activity against oncogenic BRAF signalisng. *Cancer Cell, 30*(3), 485–498. https://doi.org/10.1016/j.ccell.2016.06.024.

Knighton, D. R., Zheng, J. H., Ten Eyck, L. F., Ashford, V. A., Xuong, N. H., Taylor, S. S., et al. (1991). Crystal structure of the catalytic subunit of cyclic adenosine monophosphate-dependent protein kinase. *Science (New York, N.Y.), 253*(5018), 407–414. Retrieved from https://pubmed.ncbi.nlm.nih.gov/1862342.

Knighton, D. R., Zheng, J. H., Ten Eyck, L. F., Xuong, N. H., Taylor, S. S., & Sowadski, J. M. (1991). Structure of a peptide inhibitor bound to the catalytic subunit of cyclic adenosine monophosphate-dependent protein kinase. *Science (New York, N.Y.), 253*(5018), 414–420. Retrieved from https://pubmed.ncbi.nlm.nih.gov/1862343.

Kornev, A. P., Haste, N. M., Taylor, S. S., & Eyck, L. F. T. (2006). Surface comparison of active and inactive protein kinases identifies a conserved activation mechanism. *Proceedings of the National Academy of Sciences of the United States of America, 103*(47), 17783–17788. Retrieved from https://pubmed.ncbi.nlm.nih.gov/17095602.

Kornev, A. P., Taylor, S. S., & Ten Eyck, L. F. (2008). A helix scaffold for the assembly of active protein kinases. *Proceedings of the National Academy of Sciences of the United States of America, 105*(38), 14377–14382. https://doi.org/10.1073/pnas.0807988105.

Kufer, T. A., Sillјé, H. H. W., Körner, R., Gruss, O. J., Meraldi, P., & Nigg, E. A. (2002). Human TPX2 is required for targeting Aurora-A kinase to the spindle. *The Journal of Cell Biology, 158*(4), 617–623. Retrieved from https://pubmed.ncbi.nlm.nih.gov/12177045.

Kunkel, M. T., & Newton, A. C. (2015). Protein kinase D inhibitors uncouple phosphorylation from activity by promoting agonist-dependent activation loop phosphorylation. *Chemistry & Biology, 22*(1), 98–106. https://doi.org/10.1016/j.chembiol.2014.11.014.

Lake, E. W., Muretta, J. M., Thompson, A. R., Rasmussen, D. M., Majumdar, A., Faber, E. B., et al. (2018). Quantitative conformational profiling of kinase inhibitors reveals origins of selectivity for Aurora kinase activation states. *Proceedings of the National Academy of Sciences of the United States of America, 115*(51), E11894–E11903. https://doi.org/10.1073/pnas.1811158115.

Lamba, V., & Ghosh, I. (2012). New directions in targeting protein kinases: Focusing upon true allosteric and bivalent inhibitors. *Current Pharmaceutical Design, 18*(20), 2936–2945. Retrieved from https://pubmed.ncbi.nlm.nih.gov/22571662.

Lavoie, H., Li, J. J., Thevakumaran, N., Therrien, M., & Sicheri, F. (2014). Dimerization-induced allostery in protein kinase regulation. *Trends in Biochemical Sciences, 39*(10), 475–486. https://doi.org/10.1016/j.tibs.2014.08.004.

Lavoie, H., & Therrien, M. (2015). Regulation of RAF protein kinases in ERK signalling. *Nature Reviews. Molecular Cell Biology, 16*(5), 281–298. https://doi.org/10.1038/nrm3979.

Lawrie, A. M., Noble, M. E., Tunnah, P., Brown, N. R., Johnson, L. N., & Endicott, J. A. (1997). Protein kinase inhibition by staurosporine revealed in details of the molecular interaction with CDK2. *Nature Structural Biology, 4*(10), 796–801. Retrieved from https://pubmed.ncbi.nlm.nih.gov/9334743.

Leroux, A. E., & Biondi, R. M. (2020). Renaissance of allostery to disrupt protein kinase interactions. *Trends in Biochemical Sciences, 45*(1), 27–41. https://doi.org/10.1016/j.tibs.2019.09.007.

Liao, J. J.-L. (2007). Molecular recognition of protein kinase binding pockets for design of potent and selective kinase inhibitors. *Journal of Medicinal Chemistry, 50*(3), 409–424. Retrieved from https://pubmed.ncbi.nlm.nih.gov/17266192.

Liau, N. P. D., Wendorff, T. J., Quinn, J. G., Steffek, M., Phung, W., Liu, P., et al. (2020). Negative regulation of RAF kinase activity by ATP is overcome by 14-3-3-induced dimerization. *Nature Structural & Molecular Biology, 27*(2), 134–141. https://doi.org/10.1038/s41594-019-0365-0.

Lin, K., Lin, J., Wu, W. I., Ballard, J., Lee, B. B., Gloor, S. L., et al. (2012). An ATP-site on-off switch that restricts phosphatase accessibility of Akt. *Science Signaling, 5*(223), ra37. https://doi.org/10.1126/scisignal.2002618.

Lu, S., Deng, R., Jiang, H., Song, H., Li, S., Shen, Q., et al. (2015). The mechanism of ATP-dependent allosteric protection of Akt kinase phosphorylation. *Structure (London, England: 1993), 23*(9), 1725–1734. https://doi.org/10.1016/j.str.2015.06.027.

Lu, S., He, X., Ni, D., & Zhang, J. (2019). Allosteric modulator discovery: From serendipity to structure-based design. *Journal of Medicinal Chemistry, 62*(14), 6405–6421. https://doi.org/10.1021/acs.jmedchem.8b01749.

Lu, S., Huang, W., Wang, Q., Shen, Q., Li, S., Nussinov, R., et al. (2014). The structural basis of ATP as an allosteric modulator. *PLoS Computational Biology, 10*(9). https://doi.org/10.1371/journal.pcbi.1003831, e1003831.

Magnaghi, P., Salom, B., Cozzi, L., Amboldi, N., Ballinari, D., Tamborini, E., et al. (2018). Afatinib is a new therapeutic approach in chordoma with a unique ability to target EGFR and brachyury. *Molecular Cancer Therapeutics, 17*(3), 603–613. https://doi.org/10.1158/1535-7163.MCT-17-0324.

Manning, B. D., & Toker, A. (2017). AKT/PKB signaling: Navigating the network. *Cell, 169*(3), 381–405. https://doi.org/10.1016/j.cell.2017.04.001.

Manning, G., Whyte, D. B., Martinez, R., Hunter, T., & Sudarsanam, S. (2002). The protein kinase complement of the human genome. *Science (New York, N.Y.), 298*(5600), 1912–1934. Retrieved from https://pubmed.ncbi.nlm.nih.gov/12471243.

McIntyre, P. J., Collins, P. M., Vrzal, L., Birchall, K., Arnold, L. H., Mpamhanga, C., et al. (2017). Characterization of three druggable hot-spots in the Aurora-A/TPX2 interaction using biochemical, biophysical, and fragment-based approaches. *ACS Chemical Biology, 12*(11), 2906–2914. https://doi.org/10.1021/acschembio.7b00537.

Modi, V., & Dunbrack, R. L. (2019). Defining a new nomenclature for the structures of active and inactive kinases. *Proceedings of the National Academy of Sciences of the United States of America, 116*(14), 6818–6827. https://doi.org/10.1073/pnas.1814279116.

Oakhill, J. S., Scott, J. W., & Kemp, B. E. (2012). AMPK functions as an adenylate charge-regulated protein kinase. *Trends in Endocrinology and Metabolism, 23*(3), 125–132. https://doi.org/10.1016/j.tem.2011.12.006.

Okuzumi, T., Fiedler, D., Zhang, C., Gray, D. C., Aizenstein, B., Hoffman, R., et al. (2009). Inhibitor hijacking of Akt activation. *Nature Chemical Biology, 5*(7), 484–493. https://doi.org/10.1038/nchembio.183.

Parang, K., Till, J. H., Ablooglu, A. J., Kohanski, R. A., Hubbard, S. R., & Cole, P. A. (2001). Mechanism-based design of a protein kinase inhibitor. *Nature Structural Biology, 8*(1), 37–41. Retrieved from https://pubmed.ncbi.nlm.nih.gov/11135668.

Park, E., Rawson, S., Li, K., Kim, B.-W., Ficarro, S. B., Pino, G. G.-D., et al. (2019). Architecture of autoinhibited and active BRAF-MEK1-14-3-3 complexes. *Nature, 575*(7783), 545–550. https://doi.org/10.1038/s41586-019-1660-y.

Poulikakos, P. I., Zhang, C., Bollag, G., Shokat, K. M., & Rosen, N. (2010). RAF inhibitors transactivate RAF dimers and ERK signalling in cells with wild-type BRAF. *Nature, 464*(7287), 427–430. https://doi.org/10.1038/nature08902.

Rajakulendran, T., Sahmi, M., Lefrançois, M., Sicheri, F., & Therrien, M. (2009). A dimerization-dependent mechanism drives RAF catalytic activation. *Nature, 461*(7263), 542–545. https://doi.org/10.1038/nature08314.

Roskoski, R. (2016). Classification of small molecule protein kinase inhibitors based upon the structures of their drug-enzyme complexes. *Pharmacological Research, 103*, 26–48. https://doi.org/10.1016/j.phrs.2015.10.021.

Ross, F. A., MacKintosh, C., & Hardie, D. G. (2016). AMP-activated protein kinase: A cellular energy sensor that comes in 12 flavours. *The FEBS Journal, 283*(16), 2987–3001. https://doi.org/10.1111/febs.13698.

Rozengurt, E., Rey, O., & Waldron, R. T. (2005). Protein kinase D signaling. *The Journal of Biological Chemistry, 280*(14), 13205–13208. Retrieved from https://pubmed.ncbi.nlm.nih.gov/15701647.

Sarbassov, D. D., Guertin, D. A., Ali, S. M., & Sabatini, D. M. (2005). Phosphorylation and regulation of Akt/PKB by the rictor-mTOR complex. *Science (New York, N.Y.), 307*(5712), 1098–1101. Retrieved from https://pubmed.ncbi.nlm.nih.gov/15718470.

Schindler, T., Bornmann, W., Pellicena, P., Miller, W. T., Clarkson, B., & Kuriyan, J. (2000). Structural mechanism for STI-571 inhibition of abelson tyrosine kinase. *Science (New York, N.Y.), 289*(5486), 1938–1942. Retrieved from https://pubmed.ncbi.nlm.nih.gov/10988075.

Shah, N. P., Nicoll, J. M., Nagar, B., Gorre, M. E., Paquette, R. L., Kuriyan, J., et al. (2002). Multiple BCR-ABL kinase domain mutations confer polyclonal resistance to the tyrosine kinase inhibitor imatinib (STI571) in chronic phase and blast crisis chronic myeloid leukemia. *Cancer Cell, 2*(2), 117–125. Retrieved from https://pubmed.ncbi.nlm.nih.gov/12204532.

Sneader, W. (1996). *Drug prototypes and their exploitation*. Wiley.

Solca, F., Dahl, G., Zoephel, A., Bader, G., Sanderson, M., Klein, C., et al. (2012). Target binding properties and cellular activity of afatinib (BIBW 2992), an irreversible ErbB family blocker. *The Journal of Pharmacology and Experimental Therapeutics*, *343*(2), 342–350. https://doi.org/10.1124/jpet.112.197756.

Storz, P. (2012). PKD. In S. Choi (Ed.), *Encyclopedia of signaling molecules* (pp. 1430–1434). New York, NY: Springer New York.

Taylor, S. S., & Kornev, A. P. (2011). Protein kinases: Evolution of dynamic regulatory proteins. *Trends in Biochemical Sciences*, *36*(2), 65–77. https://doi.org/10.1016/j.tibs.2010.09.006.

Taylor, S. S., Shaw, A. S., Kannan, N., & Kornev, A. P. (2015). Integration of signaling in the kinome: Architecture and regulation of the αC Helix. *Biochimica et Biophysica Acta*, *1854*(10 Pt. B), 1567–1574. https://doi.org/10.1016/j.bbapap.2015.04.007.

Thevakumaran, N., Lavoie, H., Critton, D. A., Tebben, A., Marinier, A., Sicheri, F., et al. (2015). Crystal structure of a BRAF kinase domain monomer explains basis for allosteric regulation. *Nature Structural & Molecular Biology*, *22*(1), 37–43. https://doi.org/10.1038/nsmb.2924.

Tran, N. H., Wu, X., & Frost, J. A. (2005). B-Raf and Raf-1 are regulated by distinct autoregulatory mechanisms. *The Journal of Biological Chemistry*, *280*(16), 16244–16253. Retrieved from https://pubmed.ncbi.nlm.nih.gov/15710605.

Van Lint, J., Rykx, A., Maeda, Y., Vantus, T., Sturany, S., Malhotra, V., et al. (2002). Protein kinase D: An intracellular traffic regulator on the move. *Trends in Cell Biology*, *12*(4), 193–200. Retrieved from https://pubmed.ncbi.nlm.nih.gov/11978539.

Vulpetti, A., & Bosotti, R. (2004). Sequence and structural analysis of kinase ATP pocket residues. *Farmaco (Societa chimica italiana: 1989)*, *59*(10), 759–765. Retrieved from https://pubmed.ncbi.nlm.nih.gov/15474052.

Wu, W.-I., Voegtli, W. C., Sturgis, H. L., Dizon, F. P., Vigers, G. P. A., & Brandhuber, B. J. (2010). Crystal structure of human AKT1 with an allosteric inhibitor reveals a new mode of kinase inhibition. *PLoS One*, *5*(9), e12913. https://doi.org/10.1371/journal.pone.0012913.

Xiao, B., Sanders, M. J., Carmena, D., Bright, N. J., Haire, L. F., Underwood, E., et al. (2013). Structural basis of AMPK regulation by small molecule activators. *Nature Communications*, *4*, 3017. https://doi.org/10.1038/ncomms4017.

Xiao, B., Sanders, M. J., Underwood, E., Heath, R., Mayer, F. V., Carmena, D., et al. (2011). Structure of mammalian AMPK and its regulation by ADP. *Nature*, *472*(7342), 230–233. https://doi.org/10.1038/nature09932.

Xin, F.-J., Wang, J., Zhao, R.-Q., Wang, Z.-X., & Wu, J.-W. (2013). Coordinated regulation of AMPK activity by multiple elements in the α-subunit. *Cell Research*, *23*(10), 1237–1240. https://doi.org/10.1038/cr.2013.121.

Yan, Y., Zhou, X. E., Xu, H. E., & Melcher, K. (2018). Structure and physiological regulation of AMPK. *International Journal of Molecular Sciences*, *19*(11), 3534. https://doi.org/10.3390/ijms19113534.

Yang, J., Cron, P., Good, V. M., Thompson, V., Hemmings, B. A., & Barford, D. (2002). Crystal structure of an activated Akt/protein kinase B ternary complex with GSK3-peptide and AMP-PNP. *Nature Structural Biology*, *9*(12), 940–944. Retrieved from https://pubmed.ncbi.nlm.nih.gov/12434148.

Yang, J., Cron, P., Thompson, V., Good, V. M., Hess, D., Hemmings, B. A., et al. (2002). Molecular mechanism for the regulation of protein kinase B/Akt by hydrophobic motif phosphorylation. *Molecular Cell*, *9*(6), 1227–1240. Retrieved from https://pubmed.ncbi.nlm.nih.gov/12086620.

Zuccotto, F., Ardini, E., Casale, E., & Angiolini, M. (2010). Through the "gatekeeper door": Exploiting the active kinase conformation. *Journal of Medicinal Chemistry*, *53*(7), 2681–2694. https://doi.org/10.1021/jm901443h.

CHAPTER FIVE

Pseudokinases: Prospects for expanding the therapeutic targets armamentarium

Nayana Devang, Adyashree Pani, and G.K. Rajanikant*
School of Biotechnology, National Institute of Technology Calicut, Calicut, Kerala, India
*Corresponding author: e-mail address: rajanikant@nitc.ac.in

Contents

Abstract

Pseudokinases are pseudoenzyme variants of the protein kinase superfamily that primarily signal through non-catalytic mechanisms. The aberrant expression of pseudokinases correlates with the pathogenesis of many human diseases. However, pseudokinases remain relatively untapped as therapeutic targets due to difficulties associated with regulating their biological functions. Many protein kinase- and few pseudokinase-specific inhibitors have been reported to influence the non-catalytic functions of active kinases, giving the hope that pseudokinases can also be exploited for therapeutic purposes. This chapter presents the structural characteristics of selected pseudokinases, their known roles in human diseases, and the progress made toward developing pseudokinase-centric therapies.

Advances in Protein Chemistry and Structural Biology, Volume 124
ISSN 1876-1623
https://doi.org/10.1016/bs.apcsb.2020.09.004

Abbreviations

4HB	4-Helix bundle
AKT	AKT serine/threonine kinase 1
ATP	adenosine triphosphate
BTK	Bruton's tyrosine kinase
BiTE	bi-specific T-cell engager
C/EBPα	CCAAT/enhancer-binding protein α
CASK	calcium/calmodulin dependent serine protein kinase
Cdc37	cell division cycle 37
CDK	cyclin-dependent kinase
COP1	constitutive photomorphogenesis protein 1
DSF	differential scanning fluorimetry
DUSP3	dual specificity phosphatase 3
EGFR	epidermal growth factor receptor
ErbB	V-erb-b2 erythroblastic leukemia viral oncogene
ERBB3	Erb-B2 receptor tyrosine kinase 3
ERK	extracellular signal-regulated kinase
FAM20A	FAM20A Golgi associated secretory pathway pseudokinase
FDA	Food and Drug Administration
GSK PKIS	GlaxoSmithKline Published Kinase Inhibitor Set
HER3	human epidermal growth factor receptor 3
Hsp90	heat shock protein 90
ILK	integrin-linked protein kinase
IRAK3	interleukin-1 receptor associated kinase 3
JAK	Janus kinase
JH2	JAK homology
KSR	kinase suppressor of Ras
LNA	locked nucleic acid
MEK	mitogen-activated protein kinase kinase
MLKL	mixed lineage kinase domain-like protein
MviN	putative virulence factor MviN family
NRG1	neuregulin 1
OMIM	Online Mendelian Inheritance in Man
PAN3	poly(A) specific ribonuclease subunit 3
PEAK1	pseudopodium enriched atypical kinase 1
PI3K	phosphatidylinositol 3-kinase
PROTACs	protein degradation by proteolysis targeting chimeras
PTK7	protein tyrosine kinase 7
Raf	rapidly accelerated fibrosarcoma
RIPK3	receptor-interacting serine-threonine kinase 3
RNAi	RNA interference
RNase L	ribonuclease L
ROP2	rhoptry kinase family protein
ROR2	receptor tyrosine kinase like orphan receptor 2
SH	Src homology
STAT	signal transducer and activator of transcription
STRADα	STE20-related kinase adapter protein alpha

TIRF	total internal reflection fluorescence
TRIB1	tribbles pseudokinase 1
TYK2	tyrosine kinase 2
TYK2	tyrosine kinase 2
Ulk4	unc-51 like kinase 4
VHR	vaccinia virus phosphatase VH1-related
VRK3	vaccinia related kinase 3
WNK	WNK lysine deficient protein kinase

1. Introduction

Protein kinases (E.C.2.7.11) are phosphotransferase enzymes that catalyze phosphate transfer from adenosine triphosphate (ATP) to the serine or threonine residues of target proteins. The resulting protein phosphorylation induces a functional change in the substrate protein, cellular location, or association with other proteins, including the phosphorylation of specific relay proteins in the intracellular signal transduction pathway. The human genome comprises about 500 protein kinase genes and accounts for around 2% of all human genes (Manning, Whyte, Martinez, Hunter, & Sudarsanam, 2002). Kinase activity can alter up to 30% of all human proteins. Protein kinases coordinate and regulate the majority of signal transduction and metabolic pathways and are essential for maintaining cellular homeostasis.

However, the existence of a subset of pseudokinases in the kinome raises questions about whether protein kinases are merely catalytic enzymes. Pseudokinases account for about 10% of mammalian kinome and share the core kinase scaffold of catalytically active protein kinases, but lack essential motifs required for efficient catalytic activity (Kung & Jura, 2019; Manning et al., 2002). Pseudokinases are essentially catalytically-deficient pseudoenzyme variants of the protein kinase superfamily. Manning et al. made the first reference to the term pseudokinase in 2002, stating them as "dead kinases" (Manning et al., 2002). Their biological significance became evident when the V617F point mutation in the pseudokinase domain of human Janus kinase 2 (JAK2) was demonstrated to contribute to myeloproliferative neoplasms (Baxter et al., 2005; James et al., 2005; Kralovics et al., 2005). Pseudokinases are essential signaling proteins despite their apparent lack of catalytic activity. They perform several important cell signaling functions, such as allosteric modulation of enzymes and conventional protein

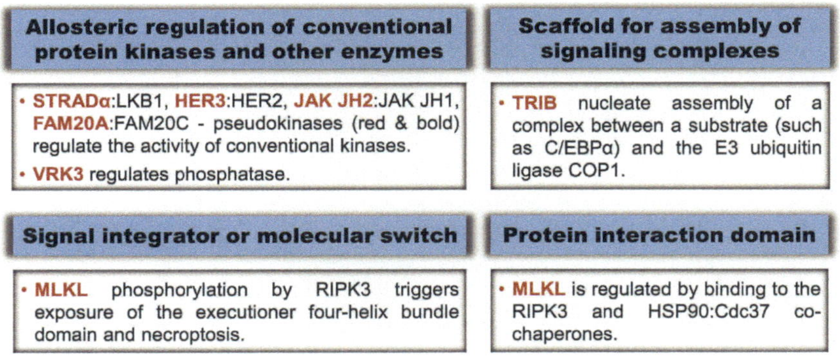

Fig. 1 Functionally diverse roles of pseudokinases. Pseudokinases are marked in red fonts.

kinases, serve as scaffolds for the assembly of signaling complexes, act as molecular switches between biochemical pathways and control of subcellular localization and trafficking (Fig. 1) (Jacobsen & Murphy, 2017; Ribeiro et al., 2019). While the functions of most pseudokinases are still poorly understood, studies have highlighted their crucial roles in development, the immune response, and metabolism. Further, dysregulation of pseudokinases is implicated in a wide range of diseases, and they are now beginning to be studied more thoroughly as potential therapeutic targets (Byrne, Foulkes, & Eyers, 2017; Jacobsen & Murphy, 2017; Kannan & Taylor, 2008; Kung & Jura, 2019; Reiterer, Eyers, & Farhan, 2014). This chapter highlights the recent progress made in understanding the structure and functions of selected pseudokinases and discusses their therapeutic implications for human diseases, including cancer.

2. General characteristics of pseudokinases

The occurrence of a significant number of pseudokinases in the human kinome suggests that they have specific mechanistic functions. Some pseudokinases are true catalysts with specialized substrates, while others act as inert scaffolds with no affinity to ATP. Few induce conformational changes when bound to ATP, but do not require phosphate transfer for their function (Shaw, Kornev, Hu, Ahuja, & Taylor, 2014). Besides, several pseudokinases act non-catalytically as allosteric modulators of the same protein or different protein complexes. The Janus kinase (JAK) family is a well-studied example of how pseudokinase and kinase activity is modulated within the same protein. JAKs are non-receptor kinases that contain both pseudokinase and canonical tyrosine kinase domains and coordinate downstream cytokine signaling. The pseudokinase domain of JAK2 has residual catalytic activity but functions primarily as a negative regulator of the kinase domain (Saharinen, Takaluoma, & Silvennoinen, 2000). Pseudokinases modulate the activity of other classes of enzymes; vaccinia related kinase 3 (VRK3) allosterically regulates the extracellular signal-regulated kinase (Erk) phosphatase dual specificity phosphatase 3/vaccinia virus phosphatase VH1-related (DUSP3/VHR), and poly(A) specific ribonuclease subunit 3 (PAN3) (Kang & Kim, 2006; Wolf et al., 2014).

Pseudokinases demonstrate a high degree of three-dimensional structural similarity to protein kinases. The pseudoactive site is composed of the β-sheet dominated N-terminal lobe, which is attached to the C-terminal lobe through the hinge region, with the ATP-binding site (pseudoactive site) buried in between the two lobes (Fig. 2) (Tomoni, Lees, Santana, Bolanos-Garcia, & Bastida, 2019).

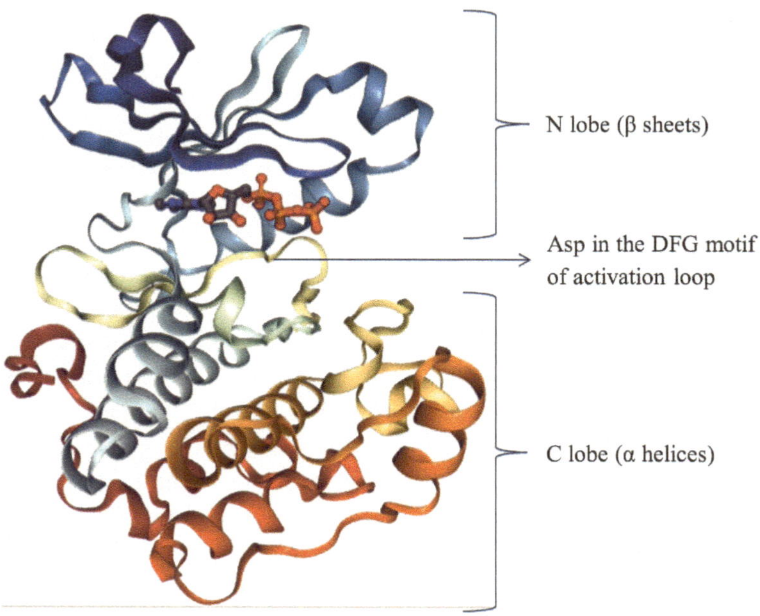

Fig. 2 General structure of a pseudokinase. Structurally, pseudokinases have the same kinase domain fold as catalytically active protein kinase counterparts. It is composed of the amino-terminal lobe (N-lobe) and the C-lobe, both of which are connected through a flexible hinge region.

Pseudokinases lack one or more conserved amino acid residues present in their catalytically active kinase counterparts: the lysine within the VAIK (Val-Ala-Ile-Lys) motif, the aspartate within the HRD (His-Arg-Asp) motif and the aspartate within the DFG (Asp-Phe-Gly) motif (Kung & Jura, 2019). The VAIK motif in active kinases, which is positioned in the β3 strand of the N-lobe, places the catalytic lysine in such a way that it coordinates with the α and β phosphates of ATP. Alteration in the lysine within the VAIK motif is detected in the kinase suppressor of Ras (KSR) and STE20-related kinase adapter protein alpha (STRADα). The HRD motif is located in the catalytic loop of active kinases. The aspartate residue in the HRD motif functions as the catalytic base during ATP hydrolysis. Aspartate mutation leads to catalytically inactive kinases. Pseudokinases like Erb-B2 receptor tyrosine kinase 3/human epidermal growth factor receptor 3 (ERBB3/HER3), integrin-linked protein kinase (ILK) and mixed lineage kinase domain-like protein (MLKL) carry HRD mutations (Asp) or lack the motif itself. Pseudokinases such as STRADα, MLKL, and calcium/calmodulin-dependent serine protein kinase (CASK) carry mutations in their DFG motifs or lack this motif entirely. Those who retain an intact DFG motif exhibit alterations such as the unusually short activation loops as in ribonuclease L (RNase L) and tyrosine kinase 2 (TYK2), which is likely to prevent the changes in canonical conformation. Recently, attention has been given to the structural integrity of the flexible glycine-rich loop, which facilitates ATP binding (Murphy et al., 2013; Scheeff, Eswaran, Bunkoczi, Knapp, & Manning, 2009). The substitution of glycine residues by bulky, negatively charged amino acids interferes with the binding of ATP to the glycine-rich loop. For instance, Asp175 and Gln177 within the glycine-rich loop protrude into the ATP-binding pocket of VRK3 (Scheeff et al., 2009). Similarly, Glu101 and Glu103 replace glycines in the glycine-rich loop of tribbles pseudokinase 1 (TRIB1) to interfere with ATP binding. The loop is shorter than usual, making the ATP-binding pocket relatively shallow (Murphy et al., 2015). Although pseudokinases lack one or more of the amino acids or experimentally validated motifs to facilitate the transfer of phosphoryl in catalytically active kinases, it does not prevent many of them (KSR1/2, CASK, TRIB2, JAK2, ERBB3, WNK lysine deficient protein kinase (WNK), STRADα, and MLKL) from binding to or hydrolyzing ATP *in vitro*. KSR1/2 has lysine mutation in its kinase domain (VAIK motif), yet it plays scaffolding and weak catalytic roles. Being a scaffold protein, it coordinates the formation of the oncogenic rapidly accelerated fibrosarcoma-mitogen-activated protein kinase kinase-ERK (Raf–MEK–ERK) signaling complex. A ligand-induced switching mechanism contributes to the biological functions of pseudokinases, which includes direct regulation of canonical protein kinase partners through a combination of switch-like or enzymatic mechanisms that are sometimes distinguishable by generalized mutations (Hu et al., 2011; Murphy et al., 2013; Zeqiraj et al., 2009). These findings have profound implications for the understanding of signaling mechanisms in healthy and diseased cells, which means that pseudokinases with atypical nucleotide interactions and little or no apparent catalytic activity, can still control cellular signaling, given the low ATP concentrations and the ability of kinases to cluster around interacting proteins, substrates and pseudosubstrates.

3. Classification

Pseudokinases are classified into four types based on whether they interact with nucleotides and/or divalent cations. Class 1: lack of detectable nucleotide or cation binding; class 2: nucleotide binding; class 3: cations binding; and class 4: nucleotide and cation binding (Table 1) (Kung & Jura, 2019; Tomoni et al., 2019).

A substantial number of pseudokinases including TRIB1, VRK3, receptor tyrosine kinase like orphan receptor 2 (ROR2), Pragmin and the bacterial pseudokinase putative virulence factor MviN family (MviN) fall within

Table 1 Pseudokinase classes.

Class 1: Do not bind to ATP or Mg^{2+}	Class 2: Bind to ATP but not Mg^{2+}	Class 3: Bind to Mg^{2+} but not ATP	Class 4: Bind to ATP and Mg^{2+}
TRIB1	MLKL	SgK269	JAK1JH2
VRK3	STRADα	PEAK1	JAK2JH2
SCYL1	TRIB2		TYK2JH2
CCK4	TRIB3		HER3
NRBP1	ULK4		KSR1
GCN2	CASK		KSR2
BUBR1	EPHB6		ADCK3
IRAK3	FAM20A		ANP-α
ROR1			HSER
ROR2			ILK
PTK7			IRAK2
STK40			PAN3
RYK			POMK
Pragmin			RNase L
			STKLD1
			TYK2 JH2

the non-nucleotide binder category (Class 1), indicating that pseudokinases have primarily evolved to serve non-catalytic functions such as serving as rigid scaffolds. For instance, TRIB1, as a scaffold, recruits the CCAAT-enhancer-binding proteins (C/EBPs) to the constitutive photomorphogenesis protein 1 (COP1) ubiquitin E3 ligase for ubiquitylation to regulate transcription and cell cycle activity. Highly distorted nucleotide-binding sites, often occluded by hydrophobic residues, are the reason for the lack of detectable nucleotide binding (Class 1). These pseudokinases do not have a canonical druggable pocket, posing a significant challenge for developing small molecule modulators. However, they can still accommodate physiological or experimentally developed non-conventional ligands (Scheeff et al., 2009). In addition, close counterparts of several class 1 pseudokinases, such as TRIB2 and pseudopodium enriched atypical kinase 1 (PEAK1), remain capable of binding to canonical kinase ligands despite their distorted pseudoactive site.

MLKL, TRIB2/3, STRADα, FAM20A Golgi associated secretory pathway pseudokinase (FAM20A), CASK, and unc-51 like kinase 4 (Ulk4) retain the ability to bind to ATP even in the absence of cations (Mg^{2+}). Except for FAM20A, this type of pseudokinases (Class 2) lacks the canonical metal-coordinating aspartate residue in the DFG motif. The presence of Mg^{2+} drastically decreases nucleotide binding (Murphy, Lucet, et al., 2014; Murphy, Zhang, et al., 2014). The pseudokinase domain of MLKL is so versatile that it acts as a molecular switch and as a protein interaction domain, suppressing the 4-Helix bundle (4HB) executioner domain and recruiting downstream effectors (cell division cycle 37-heat shock protein 90 (Cdc37-Hsp90)) (Jacobsen et al., 2016; Petrie et al., 2018). Whereas STRADα still binds to nucleotide despite the lack of three catalytic residues in its pseudoactive site (K residue in the VAIK motif, D residues in the HRD and DFG motifs) (Zeqiraj et al., 2009).

Toxoplasma protein, rhoptry kinase family protein (ROP2), and, a close homolog of Pragmin, PEAK1, are the only two proteins reported to bind divalent cations but not nucleotides (Class 3) or ATP-competitive small molecules (Murphy et al., 2013). Both proteins have highly occluded pseudoactive site with the DFG and YRDLKPEN motifs visible, whereas cation-binding residues are located in a region of low structural complexity (Ha & Boggon, 2018; Labesse et al., 2009).

Class 4 pseudokinases that include HER3, KSR, RNase L, and the JH2 pseudokinase domains of the JAKs, have an intact DFG aspartate (with a few exceptions) and exhibit kinase activity slightly lower than that of active kinases.

Mutations in catalytic residues do not generally interfere with ATP binding. Pseudokinase that lacks the equivalents of Lys72, Asp166 or Asp184 may still bind ATP with an affinity comparable to that of the wild-type protein, but cannot properly position the γ-phosphate for efficient transfer (Iyer, Garrod, Woods Jr, & Taylor, 2005). Pseudokinases such as KSR, CASK, and HER3 exhibit residual kinase activity sufficient for phosphotransfer to a specific co-localized kinase (Brennan et al., 2011). In other cases, the ATP binding alone causes signal transduction to the kinase, suggesting that certain pseudokinases function as molecular switches to oscillate between active and inactive conformations through ATP binding or hydrolysis (Murphy, Lucet, et al., 2014; Murphy, Zhang, et al., 2014). While it is clear that nucleotide binding stabilizes the core pseudokinase domain in an "active-like conformation," the modulation of downstream signals upon ATP binding is somewhat difficult to comprehend. It requires the full-length structure of the pseudokinase in both the apo and ATP-bound forms and the structural knowledge of the direct interacting partners in the protein complex to appreciate the dynamics that influence conformational flexibility.

4. Structural features and relevance to human diseases

A recent analysis of the pseudokinome achieved by scanning limited "knockout" and siRNA literature and mining the online Online Mendelian Inheritance in Man (OMIM) database reiterated the vital link between pseudokinases and a wide range of human diseases (Reiterer et al., 2014). Approximately two-thirds of pseudokinases have been associated with a range of human diseases, including cancer, neurological, immunological, and metabolic disorders. The finding of the somatic mutation V617F has led to the discovery of other mutations in the JAK2 pseudokinase domain, such as the exon 12 mutations, which cause myeloproliferative disorders (Baxter et al., 2005; Kratz et al., 2006). Mutations in the JAK1 pseudokinase domain are responsible for up to 20% of acute lymphoblastic leukemia, whereas somatic mutation hotspots in the pseudokinase domain of ERBB3/HER3 are oncogenic due to their ability to enhance the catalytic activity of conventional protein kinase, HER2/ERBB2 (Tables 2 and 3) (Flex et al., 2008; Jaiswal et al., 2013). Further, TRIB1 and TRIB2 overexpression increase the turnover of the C/EBP transcription factors, leading to acute myeloid leukemia (Keeshan et al., 2006). Whereas TRIB1 deficiency elevates C/EBPα levels, leading to aberrant Toll-like receptor

Table 2 The details of important disease-associated pseudokinases, altered canonical amino acids, relevant mutations in pseudokinase domain, their biological activities.

Pseudokinase	Biological roles in humans	Pseudokinase domain sequence/PDB ID/relevant mutation in pseudokinase domain	Associated diseases	References
ADCK3	Forms an electron-transferring membrane complex involved in isoprenoid lipid biosynthesis in the mitochondria	Pseudokinase domain (256–647, PDB ID: 4PED), Y514C	Cerebellar ataxia, breast tumors, glioblastoma, nephrotic syndrome, defects in coenzyme Q10 synthesis	Lanning et al. (2014)
CASK	Neural development and gene regulation	CaM pseudokinase domain (1–337, PDB ID: 3MFT), R28L, Y268H	Mental retardation and FG syndrome	Mukherjee et al. (2008), Mukherjee, Sharma, Jahn, Wahl, and Südhof (2010)
EPHA10	Mediators of cell-cell communication regulating cell attachment, shape, and mobility in neuronal and epithelial cells	Pseudokinase domain (645–900, UniprotID: Q5JZY3) V897H. In HRD motif, D770G, P704S	Overexpression is associated with prostate cancer and correlated with lymph node metastasis and stage progression in breast cancer	Shrestha, Byrne, Harris, Kannan, and Eyers (2020)
ILK	Links integrins to the actin cytoskeleton for actin polymerization	Pseudokinase domain/alpha-parvin core complex (183–452, PDB ID: 3KMU)	Overexpressed in breast, prostate, brain, lung, and colorectal cancers	Fukuda, Gupta, Chen, Wu, and Qin (2009)

JAK1	JH2 pseudokinase domain of JAK1 regulates JH1 domain catalytic activity	JH2 domain, (561–860, PDB ID: 4L01) and nonmutated (561–860, PDB IDs: 4L00), V658F, A634D, H647Y, L653F, R724H, T782M, L783F	Acute leukemia, gynecological tumors, rheumatoid arthritis, Crohn's disease, multiple sclerosis; myelofibrosis and psoriasis	Gustafson et al. (2014)
JAK2	JH2 domain is catalytically active and phosphorylates two negative regulatory sites in JAK2 (Ser523 and Tyr570) to suppress JAK-STAT signaling. The pathogenic V617F mutation promotes a confirmation in the JH2 αC helix that enables stimulatory interactions with the JH1 domain	JH2 domain (536–812, PDB ID: 4FVP), K607N, L611S, V617F, V717F	Myeloproliferative neoplasms, such as polycythemia vera, and acute leukemia, somatic erythrocytosis	Bandaranayake et al. (2012), Leroy et al. (2016)
TYK2	JH2 domain binds to ATP but is catalytically inactive as they lack HRD motif, important in JH1 domain regulation	JH2 domain bound to 5′ ATP (556–871, PDB ID: 5C03). Pseudokinase domain bound to BMS-066 (575–869, PDB ID: 4WOV), I684S	Systemic lupus erythematosus, Crohn's disease and psoriasis	Min et al. (2015), Tokarski et al. (2015)

Continued

Table 2 The details of important disease-associated pseudokinases, altered canonical amino acids, relevant mutations in pseudokinase domain, their biological activities.—cont'd

Pseudokinase	Biological roles in humans	Pseudokinase domain sequence/PDB ID/relevant mutation in pseudokinase domain	Associated diseases	References
HER3 Pseudokinase	Binds to ligand growth factor neuregulin 1 (NRG1), and forms heterodimers with other HER family members and activates PI3K/AKT signaling	Pseudokinase domain 698–1019 with Mg^{2+}-ANP (PDB ID: 3KEX) 698–1020 domain bound to bosutinib (PDB ID: 4OTW), V695M, Q790R, S827I, E909G	Increased expression and persistent phosphorylation of HER3 contribute to drug resistance in cancer types such as breast and lung cancer; mutations promote oncogenesis and are found in several tumor types including breast, lung and gastrointestinal, prostrate, ovarian, colon and gastric cancer; downregulated in dorsolateral prefrontal cortex in schizophrenia; overexpressed in Charcot–Marie–Tooth type 1 disease	Shi, Telesco, Liu, Radhakrishnan, and Lemmon (2010), Jura, Shan, Cao, Shaw, and Kuriyan (2009), Littlefield et al. (2014)

Kinase suppressor of Ras 2	Scaffolding pseudokinase that coordinates the formation of the Raf-MEK-ERK signaling complex. Phosphorylates MEK1	Heterodimeric pseudokinase (634–950) in complex with MEK1 (1–393) and APS-2-79 ligand (PDB ID: 2Y4I). KSR2 mutations: P662L, E667V, R684C, I801L, G816D, R818Q, R823H, D843N, S904L	Loss of function in KSR1 lead to tumorigenesis KSR2 mutation leads to obesity and insulin resistance, cancer	Brennan et al. (2011)
MLKL	Phosphorylated by RIP3 to mediate downstream necroptosis signaling. Pseudokinase domain controls 4-Helix bundle output domain	Pseudokinase domain (183–471, PDB ID: 4MWI)	Overexpressed in autoimmune hepatitis	Murphy, Lucet, et al. (2014), Murphy, Zhang, et al. (2014), Xie et al. (2013)
ROR1 ROR2	Maintain basal NMDAR-mediated synaptic transmission	ROR1 pseudokinase domain (453–752) (PDB ID: 6TU9), ROR2:N620K	ROR1: Overexpressed in cancer, including chronic lymphocytic leukemia, acute lymphocytic leukemia, gastric carcinoma, lung cancer, chronic myeloid leukemia ROR2: Brachydactyly type B; LOF in Robinow syndrome; overexpression in osteosarcoma and renal cell carcinoma	Huang et al. (2014), Sheetz et al. (2020)

Continued

Table 2 The details of important disease-associated pseudokinases, altered canonical amino acids, relevant mutations in pseudokinase domain, their biological activities.—cont'd

Pseudokinase	Biological roles in humans	Pseudokinase domain sequence/PDB ID/relevant mutation in pseudokinase domain	Associated diseases	References
RYK	Responsible for the establishment of axon tracts during the development of the central nervous system	Pseudokinase domain (PDB ID: 6TUA)	Neuron axon defect; overexpressed in epithelial ovarian cancer	Sheetz et al. (2020)
STRADα	STRADα and STRADβ binds to ATP and allosterically stimulate LKB1 and other AMPKs after complexing with MO25α	STRADα C-terminal pseudokinase domain (59–431) (PDB ID: 2WTK)	Allosterically activates tumor suppressor LKB1; deletion of exons 9–13 associated with polyhydramnios, megalencephaly, symptomatic epilepsy, Peutz-Jeghers syndrome and sporadic cancers	Zeqiraj et al. (2009)
TRIB (TRIB1, TRIB2, TRIB3)	TRIB lack ability to hydrolyse ATP, but has scaffolding role-TRIP-COP1 complex does ubiquitination of transcription factor C/EBPα They play role in lipoprotein metabolism, immune function and cellular differentiation and	Pseudokinase domain (83–343, PDB ID: 5CEK), TRIB1-R107L, TRIB3: Q84R DFG motif replaced by metal-binding motif (E [S/N]LED) SLE motif adopts a unique "inactive"-like conformation that obstructs ATP binding in the TRIB1 being a physical barrier to ATP-competitive	TRIB1: Overexpressed in acute lymphocytic leukemia, prostate cancer, acute megakaryocytic leukemia, fatty liver disease, atherosclerosis, and coronary artery disease TRIB2: Overexpressed in lung cancer and liver cancer, drug resistance in several cancer types	Murphy et al. (2015)

	proliferation. This is achieved by two modes: the first involving the E3-ligase (*e.g.*, COP1)-dependent ubiquitination of their protein substrates (*e.g.*, C/EBPα), and the second by modulation of MAPK and AKT signaling modules	inhibitors. In contrast, TRIB2 and 3 demonstrate very weak ATP affinity and ATP hydrolysis *in vitro* Mutation of Lys177 to alanine decreases TRIB2 autophosphorylation *in vitro* Mutation of conserved Lys90 similarly decrease autophosphorylation of all vertebrate TRIB family	including melanoma and breast and renal cancers. Causes acute myeloid leukemia by mediating degradation of the C/EBPα transcription factor TRIB3: Cancer, insulin resistance in diabetes, do not degrade C/EBPα, but degrade acetyl-CoA carboxylase. GOF lead to insulin resistance, lung, colon and breast cancers, Parkinson disease	
WNK (Ser/Thr pseudokinase) Egs: WNK1-4	Modulate blood pressure and ion homeostasis Stimulates activities of Na^+ driven, Cl^- importing NKCCs ($Na^+/K^+/Cl^-$ cotransporters)	Pseudokinase domain without ATP mimetic (210–482, PDB ID: 4PWN) Lys to Cys mutation in VAIK Though they lack Lys in VAIK motif. The Gly of DFG is replaced by Lys which compensates for the lost ATP binding ability of Lys in VAIK	Hereditary neuropathy and glioma Mutations in WNK1 and WNK4 have been linked to hereditary hypertension WNK1 and WNK3 also stimulate proliferative and invasive activity in glioma cells WNK2 predominantly plays a role as a tumor suppressor and downregulates cellular proliferation by increasing and decreasing activity of Rac1 and RhoA respectively	Piala et al. (2014)

Table 3 The details of relevant mutations in pseudokinase domain, their implications in diseases.

Pseudokinase	Links to disease	Pseudokinase domain mutations
ADCK3	Cerebellar ataxia, defects in coenzyme Q10 synthesis	Y514C
ANP-A	Essential hypertension or left ventricular hypertrophy	NA
ANP-B	Acromesomelic dysplasia	LOF's-Y708C, R776W
BUBR1	Gastric cancer and bladder cancer	R727C, R814H, I909T, L1012P overexpression
CASK	Mental retardation and FG syndrome	R28L, Y268H
DIA1R	Mental retardation and autism	R128K
EPHA10	Overexpression associated with prostate cancer and correlated with lymph node metastasis and stage progression in breast cancer	NA
EPHB6	Downregulated in metastatic melanoma, neuroblastoma and ovarian carcinoma	NA
FAM20A	Amelogenesis imperfecta, Nephrocalcinosis	L173R, G331D, D403N
GCN2	Pulmonary veno-occlusive disease	NA
HER3	Increased expression and persistent phosphorylation of HER3 contribute to drug resistance in cancer types such as breast and lung cancer; mutations promote oncogenesis and are found in several tumor types including breast, lung and gastrointestinal cancer; downregulated in dorsolateral prefrontal cortex in schizophrenia; overexpressed in Charcot–Marie–Tooth type 1 disease	V695M, Q790R, S827I, E909G
HSER	Loss of function develop meconium ileum; overexpression lead to familial diarrhea syndrome	NA

Table 3 The details of relevant mutations in pseudokinase domain, their implications in diseases.—cont'd

Pseudokinase	Links to disease	Pseudokinase domain mutations
ILK	Overexpressed in lung cancer and CRC	NA
IRAK2	Loss of function in colorectal cancer and hepatitis C	R214G, L392V
IRAK3	Overexpression leads to reduced immunity, associated with asthma	L400V, R429Q
JAK1 JH2	Mutations associated with acute leukemia	A634D, H647Y, L653F, V658F, R724H, T782M, L783F
JAK2 JH2	Mutations associated with myeloproliferative neoplasms, such as polycythemia vera, and acute leukemia	K607N, L611S, V617F, V717F
JAK3 JH2	Inhibitory mutations associated with severe combined immunodeficiency; activating mutations associated with acute leukemia	A572V, A573V, M576L, A593T, V715I, V722I
KSR1	Loss of function lead to tumorigenesis	NA
KSR2	Obesity and insulin resistance, cancer	P662L, E667V, R684C, I801L, G816D, R818Q, R823H, D843N, S904L
MLKL	Overexpressed in autoimmune hepatitis	NA
NRBP1	Downregulated in cancers, lung and colon cancer	NA
NRBP2	Downregulated in hepatocellular carcinoma; high expression levels correlated with greater chemosensitivity and better prognosis	NA
SGK269	Overexpressed in pancreatic cancer and breast cancers	NA
SGK196	Walker–Warburg syndrome, neuromuscular disorders and congenital muscular dystrophy	LOF: L137R, Q258R

Continued

Table 3 The details of relevant mutations in pseudokinase domain, their implications in diseases.—cont'd

Pseudokinase	Links to disease	Pseudokinase domain mutations
SGK223	Overexpressed in pancreatic cancer	NA
PTK7	Idiopathic scoliosis and neural tube defects; overexpressed in lung and breast cancer; downregulated in metastatic melanoma and ovarian cancer	NA
PXK	Systemic lupus erythematosus	NA
RETGC1	Leber congenital amaurosis and Cone–rod retinal dystrophy	F589S
RETGC2	Pancreatic, lung and breast cancers	K672T
RNase L	Prostate cancer, overexpression in chronic fatigue syndrome	R462Q
ROR1	Overexpressed in cancer, including chronic lymphocytic leukemia, acute lymphocytic leukemia, gastric carcinoma, lung cancer, chronic myeloid leukemia	NA
ROR2	Brachydactyly type B; LOF in Robinow syndrome; overexpression in osteosarcoma and renal cell carcinoma	N620K
RPS6KL1	Cervical carcinoma	NA
RYK	Neuron axon defect; overexpressed in epithelial ovarian cancer	NA
S6KΔ1	Breast cancer	NA
SCYL1	Spinocerebellar neurodegeneration and a hepatocerebellar neuropathy syndrome	NA
SCYL2	Neurodegeneration	NA
SCYL3	Colorectal cancer	NA
STK31	Overexpressed in colorectal cancer	NA
STK40	Cervical carcinoma	NA
STRADα	Allosterically activates tumor suppressor LKB1; deletion of exons 9–13 associated with polyhydramnios, megalencephaly, symptomatic epilepsy	NA

Table 3 The details of relevant mutations in pseudokinase domain, their implications in diseases.—cont'd

Pseudokinase	Links to disease	Pseudokinase domain mutations
STRADβ	Allosterically activates tumor suppressor LKB1	NA
STYK1	Overexpressed in prostate cancer, hepatocellular carcinoma	NA
TBCK	Infantile syndromic encephalopathy, intellectual disability and hypotonia	NA
Titin	Cardiomyopathy	NA
TRIB1	Overexpressed in acute lymphocytic leukemia, prostate cancer, acute megakaryocytic leukemia, altered lipid metabolism and cardiovascular disease	R107L
TRIB2	Overexpressed in acute lymphocytic leukemia, lung cancer and liver cancer, drug resistance in several cancer types, including melanoma and breast and renal cancers	NA
TRIB3	Gain of function lead to insulin resistance, lung, colon and breast cancers, Parkinson disease	Q84R
TRRAP	Developmental Delay with or without dysmorphic faces and Autism and Syndromic Intellectual Disability	NA
TYK2 JH2	Systemic lupus erythematosus, Crohn's disease and psoriasis	I684S
ULK4	Loss of function in schizophrenia	NA
VRK3	Neuropathy and cancer	NA

signaling (Satoh et al., 2013). Similarly, KSR has become a therapeutic target of interest in the treatment of RAS-driven cancers due to its positive effect on MAPK signaling (Hu et al., 2011; Lavoie et al., 2018; Rajakulendran, Sahmi, Lefrançois, Sicheri, & Therrien, 2009). In contrast, the loss of function mutation in KSR2 is associated with insulin resistance and obesity (Pearce et al., 2013). Such observations indicate that pseudokinases are central components of signal transduction networks and use multiple, but less well-characterized mechanisms to regulate cell signaling in normal and disease states, making them highly desirable therapeutic targets.

4.1 ERBB3/HER3

The HER3 receptor is a key member of the v-erb-b2 erythroblastic leukemia viral oncogene (ErbB) family and preferably signals through the phosphatidylinositol 3-kinase (PI3K) pathway (Olayioye, Neve, Lane, & Hynes, 2000). A histidine replaces HER3 conserved glutamate in helix αC, and helix αC is partially unwound and adopts inactive conformation. HER3 has been widely recognized as a potential drug target since its role as a driver of tumor progression, and acquired resistance to treatment has been understood (Fig. 3) (Liu et al., 2019; Mishra, Patel, Alanazi, Yuan, & Garrett, 2018). High frequency overexpression of HER3 in breast, pancreatic, and ovarian cancers has been associated with decreased patient survival, especially in combination with concomitant HER2 overexpression (Ocana et al., 2013). Numerous studies have shown that HER3 activation is often associated with cancer therapy failure (Amin, Campbell, & Moasser, 2010). HER3 signaling plays a pivotal role in the development of tumors with a phenotype of drug resistance, such as HER2-over-expressing breast cancer (Lee-Hoeflich et al., 2008), castration-resistant prostate cancer (Jathal, Chen, Mudryj, & Ghosh, 2011), platinum-resistant ovarian cancer (Mills & Yarden, 2010), and epidermal growth factor receptor (EGFR) tyrosine kinase inhibitor-resistant non-small cell lung cancer (Huang et al., 2013). HER3 adopts a dimer-competent conformation upon the binding of its cognate ligands. HER3 heterodimerization with the functional kinase coreceptor, HER2, results in a downstream transphosphorylation of six tyrosine residues on the cytosolic tail. The phosphorylated tyrosine residues serve as docking sites to initiate signaling cascades *via* the direct binding of the p85 subunit of PI3K (Lyu, Han, Polsdofer, Liu, & Liu, 2018). HER3 knockdown models have demonstrated that HER3/HER2 heterodimer is the most potent mitogenic unit and has a crucial role in HER2-mediated transformation and the continued growth of breast cancer cell lines (Holbro et al., 2003; Lee-Hoeflich et al., 2008).

HER3 is an allosteric activator and heterodimerization partner of EGFR and HER2, both of which are often deregulated in cancer. The earliest small molecules exploited a cysteine located at the roof of the ATP-binding site to perturb the ability of HER3 to heterodimerize and inhibit cell signaling and cancer growth (Lim et al., 2015; Xie et al., 2014). Subsequent studies have

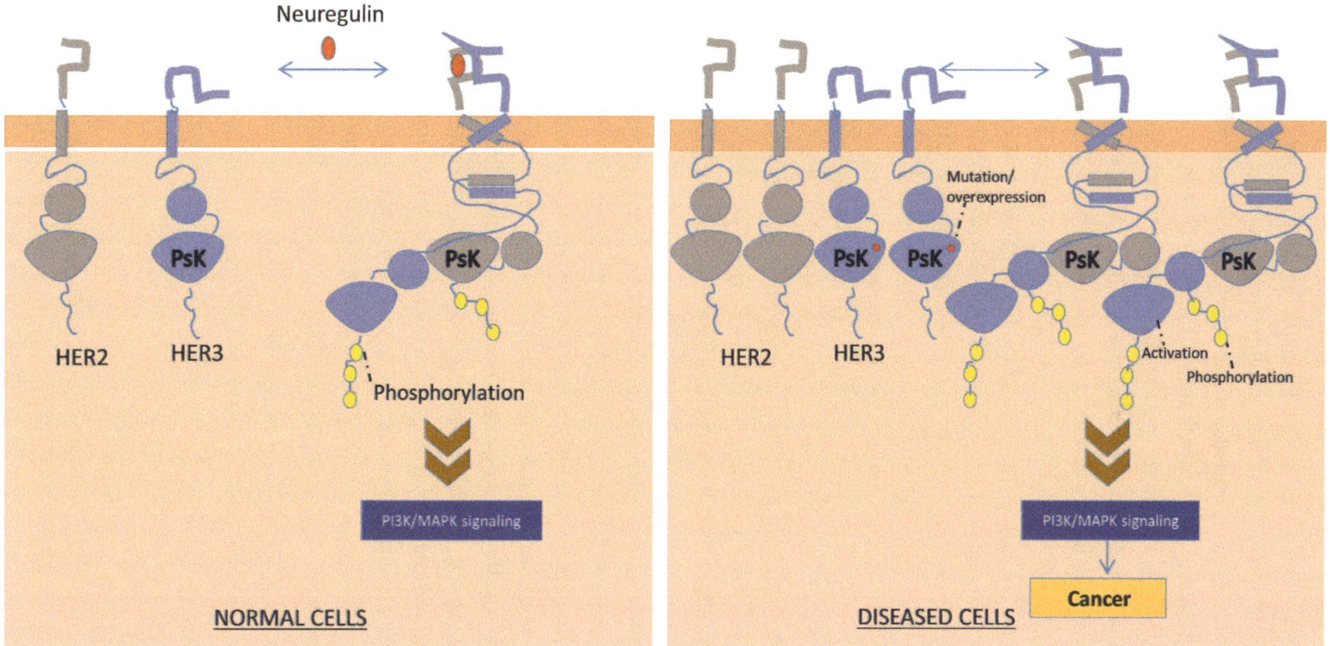

Fig. 3 In healthy cells, HER3 undergo heterodimerization to allosterically activate other HER kinases such as HER2 in a ligand-dependent manner (left panel). In diseased cells, HER3 overexpression or mutation leads to heterodimerization, independent of ligand, causing hyper-activation of signaling pathways that proceed to a cancerous growth (right panel). NRG, neuregulin; PI3K, phosphoinositide 3-kinase.

shown how clinically approved small molecules (dasatinib and bosutinib) can bind to HER3 (Davis et al., 2011). It is clear that over-expression of HER3 is correlated with a worse cancer survival rate, emphasizing the importance of developing effective HER3 targeted anticancer therapies (Gala & Chandarlapaty, 2014). Since HER2:HER3 is the most active signaling dimer in the ErbB system, pertuzumab, an antibody, has been developed to prevent dimerization (Franklin et al., 2004). Metastatic breast cancer clinical study showed that a combination of anti-HER2 (pertuzumab) and anti-HER3 (lumretuzumab) immunotherapy resulted in a positive overall response rate of 55% (Schneeweiss et al., 2018). In addition, HER3 ligand, HRG, stimulates tumor progression, and resistance to targeted therapies *via* HER2:HER3 heterodimers and is associated with worse clinical outcomes (Mota et al., 2017). Seribantumab, human immunoglobulin G2 antibody that targets HER3 by blocking HRG binding and downregulating the receptor, exhibits additive and synergistic effects on lung and breast cancer growth inhibition in combination with cetuximab and erlotinib (Curley et al., 2015; Schoeberl et al., 2017). Even though seribantumab did not improve outcomes when combined with erlotinib in patients with *EGFR* wild-type non-small cell lung cancer in phase 1/2 study, a multicenter phase II study of seribantumab, in combination with docetaxel, in patients with HRG-positive advanced adenocarcinoma has been proposed (Sequist et al., 2019). Although several anti-HER3 antibodies have been active in clinical trials, no HER3-targeted cancer therapy has been approved by the Food and Drug Administration (FDA).

HER3 is also a potential target in neurological disorders as it serves as a receptor for the growth factor neuregulin 1 (NRG1), which is essential for Schwann cells (Riethmacher et al., 1997). While downregulation of HER3 has been reported in the dorsolateral prefrontal cortex of patients with schizophrenia, HER3 overexpression has been observed in the Schwann cells of patients with Charcot–Marie–Tooth type 1 disease (Kanazawa et al., 2007; Massa et al., 2006).

4.2 Janus kinases

Janus kinases (JAK1–3 and tyrosine kinase 2, TYK2) are non-receptor tyrosine kinases involved in cell signaling *via* type I/II cytokines and interferons (Raivola, Haikarainen, & Silvennoinen, 2019). Structurally all JAKs consist of four domains at the N-terminal, one primary receptor binding Src homology 2 (SH2)-like domain, and the C-terminal JAK-homolog domains

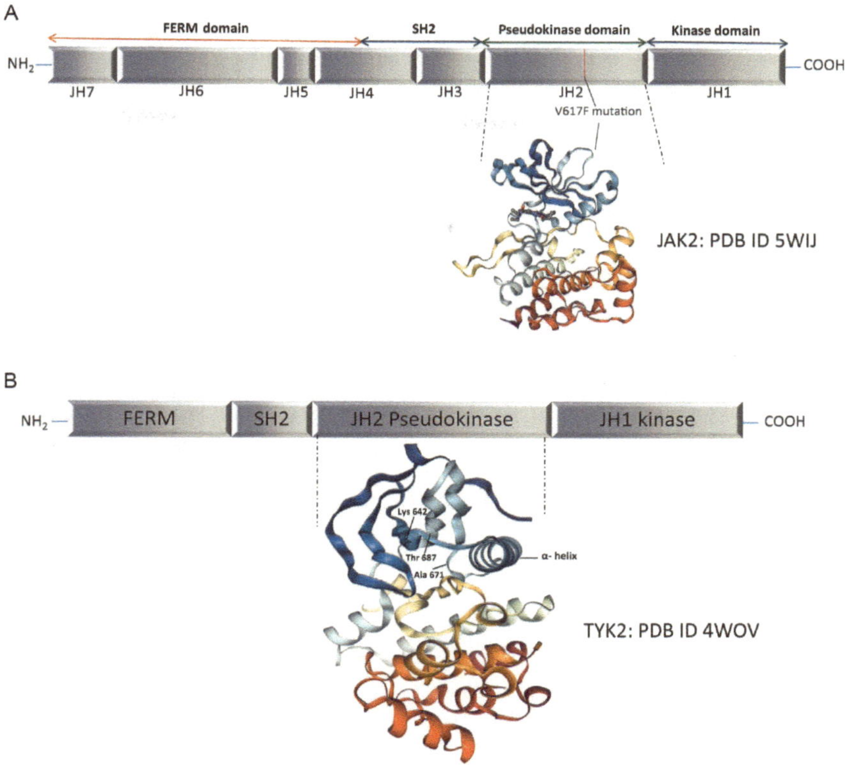

Fig. 4 Basic structural architecture of (A) JAK2, and (B) TYK2.

(JH1 and JH2) (Fig. 4A and B). While JH1 is an active domain of tyrosine kinase, JH2 is referred to as a pseudokinase domain due to altered Gly-rich region and lack of catalytic Asp. JH2 has an essential role in regulating the basal activity of JAK2 and JAK3 tyrosine kinases through allosteric mechanisms and cytokine-inducible activation of signal transduction (Saharinen & Silvennoinen, 2002). JH2 forms an auto-inhibitory interaction with JH1 by binding to the hinge region and stabilizing the inactive conformation of JH1 (Shan et al., 2014). ATP binding stabilizes JH2 and is critical to pathogenic activation of JAK2 (Hammarén et al., 2015). In addition, the SH2-JH2 linker functions as a regulatory region for the control of JAKs activities (Sanz Sanz et al., 2014). Fig. 3A shows the basic structure of the JAKs. Domain deletion studies suggested that JH2 and JH1 were required for

JAK-mediated signaling, and JH2 was also correlated with increased basal JAK activity (Saharinen et al., 2000; Velazquez et al., 1995).

The Janus kinase-signal transducer and activator of transcription protein (JAK-STAT) pathway moderates various essential biological functions, such as cytokine signaling, immune response, and hematopoiesis, to name a few. Deregulated JAK-STAT signaling causes myeloproliferative neoplasms, leukemia, and lymphomas, and autoimmune diseases (Fig. 5). JAKs have thus gained considerable importance as therapeutic targets. Multiple mutations in

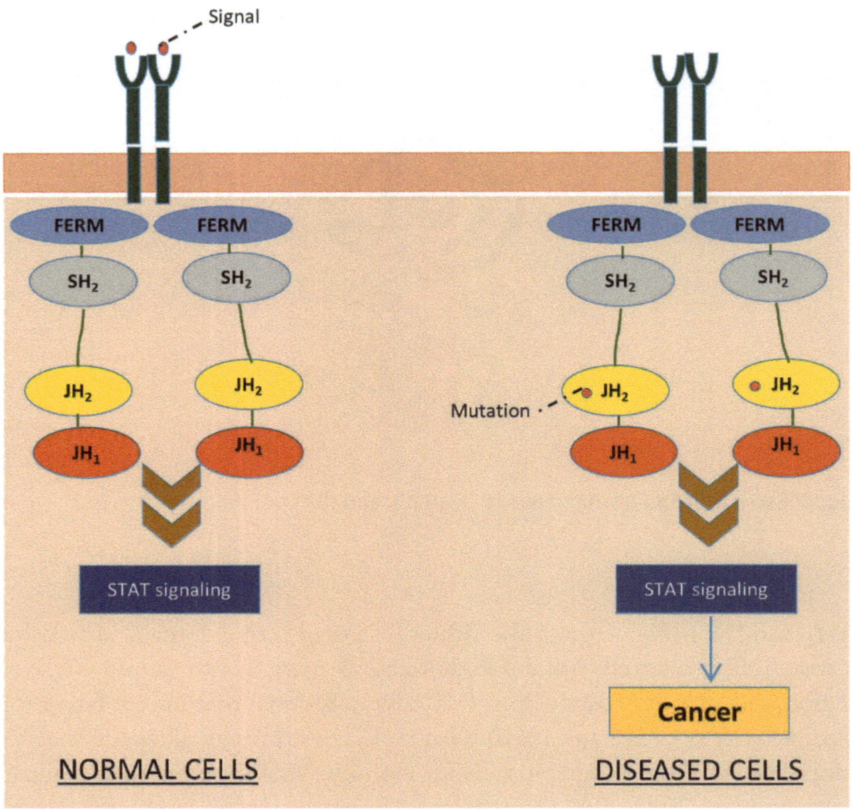

Fig. 5 In healthy cells, the JAK homology 2 (JH2) pseudokinase domain of JAK2 allosterically inhibits the JH1 kinase domain. Signaling molecule binds to the cognate receptor of JAK2 to relieve this inhibition, allowing transphosphorylation and activation of the kinase domain for STAT signaling. Mutations in the pseudokinase domain activate the kinase domain for STAT signaling even in the absence of signaling molecule, which is associated with the onset of metastasis.

the JH2 domains of JAK1, JAK2 or JAK3 contribute to the constitutive activation of the JH1 kinase domain and are associated with myeloproliferative neoplasms, polycythemia vera, and acute myeloid leukemia (Table 3) (Jeong et al., 2008; Kralovics et al., 2005). JH2 harbors both gain-of-function and loss-of-function mutations. The most common mutation, JAK2 V617F, also resides in JH2 and accounts for ~95% of patients with polycythemia vera and about 50% of patients with essential thrombocytosis and primary myelofibrosis (Levine et al., 2005; Scott, Scott, Campbell, & Green, 2006). Another gain-of-function mutation, V658F mutation in JAK1, located in the JH2 autoregulatory domain is associated with polycythemia vera, essential thrombocytosis, primary myelofibrosis and lymphoblastic leukemia (Staerk, Kallin, Demoulin, Vainchenker, & Constantinescu, 2005).

TYK2 was the first member of the JAK family to be isolated and essential for cellular responses to IL-6, -10, and -12 (Fig. 4B). Evidence suggests that target-specific inhibition of TYK2 may be a promising therapeutic strategy with an ideal safety-benefit balance for treating autoimmune disorders such as Crohn's disease, multiple sclerosis, and psoriasis. Interestingly, the distinctive catalytically inactive pseudokinase domain (JH2) has provided an ideal "allosteric" site for the discovery of a selective TYK2 small molecule inhibitor (Wrobleski et al., 2019). Bristol-Myers Squibb identified BMS-986165 as the first JH2-targeting allosteric TYK2 inhibitor candidate, which entered a phase III clinical trial for psoriasis (Moslin et al., 2019). In addition, BMS-986165 provided complete protection from colitis as determined by decreased weight loss and colonic histological scores (Burke et al., 2019). Coincidentally, TYK2 mutants, and fusion proteins have been detected in patients with leukemia, implying that TYK2 could be an oncogene (Leitner, Witalisz-Siepracka, Strobl, & Müller, 2017). Accordingly, BMS-986165 might be useful as a therapeutic agent for other inflammatory and immune disorders and cancers in the future.

However, some mutations in the JAK3 and TYK2 JH2 domains enhance the auto-inhibitory interaction between the pseudokinase and kinase domains to reduce the activity of their JH1 domains and are associated with severe combined immunodeficiency (Chen et al., 2000; Li et al., 2013). Thus, modulation of the pattern of pseudokinase–kinase interactions in either direction in the JAKs could be beneficial in such conditions. There have been attempts to develop selective inhibitors targeting the pseudokinase (JH2) domain for cancer and immunological disorders. However, the ATP-binding site of the JH2 domain has a high degree of conservation with each other and with the JH1 domain within the JAK family, making it challenging to design a selective JH2 domain inhibitor specific to one JAK kinase. Therefore, significant structural biology and medicinal chemistry efforts are needed to develop small molecule pseudokinase ATP modulators and improve their selectivity and therapeutic efficacy. Taken together, there is still a clinical need for better JAK inhibitors, and attention has been given to novel approaches targeting regions beyond the conserved kinase domain.

4.3 KSRs

Ras-Raf-MEK-ERK signal transduction is probably the most critical oncogenic pathway, as RAS mutations occur in approximately 30% of all human cancers. In this signaling cascade, KSR1/2 pseudokinase acts downstream of Ras as molecular scaffolds for the Raf/MEK/ERK kinase cascade (Roskoski Jr, 2012). KSR has two distinct functions; activation of Raf after heterodimerization and recruitment of MEK to facilitate downstream signaling (Rajakulendran et al., 2009). KSR also regulates several cellular mechanisms such as cell proliferation and survival, including participation in metabolic regulation, cell cycle re-initiation following DNA damage repair, and translational regulation of Myc (Fig. 6) (Neilsen, Frodyma, Lewis, & Fisher, 2017). While most active protein kinases possess HRD (His-Arg-Asp) or YRD (Tyr-Arg-Asp) at the beginning of the catalytic loop, KSR1/2 has HKD (His-Lys-Asp) in its place in humans. Additionally, the essential β3-lysine residue is replaced by arginine in human KSR1 and KSR2. Despite these contradictory mutations, KSR1 and KSR2 display some enzyme activity (Roskoski Jr, 2018).

The $ksr1^{-/-}$ mice have reduced Ras-driven mammary tumor burden (Lozano et al., 2003). The initiation of skin tumors with v-Ha-Ras in $ksr1^{-/-}$ mice also was lost entirely. Stable depletion of KSR1 reduces the ability of colon cancer cells expressing mutant K-Ras to grow *in vitro* and in the xenograft mouse model (Fisher et al., 2015). Further, RNA interference (RNAi)-mediated depletion of KSR1 robustly induces cell death in the Ras-mutated colon cancer cell line HCT116 cells (Fisher et al., 2015). The involvement of KSR1 in the development of Ras-driven tumors makes it an appealing target for therapeutic intervention. KSRs also play a significant role as an overall metabolic regulator in cells by regulating glucose metabolism and adipogenesis (Klutho, Costanzo-Garvey, & Lewis, 2011; Kortum et al., 2005). Although the role of KSR2 has not been established in human cancers, adequate *in vitro* evidence indicates a potential involvement of KSR2 in cancer (Dougherty et al., 2009; Fernandez, Henry, & Lewis,

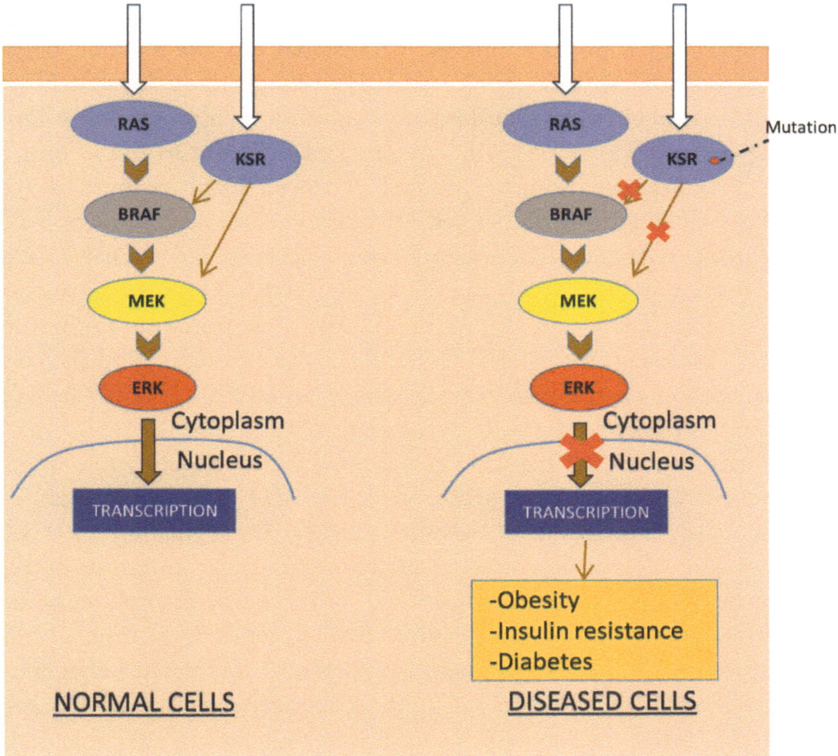

Fig. 6 In healthy cells, KSR allosterically activates BRAF that phosphorylates MEK, which activates mitogen-activated protein kinase (MAPK) signaling. KSR even directly activates MEK (left panel). In obesity and insulin resistance, associated mutations disrupt activation of MEK and BRAF by KSR, thus impairing the MAPK pathway (right panel).

2012). However, ksr2$^{-/-}$ mice have reduced fertility and become spontaneously obese (Costanzo-Garvey et al., 2009; Guo, Volle, & Lewis, 2014).

4.4 TRIBBLEs

Three human tribbles (TRIB) homologs (TRIB1, TRIB2, and TRIB3) are characterized by an N-terminal PEST region, a highly atypical bi-lobed pseudokinase domain with DFG motif replaced by metal-binding motif (E[S/N]LED) and a C-terminal COP1-binding region (Byrne et al., 2017). Helix αC is usually short in TRIB1, and they lack conserved glutamate required to form the salt bridge, resulting in a shallow nucleotide-binding pocket.

The C-terminal tail of the TRIB1 pseudokinase domain has an allosteric role. It binds to a pocket formed by helix αC in the N-lobe of the pseudokinase domain to bury the COP1-binding region located in the TRIB1 C-terminal tail so that TRIB1 would not be able to bind COP1. C/EBPα is the substrate of COP1. Binding of C/EBPα to the C-lobe of TRIB1 pseudokinase domain induces conformational changes in the TRIB1 pseudokinase domain, including ordering of the activation loop into a DFG-in-like state and displacement of the C-terminal tail (COP1 binding region) from the N-lobe (Jamieson et al., 2018). These conformation changes increase the accessibility of the TRIB1 C-terminal tail for binding to COP1. This way, the TRIB1/2-COP1 complex causes ubiquitination and degradation of the transcription factor C/EBPα. TRIB3 degrades acetyl-CoA carboxylase instead (Byrne et al., 2017).

In *Drosophila*, TRIB1 regulates cell differentiation, proliferation, migration, and interaction with the proto-oncogene AKT (Das, Sebo, Pence, & Dobens, 2014; Richmond & Keeshan, 2019). Dysregulation of TRIB expression by various cellular stresses and mitogens plays a multitude of roles in numerous cancers, and has also been associated with drug resistance through pro-survival and anti-apoptotic pathways (Table 3) (Fig. 7) (Hill et al., 2017; Richmond & Keeshan, 2019). TRIBs mediated oncogenesis is facilitated by the degradation of C/EBP transcription factors. R107L substitution in TRIB1 increases ERK phosphorylation and C/EBPα degradation in human Down syndrome-related acute megakaryocytic leukemia (Yokoyama et al., 2012). Overexpression of TRIB1 or TRIB2 induces leukemogenesis due to the upregulation of COP1 function and further depletion of C/EBPα in acute megakaryocytic leukemia (Dedhia et al., 2010; Keeshan et al., 2010; Yoshida, Kato, Nakamae, & Yoneda-Kato, 2013). Dysregulation of TRIBs is associated not only with cancer but also with liver, metabolic, cardiovascular, neurodegenerative and inflammatory disorders, highlighting its importance as a potential therapeutic target for a wide range of human diseases (Bauer, Yenilmez, & Rader, 2015; Richmond & Keeshan, 2019). For instance, overexpression of TRIB3 inhibits AKT to promote insulin resistance, apoptosis of pancreatic β-cells, and cause neuronal cell apoptosis in Parkinson's disease (Aimé et al., 2015; Du, Herzig, Kulkarni, & Montminy, 2003; Liew et al., 2010; Zareen, Biswas, & Greene, 2013) By contrast, overexpression of TRIB2 enhances AKT signaling and promotes drug resistance in cancer cells (Hill et al., 2017). TRIB2 requires an intact nucleotide-binding site to drive AML in mice, which presents a potential opportunity for pharmacological

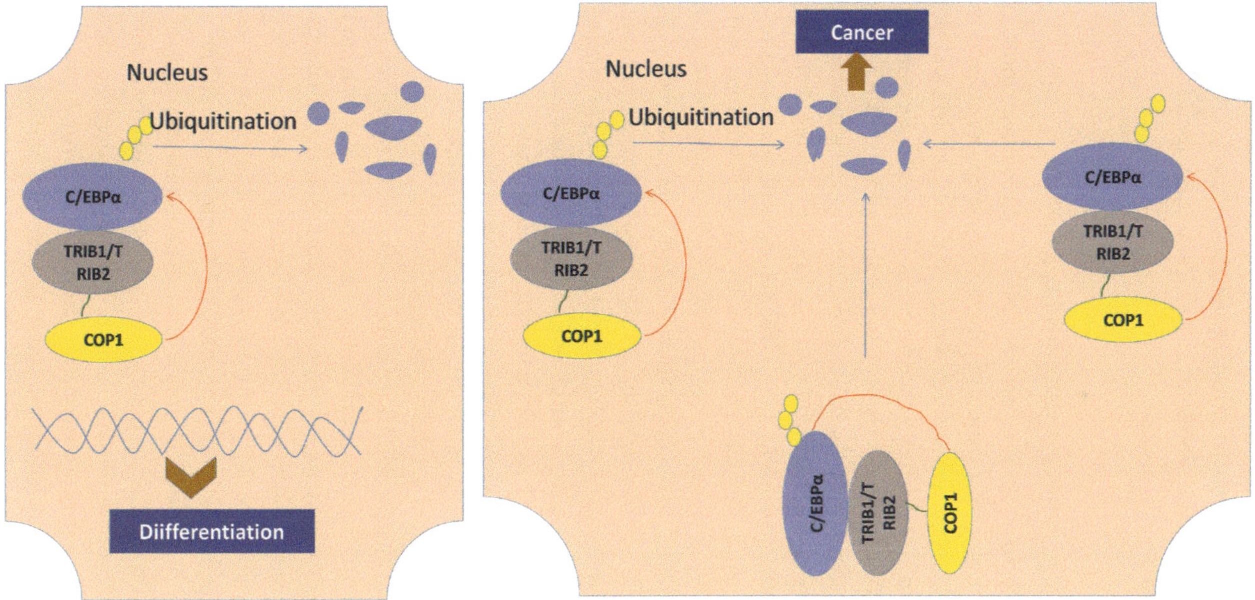

Fig. 7 The transcription factor C/EBPα promotes myeloid cell differentiation. Interaction of TRIB1 and TRIB2 with C/EBPα and the E3 ubiquitin ligase COP1 mediates ubiquitylation of C/EBPα and its proteasomal degradation (left panel). In cancer such as leukemogenesis, overexpression of TRIB1, or TRIB2 causes excessive degradation of C/EBPα (right panel).

intervention with compounds that target the TRIB2 pseudoactive site. A small molecule needs to target the TRIB2 ATP-binding site or the conserved DQLVP motif that binds with ubiquitin E3 ligases (cop1). Modulation of TRIB activity would, therefore, be therapeutically relevant to the treatment of several diseases.

4.5 Mixed lineage kinase domain-like protein

MLKL is a key effector pseudokinase in TNFα-induced necroptosis due to its role in the induction of plasma membrane permeability. Under normal physiological conditions, the pseudokinase domain of MLKL stabilizes the 4HB domain in an inhibited conformation. Pseudokinase domain phosphorylation by RIPK3 triggers the 4HB domain release, MLKL oligomerization, and its localization to the cell membrane, inducing permeabilization (Fig. 8A) (Murphy et al., 2013). Despite lacking a Glycine-rich loop, and possessing non-canonical HGK and GFE residues in place of the typical DFG and HRD motifs, MLKL binds robustly to ATP without hydrolysis. As you can see in Fig. 8A, the structure of MLKL is composed of an amino-terminal 4HB domain, a two-helix "brace" region, and a carboxy-terminal pseudokinase domain (Murphy et al., 2013). The 4HB domain has a membrane permeabilization activity (Hildebrand et al., 2014). The brace helices serve two critical functions within MLKL:MLKL oligomerization, and mediator to communicate signals from the pseudokinase domain to the 4HB domain (Davies et al., 2018). The carboxy-terminal pseudokinase domain has a regulatory role; binds the 4B domain to keep it in the dormant, inactivated state (Davies et al., 2018; Hildebrand et al., 2014; Murphy et al., 2013). Additionally, the pseudokinase domain serves as a signal integrator, in which RIPK3 phosphorylates Thr357 and Ser358 in the activation loop to trigger a conformational change within the pseudokinase domain that promotes 4HB domain exposure, enabling MLKL oligomerization, membrane translocation, permeabilize the plasma membrane and subsequent activation of necroptosis (Fig. 8B) (Byrne et al., 2017; Murphy et al., 2013).

Under normal physiological conditions, necroptosis is a part of innate defense against pathogens. Deregulation of necroptosis elicits an inflammatory response and contributes to diseases such as myocardial infarction, ischemic stroke, and neurodegeneration, highlighting MLKL as a potential novel drug target (Khoury, Gupta, Franco, & Liu, 2020). It was an exciting development to identify GW806742X (2) as a small molecule ligand of MLKL that inhibits necroptosis (Hildebrand et al., 2014).

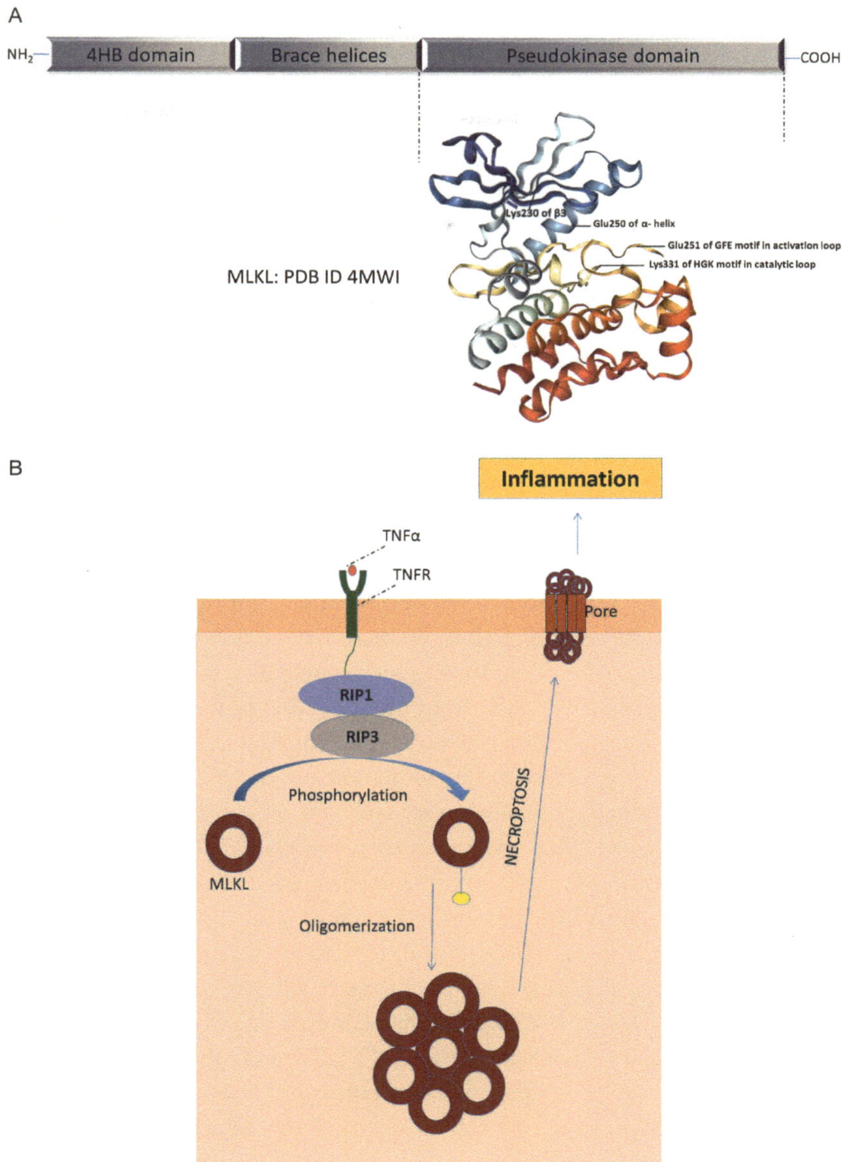

Fig. 8 (A) Basic structural architecture of MLKL. (B) In Necroptosis, on receiving a specific response from TNFα, RIPK1 interacts with RIPK3 to form a heterodimer to phosphorylate MLKL. MLKL further oligomerizes and translocates to the plasma membrane forming a pore and eliciting inflammatory response.

5. Current scenario of pseudokinase-centered therapy

Most small molecule inhibitors of canonical protein kinase are reversible ATP-competitive in nature, which can be categorized in two major classes: type 1 inhibitors act on the "active" conformation of protein kinases, bind to the ATP-binding site, and block access to ATP, and type 2 inhibitors act on the "inactive" conformation, occupy a hydrophobic groove adjacent to the ATP-binding site and lock kinases in an "inactive" state (Fabbro, Cowan-Jacob, Möbitz, & Martiny-Baron, 2012; Kufareva & Abagyan, 2008). Achieving drug target selectivity is hugely challenging in the case of protein kinases due to a high degree of sequence conservation of the active site across the kinome, which results in such inhibitors targeting multiple protein kinases (drug promiscuity) and exhibit toxicity (Byrne et al., 2017). Moreover, the development of drug resistance due to point mutations or circumvent signaling necessitates additional strategies to kinase targeting (Lu, Smaill, & Ding, 2020). In this regard, specific targeting of pseudokinases may be advantageous because these proteins have a unique amino acid sequence and structural architecture at the ATP-binding site. In addition, the targeting of allosteric sites has received increasing attention as a promising strategy for the development of new non-ATP-competitive kinase inhibitors. They are categorized as type 3 and 4 kinase inhibitors based on the distance of their allosteric sites targeted from the ATP-binding pocket (Zhang et al., 2010). Higher selectivity and low off-target toxicity are significant advantages of allosteric inhibitors over ATP-competitive kinase inhibitors. Recently, allosteric inhibitors that bind outside the catalytic kinase domain, such as those targeting the pseudokinase domain, have been identified (Kung & Jura, 2019). Inhibitors that target the pseudokinase domain to control a kinase activity allosterically are defined as type 6 (Lu et al., 2020). BMS-986165 is a classic example of a highly selective type 6 allosteric inhibitor of tyrosine kinase 2 (TYK2) (Wrobleski et al., 2019). Taken together, it is evident that significant opportunities exist to exploit each inhibitor class to test "catalytic" or "allosteric" sites of the pseudokinases, as their structural and biochemical analyses develop rapidly (Bailey, Byrne, McSkimming, Kannan, & Eyers, 2015).

However, pseudokinases pose significant challenges to drug design. For many pseudokinases, it is unknown how an ATP-competitive small molecule binding to a pseudoactive site could interfere with non-enzymatic activity. In addition, there are several pseudokinases that neither bind to nucleotide nor appear to have a structurally defined pocket that is accessible to ATP-competitive small molecules. Nevertheless, the two main strategies for the pharmacological targeting of pseudokinases continue to be investigated. While the first approach explicitly targets the ATP-binding "pseudoactive" site to modulate pseudokinase signaling functions, the second one aims the allosteric site to achieve target precision. Antibody-based approaches also target pseudokinases; however, except for the HER3 antibody, Patritumab, no other therapeutic antibodies have reached phase III clinical trials (Kung & Jura, 2019). Table 4 provides the list and mode of action of current pseudokinase inhibitors.

5.1 HER3

Over the past two decades, the importance of HER3 in cancer biology has become increasingly apparent. Increasing knowledge of the functions of HER3 has led to the development of multiple HER3-focused treatments for human cancer (Black, Longo, & Carroll, 2019). Sarup et al. (2008) developed HER ligand binding molecule, also termed RB200 that sequestered multiple EGF-like ligands and interfered with signaling by HER-1, HER-2, and HER-3. RB200 also inhibited the proliferation of a wide range of tumor cells *in vitro*, alone or in combination with tyrosine kinase inhibitors. It also inhibited lysophosphatidic acid-stimulated cell proliferation, and tumor growth in two human tumor xenograft nude mouse models. A wide variety of monoclonal antibodies that block ligand binding, promote HER3 destruction, or lock HER3 in a tethered conformation are tested clinically or currently in trials for multiple cancer types. As HER3 lacks significant kinase activity and cannot be inhibited by tyrosine kinase inhibitors, a locked nucleic acid (LNA)-based HER3 antisense oligonucleotide, EZN-3920, that specifically downregulated the expression of HER3 and exhibited antitumor activity alone or when co-administered with gefitinib or lapatinib was reported (Wu et al., 2013). Besides, ATP-competitive small molecule inhibitors (TX1-85-1, TX-121-1) of ERBB3 also effectively promoted HER3 degradation, interfered with HER3 heterodimerization, and inhibited HER3-dependent pathways (Lim et al., 2015; Xie et al., 2014). TX2-121-1 is a bi-functional small molecule containing a hydrophobic adamantane moiety responsible for inducing proteasomal degradation of HER3. A recent proof-of-concept study employed a multiplex, medium throughput thermal-shift assay screening strategy to identify selective small molecule inhibitors of HER3. A chemical probe code-named AC3573,

Table 4 Pseudokinase inhibitors, their mode of action, and implications in various diseases.

Pseudokinase	Inhibitor	Diseases	Mechanism of action	References
EPHA10	BsAb	Breast cancer therapy	Monoclonal antibody that forms a bridge between antigens expressed on tumor cells and cytotoxic T immune cells, promoting recognition of tumor cells by immune cells and redirecting T cell-mediated lysis of EPHA10-overexpressing cells	Taki et al. (2015)
	Db-1 (Heterodimeric) and Db-2 (Homodimeric)	Breast cancer and other cancers	Db-1 and Db-2 diabodies target EPHA10 expressed on cancer cells and CD3 expressed on T cells and induce T cell-mediated lysis of EPHA10-overexpressing cells	Kamada et al. (2015)
)HER3	TX1–85-1	Cancer	First selective irreversible (type V) small molecule inhibitor that forms a covalent bond with Cys721 located in the ATP-binding site of HER3 thus inhibiting HER3 signaling by partial degradation of Her3 and prevent HER2-HER3 heterodimerization	Xie et al. (2014)
	Elgemtumab (LJM716)	Cancer	Monoclonal antibody that binds to inactive conformation of extracellular domain of HER3 but does not prevent ligand binding; potent inhibitor of HER3/AKT phosphorylation and proliferation, inhibits ligand-dependent and ligand-independent HER3 signaling	Garner et al. (2013)
	Entinostat	Breast cancer	Disrupts HER2/HER3 interactions and downstream kinases; downregulates HER2 and HER3 expression by induction of repressor microRNAs	Lee et al. (2014)
	Duligotuzumab (MEHD-7945A)	Multiple tumor target; overcome inhibitor resistance in cancer cells	Monoclonal two-in-one antibody that targets EGFR- and HER3-mediated signaling; binds to inactive conformations of the extracellular domain of HER3; inhibits ligand-dependent HER3 signaling	Schaefer et al. (2011), Huang et al. (2013)

	Seribantumab (MM-121)	Cancer	Monoclonal antibody inhibits ligand-induced phosphorylation of HER3	Schoeberi et al. (2010)
	Patritumab (U3–1402)	Cancer	This is an antibody drug conjugate where patritumab is conjugated to the topoisomerase I inhibitor DXd; Inhibits HER3 signaling, the cytotoxic activity of U3–1402 in HER3-expressing cells was mediated by released DXd through DNA damage and apoptosis induction	Yonesaka et al. (2019)
	FL518	Metastatic cancer	Four in one monoclonal antibody made by combining duligotuzumab and bH1–44 recognizes EGFR, HER2, HER3 and VEGF and inhibits ligand-dependent and ligand-independent signaling.	Hu et al. (2015)
	GSK2849330	Antitumor	Monoclonal antibody disrupts ligand-dependent HER3 signaling; engineered to exhibit enhanced antibody dependent cell mediated cytotoxicity	Alsaid et al. (2017)
	KTN3379	Antitumor	Monoclonal antibody prevents HER2–HER3 heterodimerization; inhibits both ligand-dependent and ligand-independent HER3 signaling	Xiao et al. (2016)
	KIN001–051	Human	Type I reversible inhibitor whose phenyl ring forms covalent bond with the Cys 721	Lim et al. (2015)
	REGN1400	Epidermoid carcinoma, breast cancer, head and neck cancer	Monoclonal antibody inhibits binding of the HER3 ligand neuregulin 1, inhibits phosphorylation of HER3 and AKT	Zhang et al. (2014)

Continued

Table 4 Pseudokinase inhibitors, their mode of action, and implications in various diseases.—cont'd

Pseudokinase	Inhibitor	Diseases	Mechanism of action	References
	TX2–121-1	Cancer	Irreversible ATP-competitive small molecule inhibitor that forms covalent adduct with Cys721 within the ATP binding site of HER3; partial degradation of Her3 and interferes with productive heterodimerization between Her3 with Her2	Xie et al. (2014)
	Cotransin CT8	Breast cancer	Small molecule inhibitor binds the Sec61 translocon and prevents the signal peptide of the nascent HER3 protein from initiating insertion of HER3 into the ER, resulting in HER3 degradation	Ruiz-Saenz et al. (2015)
JAK1 and JAK2	Ruxolitinib (INCB018424, INC424)	Myelofibrosis, acute leukemia	Type I ATP competitive reversible inhibitor; exposure of JAK2V617F overexpressing Ba/F3 cells to ruxolitinib resulted in reduced phosphorylation of JAK2, STAT5, and ERK1/2, impaired cellular proliferation, and induced apoptosis	Quintas-Cardama et al. (2010)
	Baricitinib(reversible)	–	Reversible type I inhibitor	Calabrese et al. (2020)
JAK2 JH2	AT-9283	Polycythemia vera, essential thrombocythemia, and myelofibrosis	ATP-competitive small molecule JAK2 and JAK3 inhibitor that binds to both JAK2 JH1 and JH2	Puleo et al. (2017)
	JNJ-7706621	Chronic myeloid leukemia and myeloproliferative neoplasmas	ATP-competitive small molecule inhibitor that binds to both JAK2 JH1 and JH2	Puleo et al. (2017), Newton et al. (2017)
	Compound 2	–	Exhibited modestly improved affinity (123 nM) and selectivity for the pseudokinase domain	McNally et al. (2019)

JAK3	Ibrutinib	Rheumatoid arthritis	Type IV irreversible small molecule inhibitor binds irreversibly to a cysteine residue at an analogous position in the ATP binding site corresponding to cysteine 481 in Bruton's tyrosine kinase	Berglof et al. (2015)
	Staurosporine	Severe combined immunodeficiency	Small molecule competitive inhibitor of Mg-ATP, that disrupts Jak3 binding to γ c cytokine receptor. Staurosporine interacts with the ATP binding cleft and induces a dramatic conformational change in the N and C terminal portion of the kinase domain in JAK	Boggon, Li, Manley, and Eck (2005)
KSR2	APS-2-79	Cancer	ATP-competitive small molecule inhibitor that stabilizes an inactive conformation of the KSR2 pseudokinase domain; modulate KSR-dependent MAPK signaling by preventing heterodimerization with RAF kinases (BRAF)	Dhawan et al. (2016)
	ASC-24	Cancer	ATP-competitive small molecule inhibitor that blocks KSR kinase activity without interfering heterodimerization with RAF kinases; reduces KSR2-dependent phosphorylation of MEK1	Brennan et al. (2011)
MLKL	Compound 1	Necroptosis	ATP-competitive small molecule inhibitor of MLKL that stabilizes the pseudokinase domain in the DFG- out state; also the inhibitor of VEGFR and RIPK1; less selective for MLKL than compound 4	Ma et al. (2016)
	Compound 4	Fails to inhibit necroptosis	ATP-competitive small molecule inhibitor of MLKL that stabilizes the pseudokinase domain in the DFG-in conformation	Ma et al. (2016)
	Necrosulfonamide	Necroptosis	Type IV irreversible inhibitor covalently modifies Cys86 on the N-terminal coiled coil domain of MLKL nonpseudokinase domain	Sun et al. (2012)

Continued

Table 4 Pseudokinase inhibitors, their mode of action, and implications in various diseases.—cont'd

Pseudokinase	Inhibitor	Diseases	Mechanism of action	References
PTK7	PF-06647020	Against solid tumor	Monoclonal antibody and drug conjugate that targets PTK7 expressed on tumor cells. It is a h6M24 conjugated to the auristatin microtubule inhibitor Aur0101; upon binding PTK7 expressed on tumor cells, it's internalized and cleaved to release auristatin-0101 that further binds and inhibits tubulin polymerization, arrest G2/M phase and finally cause apoptosis of PTK7-expressing tumor cells	Katoh (2017)
RNase L	Sunitinib, flavonols (quercetin)	Suppress angiogenesis and tumor growth	ATP-competitive small molecule inhibitor. Inhibits activation of RNase L and dimerization; prevent antiviral innate immune responses mediated by RNase L and RNA dependent protein kinase namely PKR	Jha et al. (2011), Tang et al. (2020)
ROR1	1C11, 1D8, 4C10, 4A7	Chronic lymphocytic leukemia (CLL)	Monoclonal antibodies-1C11 and 1D8 act against cysteine rich extracellular domain of ROR1; monoclonal bodies- 4A7, 4C10 act against kringle region in the extracellular domain of ROR1	Daneshmanesh et al. (2012)
	Cirmtuzumab (UC-961)	Chronic lymphocytic leukemia	ROR1 is a receptor for Wnt5a, which can promote activation of Rac1 to enhance cell proliferation and survival in chronic lymphocytic leukemia. Cirmtuzumab is a monoclonal antibody that inhibits WNT5a-induced RAC1 activation	Karvonen, Niininen, Murumägi, and Ungureanu (2017)
	ROR1-CAR	Chronic lymphocytic leukemia, Mantle cell lymphoma	Chimeric antigen receptor specific (CAR) and ROR1 complex transduced into T cells selectively lyse ROR1-overexpressing cancer cells	Hudecek et al. (2010)

RYK	RWD1	Neurite outgrowth	Monoclonal antibody binds to Wnt inhibitory factor (WIF) extracellular domain of RYK and prevent Wnt ligand binding to this domain and inhibits WNT5A-induced signaling and neurite outgrowth	Halford et al. (2013)
TRIB2	Afatinib	Cancers, blood cancer	Irreversible type I ATP-competitive small molecule covalent inhibitor of EGFR and HER2; forms covalent adduct with Cys96 in TRIB2; destabilizes and degrades TRIB2 in cells	Foulkes et al. (2018)
	Neratinib, osimertinib		Irreversible type I ATP-competitive small molecule inhibitors of EGFR and HER2; destabilize TRIB2 in vitro	
	Lapatinib, TAK-285, GW693881A		Reversible ATP-competitive small molecule inhibitors of EGFR and HER2; stabilize TRIB2 in vitro	
	GW804482X		Reversible ATP-competitive small molecule inhibitor destabilizes TRIB2 in vitro	
TYK2 JH2	Compound 1, BMS-066	Treat autoimmunity	ATP-competitive small molecule inhibitor inhibit TYK2 signaling by binding to the TYK2 JH2 pseudokinase domain to cause receptor-mediated activation of the adjacent catalytic domain; blocks TYK2-mediated signaling through IL-23R in T cells	Tokarski et al. (2015)
	Compound 35 (imidazo[1,2-*b*] pyridazine (IZP) **7)**	Various immuno-inflammatory diseases	ATP-competitive small molecule inhibitor inhibit TYK2 signaling through IL-23R in T cells (more selective and stable than compound 1)	Moslin et al. (2019)
	BMS-986165	Psoriasis	ATP-competitive small molecule inhibitor that blocks TYK2-mediated signaling through type I interferon, IL-12, and IL-23	Papp et al. (2018)

Continued

Table 4 Pseudokinase inhibitors, their mode of action, and implications in various diseases.—cont'd

Pseudokinase	Inhibitor	Diseases	Mechanism of action	References
IRAK3	PROTAC 23	Modulates downstream innate immune signaling	A potent and selective degrader of IRAK3. PROTAC 23 induced proteasome-dependent degradation of IRAK3.	Degorce et al. (2020)
ULK4	Compound 1, (R788)	Deletions in ULK4 have been linked to increased susceptibility to developing schizophrenia	Although R788 would not make a suitable chemical probe for ULK4 because of the phosphate, it provides important information about optimizing the compounds further for potency, and it provides a starting point as the most potent ULK4 binder	Khamrui et al. (2020)

specifically inhibited HER3 and abrogated HER2:HER3 complex formation and downstream signaling in cells (Colomba et al., 2020). In addition, anti-HER3 antibodies conjugated to cytotoxic drugs (saponin, monomethyl auristatin F, and the topoisomerase I inhibitor DXd) have shown remarkable anticancer activity *in vitro* and *in vivo* (Capone et al., 2017, 2018; Yonesaka et al., 2019). The observation that HER3 physically interacts with other receptor tyrosine kinases has led the development of monoclonal antibodies, such as MM-111, MM-141, MCLA-128, and duligotuzumab (Camblin et al., 2019; De Pauw et al., 2017; Geuijen et al., 2018; McDonagh et al., 2012). Nevertheless, given the complex regulation of the HER3 action, a range of challenges remain to be tackled before effective ERBB3-directed therapeutics are developed.

5.2 JAK family pseudokinases

Several cytokines and growth factors are involved in the pathogenesis of hematological and immune-mediated diseases, most of which act by binding to their cognate receptor and activating one or more of the four Janus family tyrosine kinases (JAKs). Active mutant forms of JAKs (*e.g.*, JAK2 V617F in myeloproliferative disorders) and gene knockout studies provided a strong rationale for the development of JAK inhibitors. Toward these ends, in seeking molecules that bind to the JH2 ATP site, a high-throughput screening yielded several ATP-competitive small molecules, including pan-cyclin-dependent kinase (CDK) and pan-JAK inhibitor, JNJ7706621 and the Aurora, JAK2 and JAK3 inhibitor, AT-9283 (Newton et al., 2017; Puleo et al., 2017). Recently, Liosi et al. (2020) reported the design and synthesis of compounds with a diaminotriazole core that preferentially binds to the ATP-binding site in JH2 instead of JH1. A chemogenomic screen for kinase inhibitors identified JH2-specific inhibitors of TYK2, that antagonize interleukin-23 receptor (IL-23R) signaling. These molecules inhibited TYK2 autophosphorylation, and TYK2-mediated phosphorylation of STAT3 *in vitro* (Tokarski et al., 2015). In phase II clinical trial, BMS-986165, a selective JH2 domain inhibitor of TYK2 resulted in the more significant clearing of psoriasis (Papp et al., 2018). Phase 2 clinical trials of this compound are on-going for the treatment of Systemic Lupus Erythematosus, Severe Ulcerative Colitis, Crohn's Disease, and Lupus Nephritis.

5.3 KSRs

Numerous studies support the possibility of Ras-driven KSR-centric interventions for several different forms of cancers and drug resistance

(Neilsen et al., 2017). RNAi-mediated knockdown of KSR demonstrated a decrease in tumor growth *in vitro* and *in vivo* (Fisher et al., 2015). Further, antisense oligonucleotides inhibited expression KSR1 and caused regression of established tumors and metastases in Ras-driven pancreatic and non-small cell lung cancer xenografts in mice (Zhang et al., 2008). Due to its status as a pseudokinase and role as a non-catalytic regulator of core signaling enzymes, pharmacological approaches targeting KSR have so far been lacking. Guided by KSR mutations that selectively suppress oncogenic Ras signaling, Dhawan et al. (2016) have developed a class of compounds (code-named APS-2-79) that stabilize the previously unrecognized inactive state of KSR. APS-2-79 could interfere with KSR:RAF heterodimerization and inhibit oncogenic Ras signaling (Neilsen et al., 2017). Further understanding of the pathophysiological roles of KSR may facilitate the development of additional strategies for the discovery of KSR-specific inhibitors that may transform the clinical treatment of Ras-driven cancers.

5.4 RTK family pseudokinases

Pseudokinases from the receptor tyrosine kinase family such as RORs, protein tyrosine kinase 7 (PTK7), and RYK are Wnt ligand-binding receptors with essential roles in development, cell survival, migration, polarization, tissue homeostasis, and organogenesis. Although RTKs lack ATP-binding and catalytic activity, they sustain ligand-induced signaling capabilities by modulating other kinases or serving as scaffolds for signaling molecules (Mendrola, Shi, Park, & Lemmon, 2013). Enhanced expression of RTKs in hematological cancers has led to detailed studies of their oncogenic role in cancer cell survival and the development of drug resistance (Karvonen et al., 2017). ROR1-based anticancer therapies include mAb (monoclonal antibodies), antibody drug-conjugate, chimeric antigen receptor T-cells (CAR-T cells), and small molecule inhibitors (Baskar, Wiestner, Wilson, Pastan, & Rader, 2012; Daneshmanesh et al., 2012). Cirmtuzumab (UC-961) effectively reduced ROR1$^+$ leukemic cells' engraftment in the transgenic mouse model and entered phase I/II clinical trial for relapsed or refractory chronic lymphocytic leukemia (Choi et al., 2015). On the other hand, ROR1 targeted CAR-T cells, and a bi-specific T-cell engager (BiTE) effectively killed the primary ROR1$^+$ leukemia cells, while offering low off-target cytotoxicity benefits (Hudecek et al., 2010). This offers a new platform for T-cell mediated targeting of ROR1$^+$ tumors that merits further clinical studies. Further, the combination of ibrutinib (Bruton's tyrosine kinase (BTK) inhibitor) and cirmtuzumab as anti-leukemia therapy was more effective than either single agent alone *in vivo*, suggesting an additive or synergistic effect that need to be investigated in the clinic (Yu et al., 2017). These studies have collectively demonstrated that targeting Wnt-related pseudokinases (such as ROR1 or PTK7) can induce cytotoxicity on its own or improve the anticancer response of other drugs, suggesting that these proteins have to be considered in future treatment strategies. The list of reversible and irreversible FDA approved kinase inhibitors, and their pseudokinase targets are given in Table 5. Although pseudokinase-centric therapeutics have not yet been approved for clinical applications, we are better equipped to meet many challenges.

6. Challenges in developing pseudokinase-targeted therapy

Pseudokinases are distributed across the kinome and are catalytically inactive variants of canonical protein kinases. They play critical regulatory roles in signaling pathways, but have been poorly characterized, understudied and therefore merit special attention so that the scientific community can better understand the therapeutic implications of pseudokinase modulation. Pseudokinases pose a considerably higher challenge to the drug development process compared to active kinases. The lack of a solved crystal structure of many pseudokinases prevents using their druggable sites for therapeutic development. It is unlikely that the full-length, three-dimensional structures of pseudokinases that serve as membrane receptors such as Eph receptors will be resolved by traditional techniques such as NMR and X-ray crystallography. For many pseudokinases, it is not known how small molecules competing with nucleotide binding can disrupt non-enzymatic functions. While many pseudokinases do not bind to nucleotides, no structurally defined pockets appear to be available for small molecules competing with ATP in others. Though JAK2 V617F is the most common cancer-associated mutation, questions remain regarding the pathological association of JAK2 V617M, JAK2 V617W, JAK2 V617L, and JAK2 V617I mutations as well as other pseudokinase-specific mutations. Lack of reliable commercial assays hinders the development of pseudokinase-specific chemical probes, whereas the traditional kinase assays are largely incompatible due to vestigial or complete lack of kinase activity of pseudokinases (Drewry et al., 2017; Reiterer et al., 2014).

Table 5 List of reversible and irreversible FDA approved kinase inhibitors and their pseudokinase targets.

Inhibitor classification		FDA approved drugs/year of approval	Target kinases	Target kinase domain of pseudokinases	Implication (reference: https://www.fda.gov/)
Irreversible	Type I	Afatinib, 2013	EGFR	HER3, TRIB2	Carcinoma
		Neratinib, 2017	EGFR, HER2	HER3	Cancer
		Acalabrutinib, 2017	BTK	–	Mantle cell lymphoma
	Type IV	Ibrutinin, 2013		JAK3	Chronic graft *versus* host disease
Reversible	Type I	Baricitinib, 2018	–	JAK1 and JAK2	Rheumatoid arthritis
		Crizotinib, 2016	ALK/ROS1		Metastatic non-small cell lung cancer
		Ruxolitinib, 2011	–		Myelofibrosis
		Vandetanib, 2011	VEGFR/EGFR, etc.		Medullary thyroid cancer
		Vemurafenib, 2011	BRAF		Metastatic melanoma, thyroid tumors
		Tofacitinib, 2012	–	JAK1 and JAK3	Rheumatoid arthritis
		Bosutinib, 2012	ABL/SRC	HER3, TRIB2	CML (chronic myeloid leukemia)
		Dasatinib, 2006	ABL/SRC, etc.		
		Erlotinib, 2004	EGFR-HER1		Malignancy
		Gefitinib, 2003	EGFR-HER1		
		Lapatinib, 2007	EGFR, HER1,2,4		Cancers, blood cancer
		Osimertinib, 2015	EGFR	TRIB2	Non-small-cell lung carcinomas
		Sunitinib, 2006	KIT/PDGFR, etc.	RNase L	Suppress angiogenesis and tumor growth
		Abemaciclib, 2017	CDK4/CDK6	–	Metastatic breast cancers

	Alectinib, 2015	ALK	–	Non-small-cell lung carcinomas
	Axitinib, 2012	VEGFR, etc.	–	Renal cell carcinoma
	Buparlisib	PI3K, HER2	–	Prostate cancer, breast cancer, lymphoma, Metastases, lung cancer, solid tumors
	Brigatinib, 2017	ALK/EGFR	–	Metastatic anaplastic lymphoma kinase (ALK)-positive non-small cell lung cancer
	Ceritinib, 2014	ALK	–	Non-small-cell lung carcinomas
	Copanlisib, 2017	PI3Ka/PI3Kd	–	Relapsed follicular lymphoma
	Dabrafenib, 2013	BRAF	–	Skin cancer-melanoma
	Encorafenib, 2018	BRAF	–	Metastatic melanoma
	Fostamatinib, 2018	SYK	–	Thrombocytopenia
	Idelalisib, 2014	Lipid kinase PI3Kd	–	Chronic lymphocytic leukemia, follicular B-cell non-Hodgkin lymphoma, small lymphocytic lymphoma
	Lenvatinib, 2015	VEGFR1/2/3	–	Hepatocellular carcinoma
	Midostaurin, 2017	FLT3, etc.	–	Acute myeloid leukemia
	Nintedanib, 2014	VEGFR, etc.	–	Liver dysfunction
	Palbocicnib, 2015	CDK4/CDK6	–	Metastatic breast cancers
	Pazopanib, 2009	c-KIT/FGFR, etc.	–	Aortic dissection
	Ribociclib, 2017	CDK4/CDK6	–	Metastatic breast cancers

Continued

Table 5 List of reversible and irreversible FDA approved kinase inhibitors and their pseudokinase targets.—cont'd

Inhibitor classification	FDA approved drugs/year of approval	Target kinases	Target kinase domain of pseudokinases	Implication (reference: https:/www.fda.gov/)
Type II	Ponatinib, 2012	ABL	JAK3	CML (chronic myeloid leukemia)
	Cabozantinib, 2012	c-MET/VEGFR2, etc.		Hepatocellular carcinoma
	Imatinib, 2001	ABL/PDGFR/c-KIT	–	Chronic myeloid leukemia (CML) and malignant gastrointestinal stromal tumors
	Nilotinib, 2007	ABL/KIT, etc.		Philadelphia chromosome positive chronic myeloid leukemia
	Regorafenib, 2012	VEGFR, etc.		Hepatocellular carcinoma
	Sorafenib, 2005	VEGFR/PDGFR, etc.		Hepatocellular carcinoma
Type III	Binimetinib, 2018	MEK	–	Metastatic melanoma
	Cobimetinib, 2015			Melanoma
	Rapamycin, 2015	AmTOR, PI3K related kinases		Lymphangioleiomyomatosis
	Trametinib, 2013	MEK1	–	Melanoma

The majority of the present pseudokinase inhibitors have been developed based on HIT compounds, which are already known to have active kinase inhibitory activity, as evidenced by the higher structural similarity between HER inhibitors. Most of the pseudokinase inhibitors developed are for cancer treatment, whereas it is well-known that they regulate diverse pathophysiological activities related to inflammation, diabetes, necroptosis, hypercholesterolemia, cardiovascular and cerebrovascular abnormalities. Further, despite their proven efficacy, none of the pseudokinase-directed antibodies have passed the phase III clinical trial. Developing ATP-competitive pseudokinase inhibitors with maximum efficacy and minimal off-target interaction is a formidable task, as their pseudokinase domain bears structural homology to the canonical protein kinase domain. These observations collectively represent the inherent difficulties in achieving specificity by targeting the pseudoactive site with ATP-competitive compounds.

7. Way forward

Our understanding of the structural and functional aspects of pseudokinases has expanded considerably over the past decade, providing new opportunities to exploit them for therapeutic benefit. Interestingly, the "Illuminating the Druggable Genome" initiative in the United States has recently been launched to explore therapeutically relevant pseudokinases (Rodgers et al., 2018; Santos et al., 2017). Advanced state-of-the-art microscopy techniques, such as total internal reflection fluorescence (TIRF) and cryo-electron microscopy, may shed light on the dynamic oligomerization process of the pseudokinase receptors in the plasma membrane and provide clues on the transmission mechanism of extracellular cues to intracellular effectors on ligand binding. Small molecule inhibitors for a handful of pseudokinases have been developed based on structural analysis and binding patterns, which can serve as a foundation for advances in more specific inhibitor discoveries (Table 4). Moreover, many commercially available recombinant pseudokinase domains allow for the unbiased compound screening, including those that interact with the ATP, even with uncertain site geometry and the absence of known biological or pathological implications. Although there is currently a lack of pseudokinase-specific assays, differential scanning fluorimetry (DSF) is valuable as it has successfully identified TRIB2 binders (Elkins et al., 2016; Foulkes, Byrne, Bailey, & Eyers, 2015; Lucet & Murphy, 2017). Coupling DSF screening technology with available kinase inhibitor sets can yield chemical starting points for these kinases. Since pseudokinases typically lack detectable catalytic activity, thermal-shift assay has become the preferred primary screening methodology for small molecule inhibitors and the assessment of pseudoactive site accessibility (Bailey, Byrne, Oruganty, et al., 2015; Christie, Boland, Huntzinger, Weichenrieder, & Izaurralde, 2013; Hildebrand et al., 2014). The assay could be performed with small quantities of protein in solution without the need for tagging or immobilization, and is largely unaffected by the existence of tiny amounts of contaminating proteins when compared to fluorescently labeled ATP analog binding (tag leads to false binding), kinase-affinity chromatography (nucleotide immobilization necessary), isothermal titration calorimetry (large quantity of protein required), and surface plasmon resonance (protein immobilization necessary) (Bailey, Byrne, Oruganty, et al., 2015; Murphy, Lucet, et al., 2014; Murphy, Zhang, et al., 2014).

The wealth of structural and functional knowledge of pseudokinases is expected to shift the boundaries of science to overcome the limitations of developing pseudokinase-centric therapies for human diseases (Table 6). Antibody-based approaches, alone or combined with other established therapeutic agents, offer a unique opportunity to target pseudokinases. Since pseudokinases are thought to have evolved from a conventional protein kinase ancestor, substantial structural biology and medicinal chemistry endeavors would be needed to mitigate the inherent conservation found in the ATP-binding site of the kinase/pseudokinase pair. The targeting of a pseudoactive site depends on its suitability, and the implementation of this strategy would be limited to a site that is occluded by unconventional side chains (Lucet & Murphy, 2019). Structurally diverse small molecules with natural pharmacophore and scaffolds that do not resemble kinase inhibitors are ideal for ligand screening. Chemical probes that force a pseudokinase to adopt a non-functional conformation may be therapeutically useful in cases where the pseudokinase assumes a conformation that reflects either the on or off state of protein kinases (Kung & Jura, 2019). Simultaneous targeting of multiple deregulated signaling pathways with promiscuous inhibitors could be beneficial for polygenic disorders, although it could increase the risk of side effects. Specific targeting of pseudokinase would be beneficial in non-cancer diseases where mechanistic approaches have historically been defined. Another viable approach to pharmacologically target pseudokinases focuses on induced protein degradation by proteolysis targeting chimeras (PROTACs) or hydrophobic tagging, as in the case of IRAK3 (Table 4) (Degorce et al., 2020). PROTAC is a heterobifunctional compound

Table 6 Diverse online resources and databases for the kinase and pseudokinase research community.

• UniProt: A comprehensive, high-quality and freely accessible resource of protein sequence and functional information • https://www.uniprot.org/
• Protein Data Bank: Archive-information about the 3D shapes of proteins, nucleic acids, and complex assemblies • https://www.rcsb.org/
• KinHub: Resources and tools for human protein kinases • http://www.kinhub.org/index.html
• kinase.com: Genomics, evolution and function of protein kinases • http://kinase.com/
• Kinweb: A collection of protein kinases encoded in the human genome • https://www.itb.cnr.it/kinweb/
• Kinase profiling inhibitor database • http://www.kinase-screen.mrc.ac.uk/kinase-inhibitors
• KinBase: The kinase database at Sugen/Salk • http://www.kinase.com/kinbase/
• DrugKiNET: An open-access, online resource to foster the identification and characterization of inhibitors of protein kinases • http://www.drugkinet.ca/
• PKIDB: A curated, annotated and updated database of protein kinase inhibitors in clinical trials • https://www.icoa.fr/pkidb/
• RegPhos: A system to explore the protein kinase-substrate phosphorylation network in humans • http://140.138.144.141/~RegPhos/
• GPS 5.0: Computational prediction of phosphorylation sites (p-sites) with their cognate protein kinases (PKs) • http://gps.biocuckoo.cn/
• HPRD: Human protein reference database • http://hprd.org/
• KLIFS: Kinase-ligand interaction fingerprints and structures database • https://klifs.vu-compmedchem.nl/
• KinomeXplorer: An integrated framework for modeling kinase-substrate interactions and aid in the design of inhibitor-based follow-up perturbation experiments • https://kinomexplorer.info/

Continued

Table 6 Diverse online resources and databases for the kinase and pseudokinase research community.—cont'd

• KinG: Kinases encoded in genomes • http://king.mbu.iisc.ernet.in/
• PhosphoNetworks: A database for experimentally determined kinase-substrate relationships • https://phosphonetworks.org/
• KinomeX: A web application for predicting kinome-wide polypharmacology effect of small molecules • https://kinome.dddc.ac.cn/en/
• Kinomer: A multilevel HMM library that models protein kinase groups • http://www.compbio.dundee.ac.uk/kinomer/index.html
• KKB: The Kinase Knowledgebase (KKB) is Eidogen-Sertanty's database of kinase structure-activity and chemical synthesis data • https://eidogen-sertanty.com/kinasekb.php
• International Centre for Kinase Profiling • http://www.kinase-screen.mrc.ac.uk/
• HKPocket: human kinase pocket database for drug design • http://zhaoserver.com.cn/HKPocket/HKPocket.html
• ProKinO: Protein kinase ontology browser • http://vulcan.cs.uga.edu/prokino/about/browser
• KINOMEscan: A biochemical kinase profiling assay that measures drug binding using a panel of $\sim$440 purified kinases • http://lincs.hms.harvard.edu/kinomescan/
• ChEMBL: A manually curated database of bioactive molecules with drug-like properties • https://www.ebi.ac.uk/chembldb/
• DrugBank: A database resource that combines detailed drug data with comprehensive drug target information • https://www.drugbank.ca/
• KinasePhos: A web tool for identifying protein kinase-specific phosphorylation sites • http://kinasephos.mbc.nctu.edu.tw/
• KinaMetrix: A web resource to investigate kinase conformations and inhibitor space • http://KinaMetrix.com

Table 6 Diverse online resources and databases for the kinase and pseudokinase research community.—cont'd

• Kinase Atlas: Druggability analysis of potential allosteric sites in kinases • https://kinase-atlas.bu.edu
• KinMap: A web-based tool for interactive navigation through human kinome data • http://www.kinhub.org/kinmap/
• eXpression2Kinases (X2K) Web: Linking expression signatures to upstream cell signaling networks • http://amp.pharm.mssm.edu/X2K/
• Kinact: A computational approach for predicting activating missense mutations in protein kinases • http://biosig.unimelb.edu.au/kinact/
• MOKCa database: Mutations, oncogenes, knowledge and cancer has been developed to structurally and functionally annotate, and where possible predict, the phenotypic consequences of mutations implicated in cancer • http://strubiol.icr.ac.uk/extra/mokca/index.html
• Pfam: A large collection of protein families, each represented by multiple sequence alignments and hidden Markov models (HMMs) • https://pfam.xfam.org/

functions by inducing selective intracellular proteolysis. There is also considerable scope for developing inhibitors that could interfere sterically with protein-protein interactions driven by pseudokinase. Alternatively, allosteric targeting or stabilization of a pseudokinase's inactive conformation is particularly attractive, albeit less explored. Moving forward, joint collaborative research efforts between industry and academia have tremendous potential to accelerate the process of computational and structural analysis, chemical biology, high-throughput screening and proteomics, and elucidation of pseudokinase loci in known (or new) signaling networks. The GlaxoSmithKline Published Kinase Inhibitor Set (GSK PKIS) is an illuminating example of open access annotated set of small molecule kinase inhibitors with known and unknown affinities for multiple kinases and pseudokinases detected in targeted biochemical or cell screens (Drewry, Willson, & Zuercher, 2014). The collection has been widely distributed to academic collaborators to establish chemical starting points to discover small molecule modulators of kinases and investigate their signaling mechanisms in high content phenotypic assays.

8. Conclusion

Emergence and conservation of pseudokinases during evolution indicates that they are adaptable templates for new biological functions when needed. They are fundamental components of signal transduction networks and employ relatively poorly-characterized mechanisms to control signaling pathways in normal and disease-related states, making them highly attractive therapeutic targets. Few pseudokinase-specific inhibitor scaffolds have been identified, but none have made their way to the market, emphasizing the need for more robust scientific efforts. The journey toward discovering new pseudokinase-centric drugs for treating human diseases is daunting, but we hope that the near future will bring us good news.

References

Aimé, P., Sun, X., Zareen, N., Rao, A., Berman, Z., Volpicelli-Daley, L., et al. (2015). Trib3 is elevated in Parkinson's disease and mediates death in Parkinson's disease models. *The Journal of Neuroscience: The Official Journal of the Society for Neuroscience*, *35*(30), 10731–10749. https://doi.org/10.1523/JNEUROSCI.0614-15.2015.

Alsaid, H., Skedzielewski, T., Rambo, M. V., Hunsinger, K., Hoang, B., Fieles, W., et al. (2017). Non invasive imaging assessment of the biodistribution of GSK2849330, an ADCC and CDC optimized anti HER3 mAb, and its role in tumor macrophage recruitment in human tumor-bearing mice. *PLoS One*, *12*(4), e0176075. https://doi.org/10.1371/journal.pone.0176075.

Amin, D. N., Campbell, M. R., & Moasser, M. M. (2010). The role of HER3, the unpretentious member of the HER family, in cancer biology and cancer therapeutics. *Seminars in Cell & Developmental Biology*, *21*(9), 944–950. https://doi.org/10.1016/j.semcdb.2010.08.007.

Bailey, F. P., Byrne, D. P., McSkimming, D., Kannan, N., & Eyers, P. A. (2015). Going for broke: Targeting the human cancer pseudokinome. *The Biochemical Journal*, *465*(2), 195–211. https://doi.org/10.1042/BJ20141060.

Bailey, F. P., Byrne, D. P., Oruganty, K., Eyers, C. E., Novotny, C. J., Shokat, K. M., et al. (2015). The tribbles 2 (TRB2) pseudokinase binds to ATP and autophosphorylates in a metal-independent manner. *The Biochemical Journal*, *467*(1), 47–62. https://doi.org/10.1042/BJ20141441.

Bandaranayake, R. M., Ungureanu, D., Shan, Y., Shaw, D. E., Silvennoinen, O., & Hubbard, S. R. (2012). Crystal structures of the JAK2 pseudokinase domain and the pathogenic mutant V617F. *Nature Structural & Molecular Biology*, *19*(8), 754–759. https://doi.org/10.1038/nsmb.2348.

Baskar, S., Wiestner, A., Wilson, W. H., Pastan, I., & Rader, C. (2012). Targeting malignant B cells with an immunotoxin against ROR1. *MAbs*, *4*(3), 349–361. https://doi.org/10.4161/mabs.19870.

Bauer, R. C., Yenilmez, B. O., & Rader, D. J. (2015). Tribbles-1: A novel regulator of hepatic lipid metabolism in humans. *Biochemical Society Transactions*, *43*(5), 1079–1084. https://doi.org/10.1042/BST20150101.

Baxter, E. J., Scott, L. M., Campbell, P. J., East, C., Fourouclas, N., Swanton, S., et al. (2005). Acquired mutation of the tyrosine kinase JAK2 in human myeloproliferative disorders. *Lancet (London, England)*, *365*(9464), 1054–1061. https://doi.org/10.1016/S0140-6736(05)71142-9.

Berglöf, A., Hamasy, A., Meinke, S., Palma, M., Krstic, A., Månsson, R., et al. (2015). Targets for ibrutinib beyond B cell malignancies. *Scandinavian Journal of Immunology*, *82*(3), 208–217. https://doi.org/10.1111/sji.12333.

Black, L. E., Longo, J. F., & Carroll, S. L. (2019). Mechanisms of receptor tyrosine-protein kinase ErbB-3 (ERBB3) action in human neoplasia. *The American Journal of Pathology*, *189*(10), 1898–1912. https://doi.org/10.1016/j.ajpath.2019.06.008.

Boggon, T. J., Li, Y., Manley, P. W., & Eck, M. J. (2005). Crystal structure of the Jak3 kinase domain in complex with a staurosporine analog. *Blood*, *106*(3), 996–1002. https://doi.org/10.1182/blood-2005-02-0707.

Brennan, D. F., Dar, A. C., Hertz, N. T., Chao, W. C., Burlingame, A. L., Shokat, K. M., et al. (2011). A Raf-induced allosteric transition of KSR stimulates phosphorylation of MEK. *Nature*, *472*(7343), 366–369. https://doi.org/10.1038/nature09860.

Burke, J. R., Cheng, L., Gillooly, K. M., Strnad, J., Zupa-Fernandez, A., Catlett, I. M., et al. (2019). Autoimmune pathways in mice and humans are blocked by pharmacological stabilization of the TYK2 pseudokinase domain. *Science Translational Medicine*, *11*(502), eaaw1736. https://doi.org/10.1126/scitranslmed.aaw1736.

Byrne, D. P., Foulkes, D. M., & Eyers, P. A. (2017). Pseudokinases: Update on their functions and evaluation as new drug targets. *Future Medicinal Chemistry*, *9*(2), 245–265. https://doi.org/10.4155/fmc-2016-0207.

Calabrese, L., Malvaso, D., Chiricozzi, A., Tambone, S., D'Urso, D. F., Guerriero, C., et al. (2020). Baricitinib: Therapeutic potential for moderate to severe atopic dermatitis. *Expert Opinion on Investigational Drugs*, 1–10. Advance online publication https://doi.org/10.1080/13543784.2020.1800639.

Camblin, A. J., Tan, G., Curley, M. D., Yannatos, I., Iadevaia, S., Rimkunas, V., et al. (2019). Dual targeting of IGF-1R and ErbB3 as a potential therapeutic regimen for ovarian cancer. *Scientific Reports*, *9*(1), 16832. https://doi.org/10.1038/s41598-019-53322-y.

Capone, E., Giansanti, F., Ponziani, S., Lamolinara, A., Iezzi, M., Cimini, A., et al. (2017). EV20-Sap, a novel anti-HER-3 antibody-drug conjugate, displays promising antitumor activity in melanoma. *Oncotarget*, *8*(56), 95412–95424. https://doi.org/10.18632/oncotarget.20728.

Capone, E., Lamolinara, A., D'Agostino, D., Rossi, C., De Laurenzi, V., Iezzi, M., et al. (2018). EV20-mediated delivery of cytotoxic auristatin MMAF exhibits potent therapeutic efficacy in cutaneous melanoma. *Journal of Controlled Release: Official Journal of the Controlled Release Society*, *277*, 48–56. https://doi.org/10.1016/j.jconrel.2018.03.016.

Chen, M., Cheng, A., Candotti, F., Zhou, Y. J., Hymel, A., Fasth, A., et al. (2000). Complex effects of naturally occurring mutations in the JAK3 pseudokinase domain: Evidence for interactions between the kinase and pseudokinase domains. *Molecular and Cellular Biology*, *20*(3), 947–956. https://doi.org/10.1128/mcb.20.3.947-956.2000.

Choi, M. Y., Widhopf, G. F., 2nd, Wu, C. C., Cui, B., Lao, F., Sadarangani, A., et al. (2015). Pre-clinical specificity and safety of UC-961, a first-in-class monoclonal antibody targeting ROR1. *Clinical Lymphoma, Myeloma & Leukemia*, *15*(Suppl. 0), S167–S169. https://doi.org/10.1016/j.clml.2015.02.010.

Christie, M., Boland, A., Huntzinger, E., Weichenrieder, O., & Izaurralde, E. (2013). Structure of the PAN3 pseudokinase reveals the basis for interactions with the PAN2 deadenylase and the GW182 proteins. *Molecular Cell*, *51*(3), 360–373. https://doi.org/10.1016/j.molcel.2013.07.011.

Colomba, A., Fitzek, M., George, R., Weitsman, G., Roberts, S., Zanetti-Domingues, L., et al. (2020). A small molecule inhibitor of HER3: A proof-of-concept study. *The Biochemical Journal*, *477*(17), 3329–3347. BCJ20200496. Advance online publication https://doi.org/10.1042/BCJ20200496.

Costanzo-Garvey, D. L., Pfluger, P. T., Dougherty, M. K., Stock, J. L., Boehm, M., Chaika, O., et al. (2009). KSR2 is an essential regulator of AMP kinase, energy expenditure, and insulin sensitivity. *Cell Metabolism*, *10*(5), 366–378. https://doi.org/10.1016/j.cmet.2009.09.010.

Curley, M. D., Sabnis, G. J., Wille, L., Adiwijaya, B. S., Garcia, G., Moyo, V., et al. (2015). Seribantumab, an anti-ERBB3 antibody, delays the onset of resistance and restores sensitivity to letrozole in an estrogen receptor-positive breast cancer model. *Molecular Cancer Therapeutics*, *14*(11), 2642–2652. https://doi.org/10.1158/1535-7163.MCT-15-0169.

Daneshmanesh, A. H., Hojjat-Farsangi, M., Khan, A. S., Jeddi-Tehrani, M., Akhondi, M. M., Bayat, A. A., et al. (2012). Monoclonal antibodies against ROR1 induce apoptosis of chronic lymphocytic leukemia (CLL) cells. *Leukemia*, *26*(6), 1348–1355. https://doi.org/10.1038/leu.2011.362.

Das, R., Sebo, Z., Pence, L., & Dobens, L. L. (2014). Drosophila tribbles antagonizes insulin signaling-mediated growth and metabolism via interactions with Akt kinase. *PLoS One*, *9*(10), e109530. https://doi.org/10.1371/journal.pone.0109530.

Davies, K. A., Tanzer, M. C., Griffin, M., Mok, Y. F., Young, S. N., Qin, R., et al. (2018). The brace helices of MLKL mediate interdomain communication and oligomerisation to regulate cell death by necroptosis. *Cell Death and Differentiation*, *25*(9), 1567–1580. https://doi.org/10.1038/s41418-018-0061-3.

Davis, M. I., Hunt, J. P., Herrgard, S., Ciceri, P., Wodicka, L. M., Pallares, G., et al. (2011). Comprehensive analysis of kinase inhibitor selectivity. *Nature Biotechnology*, *29*(11), 1046–1051. https://doi.org/10.1038/nbt.1990.

De Pauw, I., Wouters, A., Van den Bossche, J., Deschoolmeester, V., Baysal, H., Pauwels, P., et al. (2017). Dual targeting of epidermal growth factor receptor and HER3 by MEHD7945A as monotherapy or in combination with cisplatin partially overcomes cetuximab resistance in head and neck squamous cell carcinoma cell lines. *Cancer Biotherapy & Radiopharmaceuticals*, *32*(7), 229–238. https://doi.org/10.1089/cbr.2017.2216.

Dedhia, P. H., Keeshan, K., Uljon, S., Xu, L., Vega, M. E., Shestova, O., et al. (2010). Differential ability of Tribbles family members to promote degradation of C/EBPalpha and induce acute myelogenous leukemia. *Blood*, *116*(8), 1321–1328. https://doi.org/10.1182/blood-2009-07-229450.

Degorce, S. L., Tavana, O., Banks, E., Crafter, C., Gingipalli, L., Kouvchinov, D., et al. (2020). Discovery of PROTAC molecules that selectively degrade the IRAK3 pseudokinase. *Journal of Medicinal Chemistry*. https://doi.org/10.1021/acs.jmedchem.0c01125. Advance online publication https://doi.org/10.1021/acs.jmedchem.0c01125.

Dhawan, N. S., Scopton, A. P., & Dar, A. C. (2016). Small molecule stabilization of the KSR inactive state antagonizes oncogenic Ras signalling. *Nature*, *537*(7618), 112–116. https://doi.org/10.1038/nature19327.

Dougherty, M. K., Ritt, D. A., Zhou, M., Specht, S. I., Monson, D. M., Veenstra, T. D., et al. (2009). KSR2 is a calcineurin substrate that promotes ERK cascade activation in response to calcium signals. *Molecular Cell*, *34*(6), 652–662. https://doi.org/10.1016/j.molcel.2009.06.001.

Drewry, D. H., Willson, T. M., & Zuercher, W. J. (2014). Seeding collaborations to advance kinase science with the GSK Published Kinase Inhibitor Set (PKIS). *Current Topics in Medicinal Chemistry*, *14*(3), 340–342. https://doi.org/10.2174/1568026613666131127160819.

Drewry, D. H., Wells, C. I., Andrews, D. M., Angell, R., Al-Ali, H., Axtman, A. D., et al. (2017). Progress towards a public chemogenomic set for protein kinases and a call for contributions. *PLOS One*, *12*(8), e0181585. https://doi.org/10.1371/journal.pone.0181585.

Du, K., Herzig, S., Kulkarni, R. N., & Montminy, M. (2003). TRB3: A tribbles homolog that inhibits Akt/PKB activation by insulin in liver. *Science (New York, N.Y.)*, *300*(5625), 1574–1577. https://doi.org/10.1126/science.1079817.
Elkins, J. M., Fedele, V., Szklarz, M., Abdul Azeez, K. R., Salah, E., Mikolajczyk, J., et al. (2016). Comprehensive characterization of the Published Kinase Inhibitor Set. *Nature Biotechnology*, *34*(1), 95–103. https://doi.org/10.1038/nbt.3374.
Fabbro, D., Cowan-Jacob, S. W., Möbitz, H., & Martiny-Baron, G. (2012). Targeting cancer with small-molecular-weight kinase inhibitors. *Methods in Molecular Biology (Clifton, N.J.)*, *795*, 1–34. https://doi.org/10.1007/978-1-61779-337-0_1.
Fernandez, M. R., Henry, M. D., & Lewis, R. E. (2012). Kinase suppressor of Ras 2 (KSR2) regulates tumor cell transformation via AMPK. *Molecular and Cellular Biology*, *32*(18), 3718–3731. https://doi.org/10.1128/MCB.06754-11.
Fisher, K. W., Das, B., Kim, H. S., Clymer, B. K., Gehring, D., Smith, D. R., et al. (2015). AMPK promotes aberrant PGC1β expression to support human colon tumor cell survival. *Molecular and Cellular Biology*, *35*(22), 3866–3879. https://doi.org/10.1128/MCB.00528-15.
Flex, E., Petrangeli, V., Stella, L., Chiaretti, S., Hornakova, T., Knoops, L., et al. (2008). Somatically acquired JAK1 mutations in adult acute lymphoblastic leukemia. *The Journal of Experimental Medicine*, *205*(4), 751–758. https://doi.org/10.1084/jem.20072182.
Foulkes, D. M., Byrne, D. P., Bailey, F. P., & Eyers, P. A. (2015). Tribbles pseudokinases: Novel targets for chemical biology and drug discovery? *Biochemical Society Transactions*, *43*(5), 1095–1103. https://doi.org/10.1042/BST20150109.
Foulkes, D. M., Byrne, D. P., Yeung, W., Shrestha, S., Bailey, F. P., Ferries, S., et al. (2018). Covalent inhibitors of EGFR family protein kinases induce degradation of human Tribbles 2 (TRIB2) pseudokinasse in cancer cells. *Science Signaling*, *11*(549), eaat7951. https://doi.org/10.1126/scisignal.aat7951.
Franklin, M. C., Carey, K. D., Vajdos, F. F., Leahy, D. J., de Vos, A. M., & Sliwkowski, M. X. (2004). Insights into ErbB signaling from the structure of the ErbB2-pertuzumab complex. *Cancer Cell*, *5*(4), 317–328. https://doi.org/10.1016/s1535-6108(04)00083-2.
Fukuda, K., Gupta, S., Chen, K., Wu, C., & Qin, J. (2009). The pseudoactive site of ILK is essential for its binding to alpha-Parvin and localization to focal adhesions. *Molecular Cell*, *36*(5), 819–830. https://doi.org/10.1016/j.molcel.2009.11.028.
Gala, K., & Chandarlapaty, S. (2014). Molecular pathways: HER3 targeted therapy. *Clinical Cancer Research: An Official Journal of the American Association for Cancer Research*, *20*(6), 1410–1416. https://doi.org/10.1158/1078-0432.CCR-13-1549.
Garner, A. P., Bialucha, C. U., Sprague, E. R., Garrett, J. T., Sheng, Q., Li, S., et al. (2013). An antibody that locks HER3 in the inactive conformation inhibits tumor growth driven by HER2 or neuregulin. *Cancer Research*, *73*(19), 6024–6035. https://doi.org/10.1158/0008-5472.CAN-13-1198.
Geuijen, C., De Nardis, C., Maussang, D., Rovers, E., Gallenne, T., Hendriks, L., et al. (2018). Unbiased combinatorial screening identifies a bispecific IgG1 that potently inhibits HER3 signaling via HER2-guided ligand blockade. *Cancer Cell*, *33*(5), 922–936.e10. https://doi.org/10.1016/j.ccell.2018.04.003.
Guo, L., Volle, D. J., & Lewis, R. E. (2014). Identification of a truncated kinase suppressor of Ras 2 mRNA in sperm. *FEBS Open Bio*, *4*, 420–425. https://doi.org/10.1016/j.fob.2014.04.004.
Gustafson, W. C., Meyerowitz, J. G., Nekritz, E. A., Chen, J., Benes, C., Charron, E., et al. (2014). Drugging MYCN through an allosteric transition in Aurora kinase A. *Cancer Cell*, *26*(3), 414–427. https://doi.org/10.1016/j.ccr.2014.07.015.

Ha, B. H., & Boggon, T. J. (2018). The crystal structure of pseudokinase PEAK1 (Sugen kinase 269) reveals an unusual catalytic cleft and a novel mode of kinase fold dimerization. *The Journal of Biological Chemistry*, *293*(5), 1642–1650. https://doi.org/10.1074/jbc.RA117.000751.

Halford, M. M., Macheda, M. L., Parish, C. L., Takano, E. A., Fox, S., Layton, D., et al. (2013). A fully human inhibitory monoclonal antibody to the Wnt receptor RYK. *PLoS One*, *8*(9), e75447. https://doi.org/10.1371/journal.pone.0075447.

Hammarén, H. M., Ungureanu, D., Grisouard, J., Skoda, R. C., Hubbard, S. R., & Silvennoinen, O. (2015). ATP binding to the pseudokinase domain of JAK2 is critical for pathogenic activation. *Proceedings of the National Academy of Sciences of the United States of America*, *112*(15), 4642–4647. https://doi.org/10.1073/pnas.1423201112.

Hildebrand, J. M., Tanzer, M. C., Lucet, I. S., Young, S. N., Spall, S. K., Sharma, P., et al. (2014). Activation of the pseudokinase MLKL unleashes the four-helix bundle domain to induce membrane localization and necroptotic cell death. *Proceedings of the National Academy of Sciences of the United States of America*, *111*(42), 15072–15077. https://doi.org/10.1073/pnas.1408987111.

Hill, R., Madureira, P. A., Ferreira, B., Baptista, I., Machado, S., Colaço, L., et al. (2017). TRIB2 confers resistance to anti-cancer therapy by activating the serine/threonine protein kinase AKT. *Nature Communications*, *8*, 14687. https://doi.org/10.1038/ncomms14687.

Holbro, T., Beerli, R. R., Maurer, F., Koziczak, M., Barbas, C. F., 3rd, et al. (2003). The ErbB2/ErbB3 heterodimer functions as an oncogenic unit: ErbB2 requires ErbB3 to drive breast tumor cell proliferation. *Proceedings of the National Academy of Sciences of the United States of America*, *100*(15), 8933–8938. https://doi.org/10.1073/pnas.1537685100.

Hu, S., Fu, W., Xu, W., Yang, Y., Cruz, M., Berezov, S. D., et al. (2015). Four-in-one antibodies have superior cancer inhibitory activity against EGFR, HER2, HER3, and VEGF through disruption of HER/MET crosstalk. *Cancer Research*, *75*(1), 159–170. https://doi.org/10.1158/0008-5472.CAN-14-1670.

Hu, J., Yu, H., Kornev, A. P., Zhao, J., Filbert, E. L., Taylor, S. S., et al. (2011). Mutation that blocks ATP binding creates a pseudokinase stabilizing the scaffolding function of kinase suppressor of Ras, CRAF and BRAF. *Proceedings of the National Academy of Sciences of the United States of America*, *108*(15), 6067–6072. https://doi.org/10.1073/pnas.1102554108.

Huang, D., Jiang, S., Zhang, Y., Liu, X., Zhang, J., & He, R. (2014). A new mutation in the gene ROR2 causes brachydactyly type B1. *Gene*, *547*(1), 106–110. https://doi.org/10.1016/j.gene.2014.06.035.

Huang, S., Li, C., Armstrong, E. A., Peet, C. R., Saker, J., Amler, L. C., et al. (2013). Dual targeting of EGFR and HER3 with MEHD7945A overcomes acquired resistance to EGFR inhibitors and radiation. *Cancer Research*, *73*(2), 824–833. https://doi.org/10.1158/0008-5472.CAN-12-1611.

Hudecek, M., Schmitt, T. M., Baskar, S., Lupo-Stanghellini, M. T., Nishida, T., Yamamoto, T. N., et al. (2010). The B-cell tumor-associated antigen ROR1 can be targeted with T cells modified to express a ROR1-specific chimeric antigen receptor. *Blood*, *116*(22), 4532–4541. https://doi.org/10.1182/blood-2010-05-283309.

Iyer, G. H., Garrod, S., Woods, V. L., Jr., & Taylor, S. S. (2005). Catalytic independent functions of a protein kinase as revealed by a kinase-dead mutant: Study of the Lys72His mutant of cAMP-dependent kinase. *Journal of Molecular Biology*, *351*(5), 1110–1122. https://doi.org/10.1016/j.jmb.2005.06.011.

Jacobsen, A. V., Lowes, K. N., Tanzer, M. C., Lucet, I. S., Hildebrand, J. M., Petrie, E. J., et al. (2016). HSP90 activity is required for MLKL oligomerisation and membrane translocation and the induction of necroptotic cell death. *Cell Death & Disease*, 7(1), e2051. https://doi.org/10.1038/cddis.2015.386.

Jacobsen, A. V., & Murphy, J. M. (2017). The secret life of kinases: Insights into non-catalytic signalling functions from pseudokinases. *Biochemical Society Transactions*, *45*(3), 665–681. https://doi.org/10.1042/BST20160331.

Jaiswal, B. S., Kljavin, N. M., Stawiski, E. W., Chan, E., Parikh, C., Durinck, S., et al. (2013). Oncogenic ERBB3 mutations in human cancers. *Cancer Cell*, *23*(5), 603–617. https://doi.org/10.1016/j.ccr.2013.04.012.

James, C., Ugo, V., Le Couédic, J. P., Staerk, J., Delhommeau, F., Lacout, C., et al. (2005). A unique clonal JAK2 mutation leading to constitutive signalling causes polycythaemia vera. *Nature*, *434*(7037), 1144–1148. https://doi.org/10.1038/nature03546.

Jamieson, S. A., Ruan, Z., Burgess, A. E., Curry, J. R., McMillan, H. D., Brewster, J. L., et al. (2018). Substrate binding allosterically relieves autoinhibition of the pseudokinase TRIB1. *Science Signaling*, *11*(549), eaau0597. https://doi.org/10.1126/scisignal.aau0597.

Jathal, M. K., Chen, L., Mudryj, M., & Ghosh, P. M. (2011). Targeting ErbB3: The new RTK(id) on the prostate cancer block. *Immunology, Endocrine & Metabolic Agents in Medicinal Chemistry*, *11*(2), 131–149. https://doi.org/10.2174/187152211795495643.

Jeong, E. G., Kim, M. S., Nam, H. K., Min, C. K., Lee, S., Chung, Y. J., et al. (2008). Somatic mutations of JAK1 and JAK3 in acute leukemias and solid cancers. *Clinical Cancer Research: An Official Journal of the American Association for Cancer Research*, *14*(12), 3716–3721. https://doi.org/10.1158/1078-0432.CCR-07-4839.

Jha, B. K., Polyakova, I., Kessler, P., Dong, B., Dickerman, B., Sen, G. C., et al. (2011). Inhibition of RNase L and RNA-dependent protein kinase (PKR) by sunitinib impairs antiviral innate immunity. *The Journal of Biological Chemistry*, *286*(30), 26319–26326. https://doi.org/10.1074/jbc.M111.253443.

Jura, N., Shan, Y., Cao, X., Shaw, D. E., & Kuriyan, J. (2009). Structural analysis of the catalytically inactive kinase domain of the human EGF receptor 3. *Proceedings of the National Academy of Sciences of the United States of America*, *106*(51), 21608–21613. https://doi.org/10.1073/pnas.0912101106.

Kamada, H., Taki, S., Nagano, K., Inoue, M., Ando, D., Mukai, Y., et al. (2015). Generation and characterization of a bispecific diabody targeting both EPH receptor A10 and CD3. *Biochemical and Biophysical Research Communications*, *456*(4), 908–912. https://doi.org/10.1016/j.bbrc.2014.12.030.

Kanazawa, T., Glatt, S. J., Tsutsumi, A., Kikuyama, H., Koh, J., Yoneda, H., et al. (2007). Schizophrenia is not associated with the functional candidate gene ERBB3: Results from a case-control study. *American Journal of Medical Genetics. Part B, Neuropsychiatric Genetics: The Official Publication of the International Society of Psychiatric Genetics*, *144B*(1), 113–116. https://doi.org/10.1002/ajmg.b.30367.

Kang, T. H., & Kim, K. T. (2006). Negative regulation of ERK activity by VRK3-mediated activation of VHR phosphatase. *Nature Cell Biology*, *8*(8), 863–869. https://doi.org/10.1038/ncb1447.

Kannan, N., & Taylor, S. S. (2008). Rethinking pseudokinases. *Cell*, *133*(2), 204–205. https://doi.org/10.1016/j.cell.2008.04.005.

Karvonen, H., Niininen, W., Murumägi, A., & Ungureanu, D. (2017). Targeting ROR1 identifies new treatment strategies in hematological cancers. *Biochemical Society Transactions*, *45*(2), 457–464. https://doi.org/10.1042/BST20160272.

Katoh, M. (2017). Canonical and non-canonical WNT signaling in cancer stem cells and their niches: Cellular heterogeneity, omics reprogramming, targeted therapy and tumor plasticity (review). *International Journal of Oncology*, *51*(5), 1357–1369. https://doi.org/10.3892/ijo.2017.4129.

Keeshan, K., Bailis, W., Dedhia, P. H., Vega, M. E., Shestova, O., Xu, L., et al. (2010). Transformation by Tribbles homolog 2 (Trib2) requires both the Trib2 kinase domain and COP1 binding. *Blood*, *116*(23), 4948–4957. https://doi.org/10.1182/blood-2009-10-247361.

Keeshan, K., He, Y., Wouters, B. J., Shestova, O., Xu, L., Sai, H., et al. (2006). Tribbles homolog 2 inactivates C/EBPalpha and causes acute myelogenous leukemia. *Cancer Cell*, *10*(5), 401–411. https://doi.org/10.1016/j.ccr.2006.09.012.

Khamrui, S., Ung, P., Secor, C., Schlessinger, A., & Lazarus, M. B. (2020). High-resolution structure and inhibition of the schizophrenia-linked pseudokinase ULK4. *Journal of the American Chemical Society*, *142*(1), 33–37. https://doi.org/10.1021/jacs.9b10458.

Khoury, M. K., Gupta, K., Franco, S. R., & Liu, B. (2020). Necroptosis in the pathophysiology of disease. *The American Journal of Pathology*, *190*(2), 272–285. https://doi.org/10.1016/j.ajpath.2019.10.012.

Klutho, P. J., Costanzo-Garvey, D. L., & Lewis, R. E. (2011). Regulation of glucose homeostasis by KSR1 and MARK2. *PLoS One*, *6*(12), e29304. https://doi.org/10.1371/journal.pone.0029304.

Kortum, R. L., Costanzo, D. L., Haferbier, J., Schreiner, S. J., Razidlo, G. L., Wu, M. H., et al. (2005). The molecular scaffold kinase suppressor of Ras 1 (KSR1) regulates adipogenesis. *Molecular and Cellular Biology*, *25*(17), 7592–7604. https://doi.org/10.1128/MCB.25.17.7592-7604.2005.

Kralovics, R., Passamonti, F., Buser, A. S., Teo, S. S., Tiedt, R., Passweg, J. R., et al. (2005). A gain-of-function mutation of JAK2 in myeloproliferative disorders. *The New England Journal of Medicine*, *352*(17), 1779–1790. https://doi.org/10.1056/NEJMoa051113.

Kratz, C. P., Böll, S., Kontny, U., Schrappe, M., Niemeyer, C. M., & Stanulla, M. (2006). Mutational screen reveals a novel JAK2 mutation, L611S, in a child with acute lymphoblastic leukemia. *Leukemia*, *20*(2), 381–383. https://doi.org/10.1038/sj.leu.2404060.

Kufareva, I., & Abagyan, R. (2008). Type-II kinase inhibitor docking, screening, and profiling using modified structures of active kinase states. *Journal of Medicinal Chemistry*, *51*(24), 7921–7932. https://doi.org/10.1021/jm8010299.

Kung, J. E., & Jura, N. (2019). Prospects for pharmacological targeting of pseudokinases. *Nature Reviews. Drug Discovery*, *18*(7), 501–526. https://doi.org/10.1038/s41573-019-0018-3.

Labesse, G., Gelin, M., Bessin, Y., Lebrun, M., Papoin, J., Cerdan, R., et al. (2009). ROP2 from Toxoplasma gondii: A virulence factor with a protein-kinase fold and no enzymatic activity. *Structure (London, England: 1993)*, *17*(1), 139–146. https://doi.org/10.1016/j.str.2008.11.005.

Lanning, B. R., Whitby, L. R., Dix, M. M., Douhan, J., Gilbert, A. M., Hett, E. C., et al. (2014). A road map to evaluate the proteome-wide selectivity of covalent kinase inhibitors. *Nature Chemical Biology*, *10*(9), 760–767. https://doi.org/10.1038/nchembio.1582.

Lavoie, H., Sahmi, M., Maisonneuve, P., Marullo, S. A., Thevakumaran, N., Jin, T., et al. (2018). MEK drives BRAF activation through allosteric control of KSR proteins. *Nature*, *554*(7693), 549–553. https://doi.org/10.1038/nature25478.

Lee, J., Bartholomeusz, C., Mansour, O., Humphries, J., Hortobagyi, G. N., Ordentlich, P., et al. (2014). A class I histone deacetylase inhibitor, entinostat, enhances lapatinib efficacy in HER2-overexpressing breast cancer cells through FOXO3-mediated Bim1 expression. *Breast Cancer Research and Treatment*, *146*(2), 259–272. https://doi.org/10.1007/s10549-014-3014-7.

Lee-Hoeflich, S. T., Crocker, L., Yao, E., Pham, T., Munroe, X., Hoeflich, K. P., et al. (2008). A central role for HER3 in HER2-amplified breast cancer: Implications for targeted therapy. *Cancer Research*, *68*(14), 5878–5887. https://doi.org/10.1158/0008-5472.CAN-08-0380.

Leitner, N. R., Witalisz-Siepracka, A., Strobl, B., & Müller, M. (2017). Tyrosine kinase 2—Surveillant of tumours and bona fide oncogene. *Cytokine*, *89*, 209–218. https://doi.org/10.1016/j.cyto.2015.10.015.

Leroy, E., Dusa, A., Colau, D., Motamedi, A., Cahu, X., Mouton, C., et al. (2016). Uncoupling JAK2 V617F activation from cytokine-induced signalling by modulation

of JH2 αC helix. *The Biochemical Journal*, *473*(11), 1579–1591. https://doi.org/10.1042/BCJ20160085.
Levine, R. L., Loriaux, M., Huntly, B. J., Loh, M. L., Beran, M., Stoffregen, E., et al. (2005). The JAK2V617F activating mutation occurs in chronic myelomonocytic leukemia and acute myeloid leukemia, but not in acute lymphoblastic leukemia or chronic lymphocytic leukemia. *Blood*, *106*(10), 3377–3379. https://doi.org/10.1182/blood-2005-05-1898.
Li, Z., Gakovic, M., Ragimbeau, J., Eloranta, M. L., Rönnblom, L., Michel, F., et al. (2013). Two rare disease-associated Tyk2 variants are catalytically impaired but signaling competent. *Journal of Immunology (Baltimore, Md.: 1950)*, *190*(5), 2335–2344. https://doi.org/10.4049/jimmunol.1203118.
Liew, C. W., Bochenski, J., Kawamori, D., Hu, J., Leech, C. A., Wanic, K., et al. (2010). The pseudokinase tribbles homolog 3 interacts with ATF4 to negatively regulate insulin exocytosis in human and mouse beta cells. *The Journal of Clinical Investigation*, *120*(8), 2876–2888. https://doi.org/10.1172/JCI36849.
Lim, S. M., Xie, T., Westover, K. D., Ficarro, S. B., Tae, H. S., Gurbani, D., et al. (2015). Development of small molecules targeting the pseudokinase Her3. *Bioorganic & Medicinal Chemistry Letters*, *25*(16), 3382–3389. https://doi.org/10.1016/j.bmcl.2015.04.103.
Liosi, M. E., Krimmer, S. G., Newton, A. S., Dawson, T. K., Puleo, D. E., Cutrona, K. J., et al. (2020). Selective Janus kinase 2 (JAK2) pseudokinase ligands with a diaminotriazole core. *Journal of Medicinal Chemistry*, *63*(10), 5324–5340. https://doi.org/10.1021/acs.jmedchem.0c00192.
Littlefield, P., Liu, L., Mysore, V., Shan, Y., Shaw, D. E., & Jura, N. (2014). Structural analysis of the EGFR/HER3 heterodimer reveals the molecular basis for activating HER3 mutations. *Science Signaling*, 7(354), ra114. https://doi.org/10.1126/scisignal.2005786.
Liu, X., Liu, S., Lyu, H., Riker, A. I., Zhang, Y., & Liu, B. (2019). Development of effective therapeutics targeting HER3 for cancer treatment. *Biological Procedures Online*, *21*, 5. https://doi.org/10.1186/s12575-019-0093-1.
Lozano, J., Xing, R., Cai, Z., Jensen, H. L., Trempus, C., Mark, W., et al. (2003). Deficiency of kinase suppressor of Ras1 prevents oncogenic ras signaling in mice. *Cancer Research*, *63*(14), 4232–4238. https://cancerres.aacrjournals.org/content/63/14/4232.long.
Lu, X., Smaill, J. B., & Ding, K. (2020). New promise and opportunities for allosteric kinase inhibitors. *Angewandte Chemie (International Ed. in English)*, *59*(33), 13764–13776. https://doi.org/10.1002/anie.201914525.
Lucet, I. S., & Murphy, J. M. (2017). Characterization of ligand binding to pseudokinases using a thermal shift assay. *Methods in Molecular Biology (Clifton, N.J.)*, *1636*, 91–104. https://doi.org/10.1007/978-1-4939-7154-1_7.
Lucet, I. S., & Murphy, J. M. (2019). A structural perspective of the pseudokinome: Defining the targetable space. In *Kinase drug discovery: Modern approaches* (pp. 359–380). The Royal Society of Chemistry (Chapter 13) https://doi.org/10.1039/9781788013093-00359.
Lyu, H., Han, A., Polsdofer, E., Liu, S., & Liu, B. (2018). Understanding the biology of HER3 receptor as a therapeutic target in human cancer. *Acta Pharmaceutica Sinica B*, *8*(4), 503–510. https://doi.org/10.1016/j.apsb.2018.05.010.
Ma, B., Marcotte, D., Paramasivam, M., Michelsen, K., Wang, T., Bertolotti-Ciarlet, A., et al. (2016). ATP-competitive MLKL binders have no functional impact on necroptosis. *PLoS One*, *11*(11), e0165983. https://doi.org/10.1371/journal.pone.0165983.
Manning, G., Whyte, D. B., Martinez, R., Hunter, T., & Sudarsanam, S. (2002). The protein kinase complement of the human genome. *Science (New York, N.Y.)*, *298*(5600), 1912–1934. https://doi.org/10.1126/science.1075762.
Massa, R., Palumbo, C., Cavallaro, T., Panico, M. B., Bei, R., Terracciano, C., et al. (2006). Overexpression of ErbB2 and ErbB3 receptors in Schwann cells of patients with Charcot-Marie-tooth disease type 1A. *Muscle & Nerve*, *33*(3), 342–349. https://doi.org/10.1002/mus.20460.

McDonagh, C. F., Huhalov, A., Harms, B. D., Adams, S., Paragas, V., Oyama, S., et al. (2012). Antitumor activity of a novel bispecific antibody that targets the ErbB2/ ErbB3 oncogenic unit and inhibits heregulin-induced activation of ErbB3. *Molecular Cancer Therapeutics*, *11*(3), 582–593. https://doi.org/10.1158/1535-7163.MCT-11-0820.

McNally, R., Li, Q., Li, K., Dekker, C., Vangrevelinghe, E., Jones, M., et al. (2019). Discovery and structural characterization of ATP-site ligands for the wild-type and V617F mutant JAK2 pseudokinase domain. *ACS Chemical Biology*, *14*(4), 587–593. https://doi.org/10.1021/acschembio.8b00722.

Mendrola, J. M., Shi, F., Park, J. H., & Lemmon, M. A. (2013). Receptor tyrosine kinases with intracellular pseudokinase domains. *Biochemical Society Transactions*, *41*(4), 1029–1036. https://doi.org/10.1042/BST20130104.

Mills, G. B., & Yarden, Y. (2010). The rebirth of a phoenix: Ovarian cancers are addicted to ErbB-3. *Cancer Cell*, *17*(3), 217–218. https://doi.org/10.1016/j.ccr.2010.02.023.

Min, X., Ungureanu, D., Maxwell, S., Hammarén, H., Thibault, S., Hillert, E. K., et al. (2015). Structural and functional characterization of the JH2 pseudokinase domain of JAK family tyrosine kinase 2 (TYK2). *The Journal of Biological Chemistry*, *290*(45), 27261–27270. https://doi.org/10.1074/jbc.M115.672048.

Mishra, R., Patel, H., Alanazi, S., Yuan, L., & Garrett, J. T. (2018). HER3 signaling and targeted therapy in cancer. *Oncology Reviews*, *12*(1), 355. https://doi.org/10.4081/ oncol.2018.355.

Moslin, R., Zhang, Y., Wrobleski, S. T., Lin, S., Mertzman, M., Spergel, S., et al. (2019). Identification of *N*-methyl nicotinamide and *N*-methyl pyridazine-3-carboxamide pseudokinase domain ligands as highly selective allosteric inhibitors of tyrosine kinase 2 (TYK2). *Journal of Medicinal Chemistry*, *62*(20), 8953–8972. https://doi.org/10. 1021/acs.jmedchem.9b00443.

Mota, J. M., Collier, K. A., Barros Costa, R. L., Taxter, T., Kalyan, A., Leite, C. A., et al. (2017). A comprehensive review of heregulins, HER3, and HER4 as potential therapeutic targets in cancer. *Oncotarget*, *8*(51), 89284–89306. https://doi.org/10.18632/ oncotarget.18467.

Mukherjee, K., Sharma, M., Jahn, R., Wahl, M. C., & Südhof, T. C. (2010). Evolution of CASK into a Mg^{2+}-sensitive kinase. *Science Signaling*, *3*(119), ra33. https://doi.org/10. 1126/scisignal.2000800.

Mukherjee, K., Sharma, M., Urlaub, H., Bourenkov, G. P., Jahn, R., Südhof, T. C., et al. (2008). CASK functions as a Mg^{2+}-independent neurexin kinase. *Cell*, *133*(2), 328–339. https://doi.org/10.1016/j.cell.2008.02.036.

Murphy, J. M., Czabotar, P. E., Hildebrand, J. M., Lucet, I. S., Zhang, J. G., Alvarez-Diaz, S., et al. (2013). The pseudokinase MLKL mediates necroptosis via a molecular switch mechanism. *Immunity*, *39*(3), 443–453. https://doi.org/10.1016/j.immuni.2013.06.018.

Murphy, J. M., Lucet, I. S., Hildebrand, J. M., Tanzer, M. C., Young, S. N., Sharma, P., et al. (2014). Insights into the evolution of divergent nucleotide-binding mechanisms among pseudokinases revealed by crystal structures of human and mouse MLKL. *The Biochemical Journal*, *457*(3), 369–377. https://doi.org/10.1042/BJ20131270.

Murphy, J. M., Nakatani, Y., Jamieson, S. A., Dai, W., Lucet, I. S., & Mace, P. D. (2015). Molecular mechanism of CCAAT-enhancer binding protein recruitment by the TRIB1 pseudokinase. *Structure (London, England: 1993)*, *23*(11), 2111–2121. https://doi.org/10. 1016/j.str.2015.08.017.

Murphy, J. M., Zhang, Q., Young, S. N., Reese, M. L., Bailey, F. P., Eyers, P. A., et al. (2014). A robust methodology to subclassify pseudokinases based on their nucleotide-binding properties. *The Biochemical Journal*, *457*(2), 323–334. https://doi.org/10.1042/ BJ20131174.

Neilsen, B. K., Frodyma, D. E., Lewis, R. E., & Fisher, K. W. (2017). KSR as a therapeutic target for Ras-dependent cancers. *Expert Opinion on Therapeutic Targets*, *21*(5), 499–509. https://doi.org/10.1080/14728222.2017.1311325.

Newton, A. S., Deiana, L., Puleo, D. E., Cisneros, J. A., Cutrona, K. J., Schlessinger, J., et al. (2017). JAK2 JH2 fluorescence polarization assay and crystal structures for complexes with three small molecules. *ACS Medicinal Chemistry Letters*, *8*(6), 614–617. https://doi.org/10.1021/acsmedchemlett.7b00154.

Ocana, A., Vera-Badillo, F., Seruga, B., Templeton, A., Pandiella, A., & Amir, E. (2013). HER3 overexpression and survival in solid tumors: A meta-analysis. *Journal of the National Cancer Institute*, *105*(4), 266–273. https://doi.org/10.1093/jnci/djs501.

Olayioye, M. A., Neve, R. M., Lane, H. A., & Hynes, N. E. (2000). The ErbB signaling network: Receptor heterodimerization in development and cancer. *The EMBO Journal*, *19*(13), 3159–3167. https://doi.org/10.1093/emboj/19.13.3159.

Papp, K., Gordon, K., Thaçi, D., Morita, A., Gooderham, M., Foley, P., et al. (2018). Phase 2 trial of selective tyrosine kinase 2 inhibition in psoriasis. *The New England Journal of Medicine*, *379*(14), 1313–1321. https://doi.org/10.1056/NEJMoa1806382.

Pearce, L. R., Atanassova, N., Banton, M. C., Bottomley, B., van der Klaauw, A. A., Revelli, J. P., et al. (2013). KSR2 mutations are associated with obesity, insulin resistance, and impaired cellular fuel oxidation. *Cell*, *155*(4), 765–777. https://doi.org/10.1016/j.cell.2013.09.058.

Petrie, E. J., Sandow, J. J., Jacobsen, A. V., Smith, B. J., Griffin, M., Lucet, I. S., et al. (2018). Conformational switching of the pseudokinase domain promotes human MLKL tetramerization and cell death by necroptosis. *Nature Communications*, *9*(1), 2422. https://doi.org/10.1038/s41467-018-04714-7.

Piala, A. T., Moon, T. M., Akella, R., He, H., Cobb, M. H., & Goldsmith, E. J. (2014). Chloride sensing by WNK1 involves inhibition of autophosphorylation. *Science Signaling*, 7(324), ra41. https://doi.org/10.1126/scisignal.2005050.

Puleo, D. E., Kucera, K., Hammarén, H. M., Ungureanu, D., Newton, A. S., Silvennoinen, O., et al. (2017). Identification and characterization of JAK2 pseudokinase domain small molecule binders. *ACS Medicinal Chemistry Letters*, *8*(6), 618–621. https://doi.org/10.1021/acsmedchemlett.7b00153.

Quintás-Cardama, A., Vaddi, K., Liu, P., Manshouri, T., Li, J., Scherle, P. A., et al. (2010). Preclinical characterization of the selective JAK1/2 inhibitor INCB018424: Therapeutic implications for the treatment of myeloproliferative neoplasms. *Blood*, *115*(15), 3109–3117. https://doi.org/10.1182/blood-2009-04-214957.

Raivola, J., Haikarainen, T., & Silvennoinen, O. (2019). Characterization of JAK1 pseudokinase domain in cytokine signaling. *Cancers*, *12*(1), 78. https://doi.org/10.3390/cancers12010078.

Rajakulendran, T., Sahmi, M., Lefrançois, M., Sicheri, F., & Therrien, M. (2009). A dimerization-dependent mechanism drives RAF catalytic activation. *Nature*, *461*(7263), 542–545. https://doi.org/10.1038/nature08314.

Reiterer, V., Eyers, P. A., & Farhan, H. (2014). Day of the dead: Pseudokinases and pseudophosphatases in physiology and disease. *Trends in Cell Biology*, *24*(9), 489–505. https://doi.org/10.1016/j.tcb.2014.03.008.

Ribeiro, A., Das, S., Dawson, N., Zaru, R., Orchard, S., Thornton, J. M., et al. (2019). Emerging concepts in pseudoenzyme classification, evolution, and signaling. *Science Signaling*, *12*(594), eaat9797. https://doi.org/10.1126/scisignal.aat9797.

Richmond, L., & Keeshan, K. (2019). Pseudokinases: A tribble-edged sword. *The FEBS Journal*. Advance online publication https://doi.org/10.1111/febs.15096.

Riethmacher, D., Sonnenberg-Riethmacher, E., Brinkmann, V., Yamaai, T., Lewin, G. R., & Birchmeier, C. (1997). Severe neuropathies in mice with targeted mutations in the ErbB3 receptor. *Nature, 389*(6652), 725–730. https://doi.org/10.1038/39593.

Rodgers, G., Austin, C., Anderson, J., Pawlyk, A., Colvis, C., Margolis, R., et al. (2018). Glimmers in illuminating the druggable genome. *Nature Reviews. Drug Discovery, 17*(5), 301–302. https://doi.org/10.1038/nrd.2017.252.

Roskoski, R., Jr. (2012). ERK1/2 MAP kinases: Structure, function, and regulation. *Pharmacological Research, 66*(2), 105–143. https://doi.org/10.1016/j.phrs.2012.04.005.

Roskoski, R., Jr. (2018). Targeting oncogenic Raf protein-serine/threonine kinases in human cancers. *Pharmacological Research, 135*, 239–258. https://doi.org/10.1016/j.phrs.2018.08.013.

Ruiz-Saenz, A., Sandhu, M., Carrasco, Y., Maglathlin, R. L., Taunton, J., & Moasser, M. M. (2015). Targeting HER3 by interfering with its Sec61-mediated cotranslational insertion into the endoplasmic reticulum. *Oncogene, 34*(41), 5288–5294. https://doi.org/10.1038/onc.2014.455.

Saharinen, P., & Silvennoinen, O. (2002). The pseudokinase domain is required for suppression of basal activity of Jak2 and Jak3 tyrosine kinases and for cytokine-inducible activation of signal transduction. *The Journal of Biological Chemistry, 277*(49), 47954–47963. https://doi.org/10.1074/jbc.M205156200.

Saharinen, P., Takaluoma, K., & Silvennoinen, O. (2000). Regulation of the Jak2 tyrosine kinase by its pseudokinase domain. *Molecular and Cellular Biology, 20*(10), 3387–3395. https://doi.org/10.1128/mcb.20.10.3387-3395.2000.

Santos, R., Ursu, O., Gaulton, A., Bento, A. P., Donadi, R. S., Bologa, C. G., et al. (2017). A comprehensive map of molecular drug targets. *Nature Reviews. Drug Discovery, 16*(1), 19–34. https://doi.org/10.1038/nrd.2016.230.

Sanz Sanz, A., Niranjan, Y., Hammarén, H., Ungureanu, D., Ruijtenbeek, R., Touw, I. P., et al. (2014). The JH2 domain and SH2-JH2 linker regulate JAK2 activity: A detailed kinetic analysis of wild type and V617F mutant kinase domains. *Biochimica et Biophysica Acta, 1844*(10), 1835–1841. https://doi.org/10.1016/j.bbapap.2014.07.003.

Sarup, J., Jin, P., Turin, L., Bai, X., Beryt, M., Brdlik, C., et al. (2008). Human epidermal growth factor receptor (HER-1:HER-3) Fc-mediated heterodimer has broad antiproliferative activity in vitro and in human tumor xenografts. *Molecular Cancer Therapeutics*, 7(10), 3223–3236. https://doi.org/10.1158/1535-7163.MCT-07-2151.

Satoh, T., Kidoya, H., Naito, H., Yamamoto, M., Takemura, N., Nakagawa, K., et al. (2013). Critical role of Trib1 in differentiation of tissue-resident M2-like macrophages. *Nature, 495*(7442), 524–528. https://doi.org/10.1038/nature11930.

Schaefer, G., Haber, L., Crocker, L. M., Shia, S., Shao, L., Dowbenko, D., et al. (2011). A two-in-one antibody against HER3 and EGFR has superior inhibitory activity compared with monospecific antibodies. *Cancer Cell, 20*(4), 472–486. https://doi.org/10.1016/j.ccr.2011.09.003.

Scheeff, E. D., Eswaran, J., Bunkoczi, G., Knapp, S., & Manning, G. (2009). Structure of the pseudokinase VRK3 reveals a degraded catalytic site, a highly conserved kinase fold, and a putative regulatory binding site. *Structure (London, England: 1993), 17*(1), 128–138. https://doi.org/10.1016/j.str.2008.10.018.

Schneeweiss, A., Park-Simon, T. W., Albanell, J., Lassen, U., Cortés, J., Dieras, V., et al. (2018). Phase Ib study evaluating safety and clinical activity of the anti-HER3 antibody lumretuzumab combined with the anti-HER2 antibody pertuzumab and paclitaxel in HER3-positive, HER2-low metastatic breast cancer. *Investigational New Drugs, 36*(5), 848–859. https://doi.org/10.1007/s10637-018-0562-4.

Schoeberl, B., Faber, A. C., Li, D., Liang, M. C., Crosby, K., Onsum, M., et al. (2010). An ErbB3 antibody, MM-121, is active in cancers with ligand-dependent activation. *Cancer Research, 70*(6), 2485–2494. https://doi.org/10.1158/0008-5472.CAN-09-3145.

Schoeberl, B., Kudla, A., Masson, K., Kalra, A., Curley, M., Finn, G., et al. (2017). Systems biology driving drug development: From design to the clinical testing of the anti-ErbB3 antibody seribantumab (MM-121). *NPJ Systems Biology and Applications*, *3*, 16034. https://doi.org/10.1038/npjsba.2016.34.

Scott, L. M., Scott, M. A., Campbell, P. J., & Green, A. R. (2006). Progenitors homozygous for the V617F mutation occur in most patients with polycythemia vera, but not essential thrombocythemia. *Blood*, *108*(7), 2435–2437. https://doi.org/10.1182/blood-2006-04-018259.

Sequist, L. V., Gray, J. E., Harb, W. A., Lopez-Chavez, A., Doebele, R. C., Modiano, M. R., et al. (2019). Randomized phase II trial of seribantumab in combination with erlotinib in patients with EGFR wild-type non-small cell lung cancer. *The Oncologist*, *24*(8), 1095–1102. https://doi.org/10.1634/theoncologist.2018-0695.

Shan, Y., Gnanasambandan, K., Ungureanu, D., Kim, E. T., Hammarén, H., Yamashita, K., et al. (2014). Molecular basis for pseudokinase-dependent autoinhibition of JAK2 tyrosine kinase. *Nature Structural & Molecular Biology*, *21*(7), 579–584. https://doi.org/10.1038/nsmb.2849.

Shaw, A. S., Kornev, A. P., Hu, J., Ahuja, L. G., & Taylor, S. S. (2014). Kinases and pseudokinases: Lessons from RAF. *Molecular and Cellular Biology*, *34*(9), 1538–1546. https://doi.org/10.1128/MCB.00057-14.

Sheetz, J. B., Mathea, S., Karvonen, H., Malhotra, K., Chatterjee, D., Niininen, W., et al. (2020). Structural insights into pseudokinase domains of receptor tyrosine kinases. *Molecular Cell*, *79*(3), 390–405.e7. https://doi.org/10.1016/j.molcel.2020.06.018.

Shi, F., Telesco, S. E., Liu, Y., Radhakrishnan, R., & Lemmon, M. A. (2010). ErbB3/HER3 intracellular domain is competent to bind ATP and catalyze autophosphorylation. *Proceedings of the National Academy of Sciences of the United States of America*, *107*(17), 7692–7697. https://doi.org/10.1073/pnas.1002753107.

Shrestha, S., Byrne, D. P., Harris, J. A., Kannan, N., & Eyers, P. A. (2020). Cataloguing the dead: Breathing new life into pseudokinase research. *The FEBS Journal*. Advance online publication https://doi.org/10.1111/febs.15246.

Staerk, J., Kallin, A., Demoulin, J. B., Vainchenker, W., & Constantinescu, S. N. (2005). JAK1 and Tyk2 activation by the homologous polycythemia vera JAK2 V617F mutation: Cross-talk with IGF1 receptor. *The Journal of Biological Chemistry*, *280*(51), 41893–41899. https://doi.org/10.1074/jbc.C500358200.

Sun, L., Wang, H., Wang, Z., He, S., Chen, S., Liao, D., et al. (2012). Mixed lineage kinase domain-like protein mediates necrosis signaling downstream of RIP3 kinase. *Cell*, *148*(1–2), 213–227. https://doi.org/10.1016/j.cell.2011.11.031.

Taki, S., Kamada, H., Inoue, M., Nagano, K., Mukai, Y., Higashisaka, K., et al. (2015). A novel bispecific antibody against human CD3 and ephrin receptor A10 for breast cancer therapy. *PLoS One*, *10*(12), e0144712. https://doi.org/10.1371/journal.pone.0144712.

Tang, J., Wang, Y., Zhou, H., Ye, Y., Talukdar, M., Fu, Z., et al. (2020). Sunitinib inhibits RNase L by destabilizing its active dimer conformation. *The Biochemical Journal*, *477*(17), 3387–3399. BCJ20200260, Advance online publication https://doi.org/10.1042/BCJ20200260.

Tokarski, J. S., Zupa-Fernandez, A., Tredup, J. A., Pike, K., Chang, C., Xie, D., et al. (2015). Tyrosine kinase 2-mediated signal transduction in T lymphocytes is blocked by pharmacological stabilization of its pseudokinase domain. *The Journal of Biological Chemistry*, *290*(17), 11061–11074. https://doi.org/10.1074/jbc.M114.619502.

Tomoni, A., Lees, J., Santana, A. G., Bolanos-Garcia, V. M., & Bastida, A. (2019). Pseudokinases: From allosteric regulation of catalytic domains and the formation of macromolecular assemblies to emerging drug targets. *Catalysts*, *9*(9), 778. https://doi.org/10.3390/catal9090778.

Velazquez, L., Mogensen, K. E., Barbieri, G., Fellous, M., Uzé, G., & Pellegrini, S. (1995). Distinct domains of the protein tyrosine kinase tyk2 required for binding of interferon-alpha/beta and for signal transduction. *The Journal of Biological Chemistry*, *270*(7), 3327–3334. https://doi.org/10.1074/jbc.270.7.3327.

Wolf, J., Valkov, E., Allen, M. D., Meineke, B., Gordiyenko, Y., McLaughlin, S. H., et al. (2014). Structural basis for Pan3 binding to Pan2 and its function in mRNA recruitment and deadenylation. *The EMBO Journal*, *33*(14), 1514–1526. https://doi.org/10.15252/embj.201488373.

Wrobleski, S. T., Moslin, R., Lin, S., Zhang, Y., Spergel, S., Kempson, J., et al. (2019). Highly selective inhibition of tyrosine kinase 2 (TYK2) for the treatment of autoimmune diseases: Discovery of the allosteric inhibitor BMS-986165. *Journal of Medicinal Chemistry*, *62*(20), 8973–8995. https://doi.org/10.1021/acs.jmedchem.9b00444.

Wu, Y., Zhang, Y., Wang, M., Li, Q., Qu, Z., Shi, V., et al. (2013). Downregulation of HER3 by a novel antisense oligonucleotide, EZN-3920, improves the antitumor activity of EGFR and HER2 tyrosine kinase inhibitors in animal models. *Molecular Cancer Therapeutics*, *12*(4), 427–437. https://doi.org/10.1158/1535-7163.MCT-12-0838.

Xiao, Z., Carrasco, R. A., Schifferli, K., Kinneer, K., Tammali, R., Chen, H., et al. (2016). A potent HER3 monoclonal antibody that blocks both ligand-dependent and -independent activities: Differential impacts of PTEN status on tumor response. *Molecular Cancer Therapeutics*, *15*(4), 689–701. https://doi.org/10.1158/1535-7163.MCT-15-0555.

Xie, T., Lim, S. M., Westover, K. D., Dodge, M. E., Ercan, D., Ficarro, S. B., et al. (2014). Pharmacological targeting of the pseudokinase Her3. *Nature Chemical Biology*, *10*(12), 1006–1012. https://doi.org/10.1038/nchembio.1658.

Xie, T., Peng, W., Yan, C., Wu, J., Gong, X., & Shi, Y. (2013). Structural insights into RIP3-mediated necroptotic signaling. *Cell Reports*, *5*(1), 70–78. https://doi.org/10.1016/j.celrep.2013.08.044.

Yokoyama, T., Toki, T., Aoki, Y., Kanezaki, R., Park, M. J., Kanno, Y., et al. (2012). Identification of TRIB1 R107L gain-of-function mutation in human acute megakaryocytic leukemia. *Blood*, *119*(11), 2608–2611. https://doi.org/10.1182/blood-2010-12-324806.

Yonesaka, K., Takegawa, N., Watanabe, S., Haratani, K., Kawakami, H., Sakai, K., et al. (2019). An HER3-targeting antibody-drug conjugate incorporating a DNA topoisomerase I inhibitor U3-1402 conquers EGFR tyrosine kinase inhibitor-resistant NSCLC. *Oncogene*, *38*(9), 1398–1409. https://doi.org/10.1038/s41388-018-0517-4.

Yoshida, A., Kato, J. Y., Nakamae, I., & Yoneda-Kato, N. (2013). COP1 targets C/EBPα for degradation and induces acute myeloid leukemia via Trib1. *Blood*, *122*(10), 1750–1760. https://doi.org/10.1182/blood-2012-12-476101.

Yu, J., Chen, L., Cui, B., Wu, C., Choi, M. Y., Chen, Y., et al. (2017). Cirmtuzumab inhibits Wnt5a-induced Rac1 activation in chronic lymphocytic leukemia treated with ibrutinib. *Leukemia*, *31*(6), 1333–1339. https://doi.org/10.1038/leu.2016.368.

Zareen, N., Biswas, S. C., & Greene, L. A. (2013). A feed-forward loop involving Trib3, Akt and FoxO mediates death of NGF-deprived neurons. *Cell Death and Differentiation*, *20*(12), 1719–1730. https://doi.org/10.1038/cdd.2013.128.

Zeqiraj, E., Filippi, B. M., Goldie, S., Navratilova, I., Boudeau, J., Deak, M., et al. (2009). ATP and MO25alpha regulate the conformational state of the STRADalpha pseudokinase and activation of the LKB1 tumour suppressor. *PLoS Biology*, 7(6), e1000126. https://doi.org/10.1371/journal.pbio.1000126.

Zhang, J., Adrián, F. J., Jahnke, W., Cowan-Jacob, S. W., Li, A. G., Iacob, R. E., et al. (2010). Targeting Bcr-Abl by combining allosteric with ATP-binding-site inhibitors. *Nature*, *463*(7280), 501–506. https://doi.org/10.1038/nature08675.

Zhang, L., Castanaro, C., Luan, B., Yang, K., Fan, L., Fairhurst, J. L., et al. (2014). ERBB3/HER2 signaling promotes resistance to EGFR blockade in head and neck and colorectal cancer models. *Molecular Cancer Therapeutics*, *13*(5), 1345–1355. https://doi.org/10.1158/1535-7163.MCT-13-1033.

Zhang, J., Zafrullah, M., Yang, X., Yin, X., Zhang, Z., Fuks, Z., et al. (2008). Downregulation of KSR1 in pancreatic cancer xenografts by antisense oligonucleotide correlates with tumor drug uptake. *Cancer Biology & Therapy*, 7(9), 1490–1495. https://doi.org/10.4161/cbt.7.9.6472.

CHAPTER SIX

Chemogenomics and bioinformatics approaches for prioritizing kinases as drug targets for neglected tropical diseases

Joyce Villa Verde Bastos Borba[a,b,†], Arthur Carvalho Silva[a,†], Marilia Nunes Nascimento Lima[a,†], Sabrina Silva Mendonca[a], Nicholas Furnham[c], Fabio Trindade Maranhão Costa[b], and Carolina Horta Andrade[a,c,*]

[a]LabMol—Laboratory for Molecular Modeling and Drug Design, Faculty of Pharmacy, Federal University of Goiás, Goiânia, GO, Brazil
[b]Laboratory of Tropical Diseases—Prof. Luiz Jacintho da Silva, Department of Genetics, Evolution and Bioagents, University of Campinas, Campinas, SP, Brazil
[c]Department of Infection Biology, London School of Hygiene and Tropical Medicine, London, United Kingdom
*Corresponding author: e-mail address: carolina@ufg.br

Contents

† These authors contributed equally.

Advances in Protein Chemistry and Structural Biology, Volume 124
ISSN 1876-1623
https://doi.org/10.1016/bs.apcsb.2020.10.006

Abstract

Neglected tropical diseases (NTDs) are a group of twenty-one diseases classified by the World Health Organization that prevail in regions with tropical and subtropical climate and affect more than one billion people. There is an urgent need to develop new and safer drugs for these diseases. Protein kinases are a potential class of targets for developing new drugs against NTDs, since they play crucial role in many biological processes, such as signaling pathways, regulating cellular communication, division, metabolism and death. Bioinformatics is a field that aims to organize large amounts of biological data as well as develop and use tools for understanding and analyze them in order to produce meaningful information in a biological manner. In combination with chemogenomics, which analyzes chemical-biological interactions to screen ligands against selected targets families, these approaches can be used to stablish a rational strategy for prioritizing new drug targets for NTDs. Here, we describe how bioinformatics and chemogenomics tools can help to identify protein kinases and their potential inhibitors for the development of new drugs for NTDs. We present a review of bioinformatics tools and techniques that can be used to define an organisms kinome for drug prioritization, drug and target repurposing, multi-quinase inhibition approachs and selectivity profiling. We also present some successful examples of the application of such approaches in recent case studies.

1. Introduction

Neglected tropical diseases (NTDs) are a group of communicable diseases predominat in regions with tropical and subtropical climate that mostly affect the poor and disadvantaged (World Health Organization, 2020). The latest estimatives show that NTDs affect more than one billion people with half a million deaths per year and disability-adjusted life years (DALYs) of up to 62.3 million (Kyu et al., 2018; Silva-Moraes et al., 2019; World Health Organization, 2020).

Bioinformatics is a field that aims to organize large amounts of biological data, develop and use tools for understanding and analyze them in order to produce meaningful information in a biological manner (Luscombe, Greenbaum, & Gerstein, 2001). In combination with chemogenomics, which analyzes chemical-biological interactions to screen ligands against selected targets, these approaches can be used to establish a rational strategy for prioritizing new drug targets for NTDs (Kubinyi, 2006).

Among potential new drug targets for developing drugs against NTDs, protein kinases are highly attractive as they are essential to a parasite's survival (Hanks & Hunter, 1995). This class of enzymes catalyze the phosphorylation of proteins, thereby regulating a variety of cellular processes and playing crucial roles in many biological processes, such as signaling pathways, regulating cellular communication, division, metabolism and cell death (Dichiara et al., 2017; Hernandez et al., 2019).

In this chapter, we highlight the urgent need to control and eliminate NTDs and aim to describe how bioinformatics and chemogenomics tools can help in this effort though the example of finding protein kinases and their potential inhibitors for the development of new drugs for NTDs. We present a review of bioinformatics tools and techniques that can be used in kinome defining for drug prioritization, drug and target repurposing, multi-quinase inhibition approachs and selectivity profiling. We also present some examples of how these approaches can be successfully applied in recent case studies.

2. Neglected tropical diseases paradigm

NTDs are a group of infectious diseases prevalent in 149 countries in tropical and subtropical climate conditions and are intimately linked to poverty (World Health Organization, 2020). The most marginalized communities, with limited access to adequate sanitization, clean water and healthcare, and people living in proximity with animals and infective disease vectors, are the most vunerable and worst affected by these diseases (Hamill et al., 2019). Globally, there are more than 1.5 billion (Mitra & Mawson, 2017; World Health Organization, 2020) people suffering from at least one NTD and the total worldwide burden is of approximately 62.3 million disability-adjusted life years (DALY) (Silva-Moraes et al., 2019; World Health Organization, 2020).

NTDs have traditionally suffered from lack of market incentive to develop new drugs and vaccines for markets that cannot pay. Since 2003, WHO began to focus control measures away from specific diseases to the health needs of poor communities (World Health Organization, 2004). In 2007, WHO hosted the first Global Partners' Meeting, which resulted in a shared commitment to support WHO's strategies, goals and targets (World Health Organization, 2007). Hence, on 30th January 2012, WHO published a roadmap aiming to eliminate and/or control by 2020 the 17 NTDs defined by WHO (World Health Organization, 2012). Then, a community of partners—today collectively known as Uniting to Combat NTDs—were signatories to the "London Declaration on Neglected Tropical Diseases" (Uniting to Combat Neglected Tropical Diseases, 2012), committing to the goals of WHO's roadmap. Pharmaceutical companies committed to donate medicines for a value of USD17.8 billion for the 10 London Declaration Diseases through to 2020. Philanthropic foundations (Bill and Melinda Gates Foundation) and bilateral donors (UKAID and USAID) have pledged support to deliver those medicines to people in need, and governments in endemic regions have committed to the deployment of their health services to make this happen. As a result, 1.12 billion people were threated for at least 1 NTD in 2018 (World Health Organization, 2019a), 300 million fewer people required treatment for an NTD in 2017 compared to 2012 and 31 countries have eliminated at least one NTD since 2012 (World Health Organization, 2019a).

Many other poverty-associated diseases are not on WHO's list, but must be considered as NTDs and should be under surveillance by governments and international agencies. Unfortunately such diseases are only given attention when there is an outbreak as in the cases of Ebola, Zika virus and Severe Acute Respiratory Syndrome Coronavirus 2 (SARS-CoV-2) (Ekins et al., 2016; Lai, Shih, Ko, Tang, & Hsueh, 2020; Moran et al., 2012). The SARS-CoV-2 is not classified as NTD, but it has been neglected since the emergence of SARS-CoV and a study has already shown other coronaviruses that could cause symptoms worse than SARS-CoV (Cheng, Lau, Woo, & Yuen, 2007). This virus is the cause of the COVID-19 pandemia (Lai et al., 2020), leading to the rush to find new treatments through drug repurposing (Ekins et al., 2020) and also towards the development of new vaccines (Chen, Strych, Hotez, & Bottazzi, 2020). This follows the hard experience of dealing with the Ebola virus outbreak in West Africa and ZIKV in Brazil and South American countries, but there should be more attention to all NTDs and emerging diseases in order to prevent large scale outbreaks and epidemics that can affect many lives around the world. There is an urgent need to develop new drugs targeting NTDs, since many of the infective agents have already developed resistance to the limited number of currently available drugs and face many other issues, such as toxicity and pharmacokinetics problems, that prevent the eradication of these diseases. Among the first tasks in a drug discovery pipeline is the identification and validation of a drug target based on the organism's biology (Duffy, Zhu, Decornez, & Kitchen, 2012). One of the most widely targeted groups of proteins for many non-communicable as well as infectious diseases are the protein kinases and the elucidation of an organism's kinome is often a starting point for these studies (You, McManus, & Gobert, 2015).

3. Bioinformatics tools and chemogenomics approaches

Bioinformatics represents a relatively new interdisciplinary branch of science to systematically manage large and diverse amount of biological data and analyze it to extract its hidden information (Kelly, 2009). In this context, bioinformatics has three main goals: (i) organize the data and allow researchers to access the existing information; (ii) develop tools and resources through computational and biological expertise that help directly in the data analysis; (iii) the use of these tools to analyze and properly interpret the data in a biologically meaningful manner (Kelly, 2009; Ouzounis, 2012).

Chemogenomics is a term created to denote the integration of chemistry, biology, genetics, informatics, structural biology and chemical screening information. Lattely, phenotypic target-based cellular loss and gain-of-function screens have also been included (Gaither, 2007; Paolini, Shapland, van Hoorn, Mason, & Hopkins, 2006). In summary, chemogenomics handles chemical biology data through bioinformatics techniques and approaches to extract information hidden in the vast quantities of data generated in labs worldwide and to provide meaningful and useful biochemical insight that could be applied in drug discovery. Using the information gained from the "omics" tools in combinations with chemical screens as well as *in silico* and/or high throughput screens (HTS), raises the possibilities and new ways to identify new druggable targets associated to diverse diseases, in the context of this chapter, to NTDs.

In this section we present some key applications of bioinformatics tools for chemogenomics focusing on NTDs, especially involving target prioritization, protein-protein interactions, target-ligand interactions and drug discovery.

Neves and coworkers (Neves, Braga, Bezerra, Cravo, & Andrade, 2015) applied a chemogenomics approach to identify druggable targets in the different life stages of *Schistosoma mansoni*, the main causative agent of schistosimaisis, prioritizing targets found to be more expressed in the host stage of the parasite's life cycle. The heart of their analysis was the use of pairwise alignment to choose targets and find approved drugs that were not yet tested against *S. mansoni* (Neves et al., 2015). They used a combination of bioinformatics tools including: BLAST (Altschul, Gish, Miller, Myers, & Lipman, 1990), ConSurf server (Ashkenazy, Erez, Martz, Pupko, & Ben-Tal, 2010), MAFFT (Katoh & Standley, 2013) and CD-HIT algorithm (Fu, Niu, Zhu, Wu, & Li, 2012; Li & Godzik, 2006). Basic Local Alignment Search Tool (BLAST) searches for local similarity regions between two or more nucleotide or amino acid sequences and, after identifying these similarity regions, calculates a score related to matches' significance (Altschul et al., 1990). The ConSurf server is a computational tool to determine the conservation between two or more sequences (amino acid or nucleotide) accordingly to phylogenetic relations between them and homologous sequences (Ashkenazy et al., 2010). Starting from the amino acid or nucleotide sequence, ConSurf follows the following steps: (i) search for close homologs, based on user-defined criteria using BLAST; (ii) sequences clustering and removal of probably redundant sequences using the CD-HIT algorithm; (iii) a multiple sequence alignment (MSA) is built using the homologs sequences by software such as CLUSTALW (Chenna et al., 2003), T-COFFEE (Notredame, Higgins, & Heringa, 2000), MAFFT, PRANK (Loytynoja & Goldman, 2005) or MUSCLE (Edgar, 2004); (iv) generate a phylogenetic tree from the MSA output using a neighbor-joining algorithm. Finally, position-specific conservation scores are computed using a empirical Bayesian method or machine learning method, generating a discrete scale that can be displayed as nine different colors to facilitate the visual inspection. This was applied to the large amount of data available from *S. mansoni* transcriptional profiles from its different life cycle stages. Focusing on intra-host life stages of the parasite and, through chemoinformatics approaches, they found 115 drugs that lay inside the chemical space of known active anti-schistosoma compounds that could be investigated for their biological activity (Neves et al., 2015).

Taylor and coworkers (2013) were pursuing broad spectrum anti-helmintic drugs by screening DrugBank compounds that could target parasite protein kinases. To achieve this goal, multilevel HMM profiles and Markov clustering were employed to proteomic data of the parasites of interest (*Caenorhabditis elegans*, *Brugia malayi* and *Haemonchus contortus*). HMM profiles are fully probabilistic model that specifies the distribution of position-specific insertions and deletions accumulated during the evolutionary process within an entire sequence family (Eddy, 1998; Schuster-Böckler, Schultz, & Rahmann, 2004) based on pre-aligned sequence converting the observed counts of residue emissions and state transitions into probabilities (Eddy, 1998). These probabilities, as HMM profiles, are used to describe features such as a specific domain superfamily or more specifically a catalytic domain of a group of proteins. HMM profiles can be used to scan whole genomes to identify protein domains and domain architecture (Jurberg & Brindley, 2015). In the context of a chemogenomics, this approach can be used to identify common protein domains, and to shed light in the prediction of how similar domains interact with known chemicals.

From the evolutionary perspective, functionally important regions such as the the binding site of proteins are more conserved. This characteristic can be exploited in approaches such as substructure or fingerprint searches.

In this context, Konc and colleagues (2015) presented a new approach that, in contrast to molecular docking technique, predicts ligands for a protein through their transposition between similar binding sites. They introduced the ProBis algorithm (Konc & Janežič, 2010) that can detect structurally similar sites on protein surfaces using local surface structure alignment to identify with sub-residue precision similar sites as patterns of physicochemical properties on the protein surface. Likewise, another approach developed by Kuhn and colleagues (2007), named Cavbase. This program performs both a comparison and a classification of proteins *via* physicochemical properties of their respective active sites and, in contrast to ProBis, it does not rely on any similarity information and operates automatically. Cavbase detects protein binding sites using the Ligsite algorithm (Hendlich, Rippmann, & Barnickel, 1997), which places the protein into a Cartesian grid with a 0.5 Å spacing and classifies as "solvent inaccessible" any grid points falling within or less than 1.5 Å outside the van der Waals radius of any protein atom. A depth value is then determined for the other grid points in order to represent how buried it is in the protein binding pocket. A minimum size for the buried points must be specified and, hence, the algorithm is able to generate a cavity, whose surface is represented by the grid points of the cluster in contact with the protein surface. The Cavbase approach can be also applied to analyze similarity relationships between kinases and to sort them accordingly to their Cavbase similarity score (Glinca & Klebe, 2013).

4. Protein kinases and kinome

Protein kinases are among the most studied protein families (Johnson & Hunter, 2005) because of their participation in many signaling pathways that control biological processes such as cell growth, development, differentiation and death (Stroehlein et al., 2015). Much research effort has been made to find compounds targeting kinases for different diseases (Cohen, 2002; Melnikova & Golden, 2004; Paul, Kaur, Geete, & Sobhia, 2014) and compounds targeting these proteins have reached success particularly for cancer chemotherapy (Vieth et al., 2005).

Eukaryotic protein kinases (ePKs) transfere a phosphate group from ATP to serine, threonine, tyrosine residues of other proteins, regulating their activities (Ferguson & Gray, 2018; Kanev et al., 2019). Some proteins that have kinase activity do not have sequence similarity to ePKs and are called atypical protein kinases (aPKs) (Manning, Whyte, Martinez, Hunter, & Sudarsanam, 2002). The ePKs are characterized by a conserved domain with

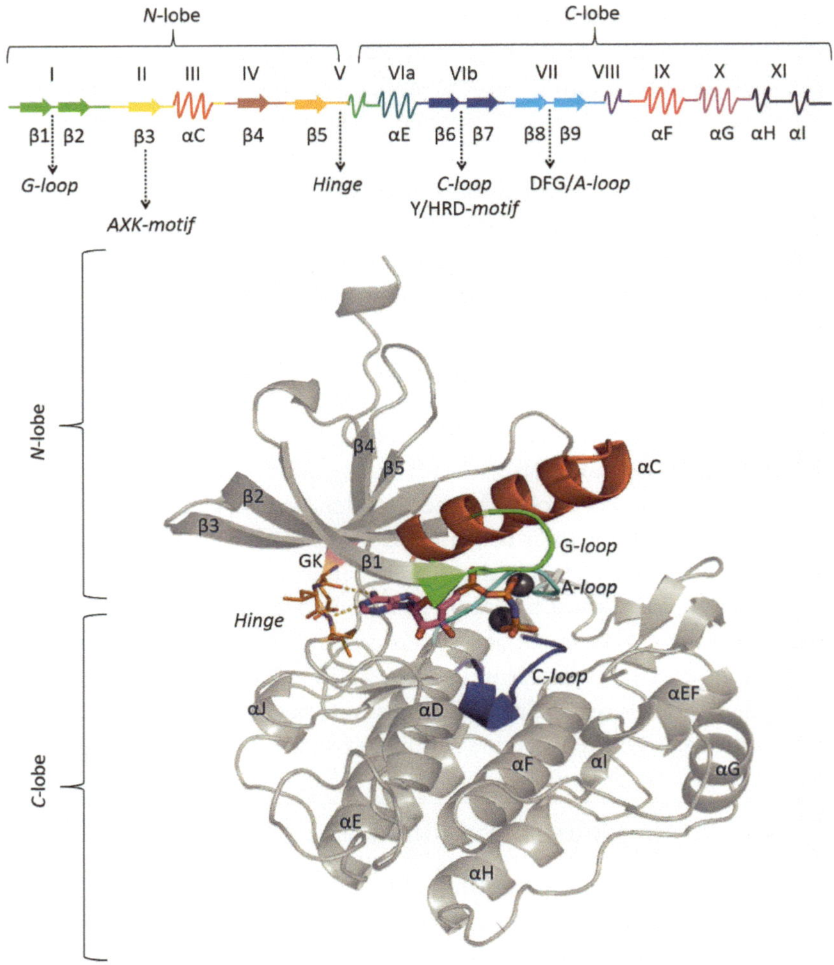

Fig. 1 2D and 3D general structure of a protein kinase. Representation of the secondary structure of a protein kinase, highlighting standard annotations of subdomains and structural elements. The structural elements α-C (C-helix), hinge region, G-loop (gly-rich loop), A-loop (activation loop), C-loop (catalytic loop) and GK (gatekeeper) are highlighted in red, orange, green, cyan, blue and pink, respectively. The magnesium ion is represented as a gray sphere. The carbon, oxygen, nitrogen and phosphorus are represented in pink, red, blue and orange, respectively.

11 sub-domains (Fig. 1) that are strongly conserved among the kinase family (Manning et al., 2002). Structurally, ePKs can be recongnized by an *N*-terminal lobe composed of antiparallel β-sheet, a *C*-terminal lobe with α-helices, and the hinge segment with 3– 5 aminoacids, connecting both domains. The ATP interacts with kinases into the slit formed between

N- and *C*- lobes (Knighton et al., 1991). The most important interacting regions of ATP and inhibitors include the hinge, the conserved residues HRD (histidine, arginine and aspartate) and DFG (aspartate, phenylalanine and glycine), the glycine loop, the activation helice αC and the activation and catalytic loops. The DFG motif can be in open or closed conformations, activating or inactivating the kinase. The HRG sustains kinase activity by its involvement into the phosphate group transference (Kanev et al., 2019).

Protein kinases are classified into super families, groups, families and subfamilies according to the amino acid they phosphorylate and the conserved sequence of their catalytic domain (Hanks & Hunter, 1995). ePKs are divided into two super families, the serine/threonine protein kinases, that phosphorylate serine or threonine resisues and tyrosine kinases, that phosphorylate tyrosine residues (Deshmukh, Anamika, & Srinivasan, 2010). There are nine main groups of protein kinases (Table 1) classified based on domain sequence similarity: (i) the AGC group—protein kinases A, G and C; (ii) CAMK group—Ca^{+}/CAM- dependent kinases; (iii) CMGC group—CDK, MAPK, GSK3 and CLK; (iv) CK1—casein kinase 1; (v) STE group—homologs of yeast sterile 7, 11 and 20; (vi) RGC group—receptor guanylate cyclases (vii) TK group—tyrosine kinase; (viii) TKL group—tyrosine kinase like; and (ix) Other group—several kinase families that do not fit within any of the other main kinase groups (Shchemelinin, Šefc, & Nečas, 2006).

The human kinome was elucidated in 2002 by Manning and co-workers. It consists of 518 protein kinases of which 478 are ePKs, distributed in the kinase groups, and 40 are aPKs (Manning et al., 2002). The elucidation of the Human kinome helped in the understanding of kinases functions, their role in biological pathways and in the development of new compounds targeting this protein family (Cohen, 2002; Melnikova & Golden, 2004). As well as being good targets for non-communicable diseases such as cancer, they can also be considered a good protein family to target in infective diseases, as they share many of the human protein kinases functions but as distant phylogenetic organisms are divergent enough to be parasite specific (Doerig, Billker, Pratt, & Endicott, 2005).

4.1 Generating a kinome: Multiple ways, same principle

The first experimentally resolved crystal structure of a protein kinase was from mouse, PKA-Cα, which markedly showed the general topology of the catalytic core of a kinase protein (Knighton et al., 1991). Later structures

Table 1 Classification of protein kinases groups.

Kinase group	Kinase activity	Kinase protein	
RTK **R**eceptor **t**yrosine **k**inase	**TPK** Tyrosine-specific protein kinase	EGFR, VEGFR, RET	**ePKs**
NRTK **N**on-**r**eceptor **t**yrosine **k**inase		Src, Abl, Syk	
TKL **T**yrosine **k**inase **l**ike	**STPK** Serine/threonine-specific protein kinase	RAF, MLK, LISK	
CMGC **C**yclin dependent kinases, **M**AP kinases, **G**lycogen synthase kinases, **C**asein kinases	**STPK** Serine/threonine-specific protein kinase	CDK family, MAPK family	
CAMK **C**alcium/cal**m**odulin dependent **k**inases	**STPK** Serine/threonine-specific protein kinase	MARK, AMPK, CHK1	
CK1 **C**asein **k**inases 1	**STPK** Serine/threonine-specific protein kinase	CK1, TTBK	
Other	**STPK** Serine/threonine-specific protein kinase	PLK1, PLK2, SAK	
STE homologs of yeast STE7, STE11 and STE20 genes	**STPK** Serine/threonine-specific protein kinase	MAP2K, MAP3K, MAP4K	
AGC Protein kinases **A**, **G** and **C**	**STPK** Serine/threonine-specific protein kinase	PKA, PKC, PKC	
APKs **A**typical **p**rotein **k**inases	Dissimilar to the abovementioned groups, but presenting kinase activity	RIO1, RIO2, ABC1	**aPKs**
PKL **P**rotein **k**inase **l**ike	Kinases that act as part of small molecule (*e.g.*, sugar, lipids) metabolism	PI3K	**None**

revlealed the role of ATP coordinated to magnesium ion (Bossemeyer, Engh, Kinzel, Ponstingl, & Huber, 1993; Zheng et al., 1993). From those studies, the highly conserved residues of kinase domain were assigned and their function could be more precisely understood.

In 1995, Hanks and Hunter built phylogenetic trees to better classifiy and display the relations between kinase proteins and, as a consequence, to facilitate the analysis and handling of this large superfamily of proteins. They classified kinases into four large groups accordingly to the similarity found in their kinase domain amino acid sequences: AGC, CaMK, CMGC and PTK (Hanks & Hunter, 1995).

With the advances in genomic sequencing more information has became available and, using the latest computational tools, nine groups of kinase proteins (Table 1) have been classified. This has increased the possibility of identification and prioritization of new drug targets, especially protein kinases involved in multiple processes, both physiological and pathological, in diverse organisms (Hanks, 2003; Katara, 2013). It also allows the better comprehension of cell-signaling pathways and, thus, an improved way of interfering in such pathways accordingly to the goal of the study.

The kinome of an organism is the complete set of protein kinases encoded in its genome, or it can be defined as the proteome of protein kinases of an organism (Manning et al., 2002). Kinome interpretation may represent a good way to better know and prioritize proteins that are associated with diseases and with lethal phenotypes of etiologic agents when inhibited for target selection (Robertson & Renslo, 2011). This is true for many NTDs as schistosomiasis, leishmaniasis, trypanosomiasis as well as for tuberculosis and malaria, where there is an urgent need to find new therapeutic targets and new drugs (Doerig & Grevelding, 2015).

Parsons, Worthey, Ward, and Mottram (2005) made a comparative analysis of kinomes of three pathogenic trypanosomatids (*L. major, T. brucei, T. cruzi*) using bioinformatics techniques, based on BLAST and Hidden Markov Model (HMM), searching for the presence of specific subdomains on the queried sequences of each parasite's proteome. Trained HMM models were used to find these subdomains and BLAST to identify atypical protein kinases (aPKs) and ePKs, in model organisms such as *H. sapiens, C. elegans, S. cerevisae* and *D. melanogaster*. Second, a phylogenetic inference was conducted at the amino acid alignment due to the relatively low degree of sequence conservation at the nucleotide level within some kinase families of the trypanosomatids under investigation. Protein kinase domains from the Pfam database (Finn et al., 2015) (http://pfam.xfam.org/) were used to identify kinase domains in sequences identified.

Ward and coworkers generated (Ward, Equinet, Packer, & Doerig, 2004) the kinome of *P. falciparum*, showing the kinome features of this pathogenic organism. They applied a combination of bioinformatics approaches, using multiple alignment techniques and HMM trained models to identify the conserved kinase domains among the parasite's proteins, as well as using phylogenetic analysis to better visualize the relationships between *P. falciparum* and Human kinome.

Likewise, Andrade and colleagues reported an initial approach attempting to identify and functionally classify potential protein kinases of *S. mansoni* into groups, families and subfamilies (Andrade et al., 2011). They also used HMM models followed by manual annotation and integration of data obtained from InterProScan and a reverse PSI-BLAST (Altschul et al., 1997) analysis and phylogenetic analysis.

A deeper approach of reporting a kinome was made recently by Stroehlein et al. (2015). They showed a powerful combination of bioinformatics tools not only to report the kinome, but to perform a target prioritization through gene expression data analysis and interrogation of chemical databases to investigate the possibility of repurposing compounds against possible targets identified in their study. In summary, the authors built a pipeline to characterize and curate kinases in *S. haematobium*, including the addition of orthologous sequences and the complementation of incomplete homologous sequences of *S. haematobium* using the well-annotated biological data of *S. mansoni*. The pipeline started with an initial prediction of kinases from the proteome of *S. haematobium* and using sequence similarity techniques, such as trematode-specific HMM models extensively trained to identify kinase domains of each kinase group, combined with multi-alignment techniques such as MAFFT and MUSCLE for phylogenetic analysis and to generate the kinome tree.

4.2 Selectivity profiling in protein kinases: A balance between efficacy and toxicity

All protein kinases share conserved amino acids at the catalytic domain and have similar structures that provide the γ-phosphate transfer from ATP to a specific target molecule (e.g., proteins, lipids, carbohydrates or nucleotides). Due to the conservation of ATP binding site, it is easy to develop ATP-mimetic inhibitors with promiscuous profiles but this often leads to toxicity issues (Cheng & Force, 2010). Conversely, inhibitors targeting only one protein can lose their efficacy due to mutations on that target, as this is the case of resistance to imatinib acquired in chronic phase myeloid leukemia patients due to mutations on BCR-ABL kinase domain (Shah et al., 2002). For those reasons, it is important to develop selective inhibitors that can target more than one kinase in order to reach the balance of efficacy and toxicity. The ATP binding site is composed of many pockets

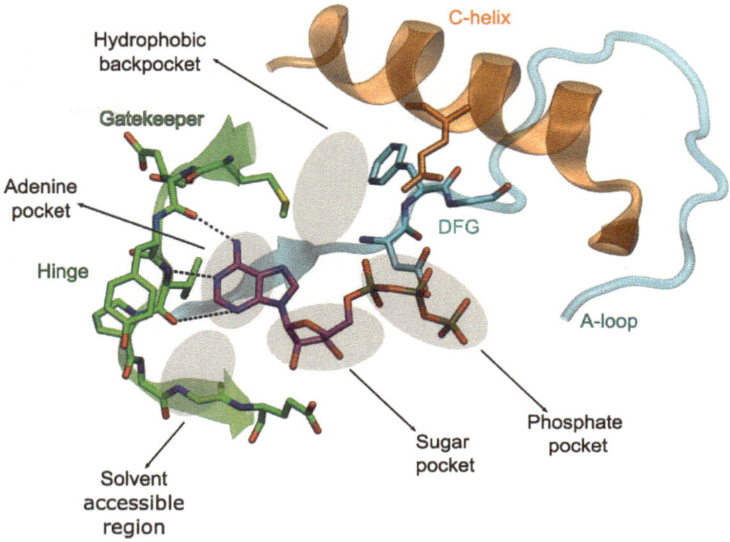

Fig. 2 ATP binding pockets in a protein kinase.

(Fig. 2): (i) the adenine binding pocket or hinge binder—where the inhibitors/ATP form hydrogen bonds with the hinge region; (ii) the sugar pocket—hydrophilic character for most kinases; (iii) hydrophobic back pocket—can be either small or large, depending on the shape of gatekeeper residue; (iv) solvent accessible region and (v) phosphate pocket—where the phosphates of ATP are placed (Fabbro, Cowan-Jacob, & Moebitz, 2015; Traxler & Furet, 1999; Zuccotto, Ardini, Casale, & Angiolini, 2010).

Kinase inhibitors developed to date bind kinases in different binding modes. They can be categorized as covalent (irreversible) inhibitors or reversible inhibitors. Reversible inhibitors can be classified in four types (Fig. 3), based on their binding sites and the kinase conformation (Fabbro et al., 2015; Wu, Nielsen, & Clausen, 2015). Type 1 inhibitors bind to the ATP site with DFG-in conformation (Fig. 3A). The active site of kinases in active conformation is usually rigid and has a great complementarity with its inhibitors. For this reason, it is easier to develop ATP-mimetic type 1 inhibitors. Type 1 kinase pharmacophore consists of one hydrogen acceptor, two hydrogen donors and one hydrophobic moiety (Zuccotto et al., 2010). Selectivity can be reached by targeting poorly conserved residues, by developing inhibitors that lead to structural penalties, binding to additional sites near the substrate binding site or at the back pocket (Müller, Chaikuad, Gray, & Knapp, 2015; Zuccotto et al., 2010). Inhibitors that reach the back pocket are also called type 1.5 inhibitors (Fig. 3B). The back pocket of

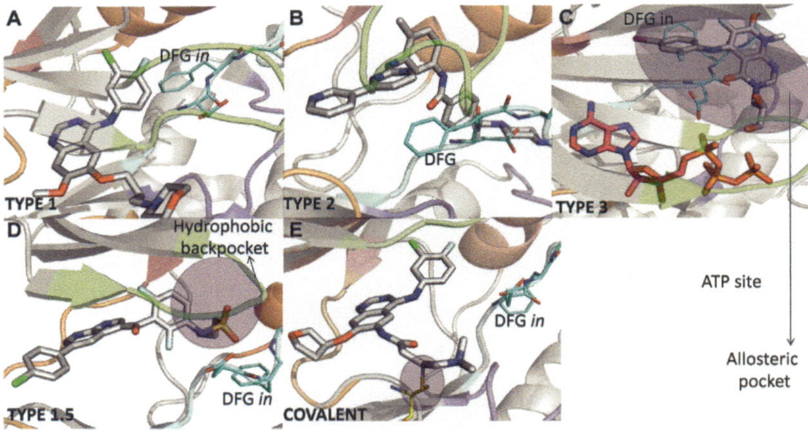

Fig. 3 Type of kinase inhibitors. (A) Type 1 inhibitor—"DFG in" [PDB ID: 2ITY]; (B) Type 1.5 inhibitor—"DFG in" reaching the hydrophobic backpocket [PDB ID: 3OG7]; (C) Type 2 inhibitor—"DFG out" [PDB ID: 1IEP]; (D) Covalent inhibitor [PDB ID: 4G5J]; (E) Type 3 inhibitor—allosteric pocket close to ATP site [PDB ID: 3PP1]. In this figure, the structural elements α-C (C-helix), hinge region, G-loop (gly-rich loop), A-loop (activation loop), C-loop (catalytic loop) and GK (gatekeeper) are depicted in red, orange, green, cyan, blue and pink, respectively. The carbon atoms of ATP e are represented in pink and of the inhibitors in white. The oxygen, nitrogen, chlorine, fluorine, iodine, phosphorus and sulfur atoms are shown in red, blue, green, cyan, purple, orange and yellow, respectively.

kinases has differences on their sizes depending on the gatekeeper's shape and for this reason, selectivity can increase depending on the target kinase if the inhibitor occupies the back pocket (Fabbro et al., 2015; Müller et al., 2015; Zuccotto et al., 2010).

Type 2 inhibitors bind to the ATP site in its inactive conformation (Fig. 3C). The DFG-out conformation exposes a hydrophobic pocket that is also called the allosteric pocket (Fabbro et al., 2015; Wu et al., 2015). The pharmacophore of type 2 inhibitors is the same as type 1 with the addition of a linker, a hydrogen donor that interacts with a conserved Glu from the c-helix, a hydrogen acceptor that forms a hydrogen bond with Asp from the DFG-motif (Zuccotto et al., 2010). Usually, this kind of inhibitor is less promiscuous because of its level of conserved residues at the allosteric pocket sequence is smaller than other pockets. Also, the less rigid conformation of DFG-out kinases can accommodate a higher structural diversity of inhibitors. Unfortunately, many kinases can assume the inactive state and having the type 2 conformations as targets not necessarily generates selective compounds. Type 2 inhibitors also have higher molecular weight, leading to a negative impact in cell penetration/biodistribution, resulting in a lower efficacy (Müller et al., 2015; Wu et al., 2015).

Covalent inhibitors usually bind irreversibly to the nucleophilic thiol group of Cys close to the ATP binding site (Fig. 3D). The potential advantages of this binding mode are prolonged pharmacodynamics, suitability for rational design, high potency and selectivity. The disadvantages are that mutations on the covalent binding amino acid can lead to loss of potency or activity (Liu et al., 2013).

Type 3 and type 4 inhibitors bind in allosteric pockets. Type 3 binds to a pocket close to the ATP site and type 4 binds to pockets far from the ATP site (Fig. 3E). These type of inhibitors are more selective because they bind to pockets that are unique to specific kinases but they are hard to develop because of the lack of inhibition assays for this type of inhibitor. There are no general strategies for developing allosteric inhibitors since unique sites were discovered in crystalized structures after a specific inhibitor's selectivity was determined (Fabbro et al., 2015; Müller et al., 2015).

Protein kinases are selective to the substrate peptides they phosphorylate. These peptides have a conserved sequence of amino acids (close to the Ser/Thr/Tyr that will be phosphorylated) that interact at the kinase substrate site. Each kinase has a different electrostatic potential and is selective to its substrates (Ubersax & Ferrell, 2007). Bivalent kinase inhibitors, containing a peptide tag ligand that binds to a signaling interaction site of the targeted protein and an ATP-competitive inhibitor, were developed and are able to increase selectivity of the ATP-competitive inhibitors alone (Hill, Perera, Andrews, & Maly, 2012).

As the kinases of organisms that are etiologic agents of NTDs share human kinase's catalytic domain, those types of inhibition can also be exploited for them. A comparison between sequence similarity of protein kinases and SAR (structure activity-relationship) analysis of kinase inhibitors showed that kinases with more than 60% of sequence similarity tend to have high SAR similarity with their inhibitors (Vieth et al., 2005). This relationship can be exploited knowing that many microorganisms' kinases have little sequence similarity with human kinases. Some of the important residues in the ATP pocket are conserved but some are divergent and maintain kinase activity (Kannan, Taylor, Zhai, Venter, & Manning, 2007).

4.3 Multi-kinase inhibition: A powerful approach to overcome drug resistance and lack of efficacy

Single-target-based drug discovery approaches developed with the progresses in molecular and structural biology knowledge (Macarron et al., 2011). This idea was built on the concept of "*magic bullets*", founded by Paul Ehrlich (Kaufmann, 2008), stating that: "*one molecule*" is capable of interacting with "*one target*" and thereby eradicates "*one disease*" with superior efficiency and selectivity. But lack of efficacy and safety in clinical trials made this concept less secure as it appeared, attributing the lack of effectiveness mainly to the fact that it is impossible to find a single target for effective treatment of complex diseases (Chen, Han, Xu, Xu, & Qian, 2013; Kola & Landis, 2004; Viayna, Sola, Di Pietro, & Muñoz-Torrero, 2013). Network pharmacology or polypharmacological therapeutic approaches have been developed to overcome these problems, and to revolutionize the process of drug discovery, designing molecules that are capable of interacting with multiple targets (Hopkins, 2008). This new paradigm uses one drug to interact with appropriate protein targets with the aim to achieve the expected pharmacological activity (Viayna et al., 2013). In addition, the use of drug cocktails or fixed-dose combinations has disadvantages regarding drug-drug interactions and dosing complexity (Baumert & Hilgeroth, 2009), and the difficulty to co-formulate theses products (Zimmermann, Lehár, & Keith, 2007). Thus, the best therapeutic approach could be the multi-target drug (MTD). The MTD approach can be applied using multi-target screening of chemical libraries (*Screening-approach*) or rational design with hybridization of compounds framework or pharmacophore combination approach (Morphy & Rankovic, 2005; Viayna et al., 2013).

Many MTDs have been approved by the FDA, like sorafenib, sunitinib, ponatinib (anti-cancer); paliperidone, asenapine, lurasidone, (for psychotic disease); and vilazodone (antidepressant) (Peters, 2013). This has enhanced the reliability of this method, and has aroused the interest of researchers working in NTDs to overcome obstacles that have hindered effective treatment. Cavalli and Bolognesi (2009) proposed MTDs as a new and interesting approach for NTD drug discovery by improving potency and efficacy, decreasing drug resistance and expenses, and encouraging treatment adhesion compared to cocktails and fixed-drug combination (Cavalli & Bolognesi, 2009).

Multi-kinase Inhibition (MKI) emerged from the concept of MTD and is a useful technique in which a single molecule is able to inhibit more than one kinase target at the same time. The synergic effect of MKI makes it more effective and avoids the appearance of resistant cells or strains (Garuti, Roberti, & Bottegoni, 2015). As the ATP kinases share a similar ATP domain, this makes them targets for potent inhibition by a single molecule. However, the inhibitor promiscuity also makes it disadvantageous causing adverse effects. To avoid such toxic effects, researchers must pay attention on the development of molecules that are not capable of interacting with proteins unrelated to the disease and/or target organism, as highlighted by researches at University of Dundee (Bain et al., 2007; Bain, McLauchlan, Elliott, & Cohen, 2003; Davies, Reddy, Caivano, & Cohen, 2000).

A variety of *in silico* methods can predict, with accuracy and robustness, molecules with activity against multi-targets. From the knowledge of the target structure, in *structure-based methods*, as one example of an *in silico* technique, potent and selective ligands can be developed using methods such as molecular docking and receptor-based pharmacophore. Alternatively, *ligand based method* uses known ligand features necessary for interaction with the target to select molecules with similar properties (structural, physicochemical and electronic) using QSAR models and machine learning methods. Another approach is *fragment-based methods*, that screen small pharmacophore like fragments in co-crystallization experiments whoseresults are used to build a framework to combine the fragments and generate a potent compound. Finally *expression-based methods* use gene expression profile to assist in the prediction of new targets (Liu et al., 2013). An integrative computational approach to select inhibitors for *Plasmodium falciparum* calcium-dependent protein kinases 1 and 4 (CDPK1 and CDPK4) and protein kinase 6 (PK6) led to the identification of three potential multi-kinase inhibitors with nanomolar activities against bood stage *P. falciparum* using computational shape-based and machine learning models (Lima et al., 2019).

MKI can circumvent the pharmacologically resistance, one of the main problems faced in diseases such as malaria, leishmaniasis, schistosomiasis, and other NTDs. It proves to be more attractive for patient adherence to treatment, whereas one drug can act on several targets, avoiding more complex pharmacotherapies as Artemisinin-based Combination Therapy (ACT) for malaria. Furthermore, the availability of *in silico* methodologies can help getting promising and cost-effective results.

5. Drug and target repurposing

Due to the low investments by private companies to discover and develop drugs for NTDs, the drug discovery process for these diseases is mainly led by academic laboratories. Aiming to reduce cost and time in this process, academic groups have used "*repurposing*" strategies. Repuposing strategies are based on data already known about a certain organism that can be used as a starting point for the development of drugs for homologous targets in other organisms. It involves four main categories: drug repurposing, target repurposing, target class repurposing and lead repurposing

(Klug, Gelb, & Pollastri, 2016). The "*Drug repurposing*" approach often uses the FDA-approved drugs to identify an alternative indication that was not originally approved, in order to reduce research time and pharmaceutical development costs (Ashburn & Thor, 2004; Klug et al., 2016). "*Target repurposing*" starts with the search of homologs of specific target, with the aim of beginning drug design of this specific target based on approved drugs or clinical candidates of homologous target. Meanwhile, "*target class repurposing*" differ from "*target repurposing*" by not identifying an exact target, but using chemical information based on recognizing cellular features (for example) of a determined target class. Unlike the other repurposing strategies, "*lead repurpusing*" has as the starting point lead candidates rather than an approved drug or clinical candidates (Klug et al., 2016).

Drug and target repurposing have emmerged as promising strategies for drug design and discovery against NTDs, since only 1% of drug candidates that reached clinical phases in the last 10 years aim treating NTDs. In leishmaniasis context, drug repurposing is the main strategy applyed and most of the currently clinically available drugs for treatment of this disease were obtained from repositioning: (i) Miltefosine, that was initially developed for cancer treatment (Berman, 2015), (ii) Paromomycin, was designed to treat amebiasis (Ben Salah et al., 2013), (iii) amphotericin B, which is an antifungal drug (Sundar, Chakravarty, Agarwal, Rai, & Murray, 2010). There are many experimental and computational (Hernandez et al., 2019) strategies for drug repositioning (Andrade et al., 2019; Pushpakom et al., 2019). Experimental strategies like phenotypic HTS screenings (Moffat, Vincent, Lee, Eder, & Prunotto, 2017), chemoproteomics (Sleno & Emili, 2008), transcriptomics, proteomics, and metabolomics have being largely used (Hieronymus et al., 2006). Among computational strategies, similarity searches (Muegge & Mukherjee, 2016), shape-based and pharmacophore models (Grant, Gallardo, & Pickup, 1996), protein-domain alignment (Rognan, 2007), searching for common protein motifs (Pollastri & Campbell, 2011), and molecular docking have been also used (Ehrt, Brinkjost, & Koch, 2016).

Genetic sequencing of numerous parasite species has assisted the selection of targets using similarity studies to detect homologous Human sequences and structures that are already targets of known active agents. This strategy is useful due to a large number of drug-like compounds identified for human targets, which already have some toxicological and pharmacological assessment, as well as, pharmacokinetics studies. Thus, drug repurposing can speed up the process to discover drug candidates for other diseases, and therefore reduce costs in the process of drug discovery (Amata et al., 2016; Pollastri & Campbell, 2011). As already discussed, kinases are the targets in many human pathological diseases, such as, endocrine disorders, cancer, immunodeficiency and cardiovascular diseases (Stenberg, Riikonen, & Vihinen, 2000). This large diversity of human kinase inhibitors appears to be a rich source for hit-to-lead optimization for parasitic kinases inhibitors, with the advantage that by repurposing between evolutionary quite distinct organism differences can be exploited for improving selectivity (Rotella, 2012). The relatively faster and cost effective approach of repurposing strategies (Pollastri & Campbell, 2011) is highly attractive for NTDs. Based on the use of bioinformatics tools and the availability of a range information about human kinase inhibitors, this class of drugs could yield effective treatments for many NTDs.

6. Succesfull applications of target reporpusing for discovering kinase inhibitors in NTDs and malaria

6.1 Malaria

Malaria is one of the most impacting public health problems in tropical and subtropical areas of the globe, with approximately 200 million cases worldwide annually (World Health Organization, 2019b). The Phase III trials of the malaria vaccine recently identified three safety concerns: higher risks of meningitis, cerebral malaria, and doubled female mortality (Aaby, Fisker, Björkman, & Benn, 2020). In the absence of an effective vaccine, rapid treatment is crucial for effective malaria control. However, parasite resistance to currently available drugs, including the Artemisinin-Based Combination therapies (ACTs) (Sibley, 2013; World Health Organization, 2019b) underscores the urgent need for identifying new antimalarial drugs.

The human malaria parasite *Plasmodium falciparum* had its kinome identified and classified after its genome was sequenced (Ward et al., 2004). Using a bioinformatics strategy, the authors found 65 ePK sequences classified into seven of the main protein kinase groups and 20 ePK-related sequences classified into a new group, restricted to Apicomplexan species, termed the FIKK group (Ward et al., 2004). There were not any members of the STE and TK groups in malaria kinome. The human and parasite kinomes show great divergence and there is a significant fraction, the FIKK, of plasmodial kinome without orthologs in mammalians. For this reason, it's likely that inhibitors already tested for human kinases can be optimized to malaria kinases and that this protein family is a rich source of potential targets (Lucet, Tobin, Drewry, Wilks, & Doerig, 2012).

P. falciparum cyclin-dependent kinase (CDK) is an example of target repurposing, with three proteins *Pf*PK5, *Pf*PK6 and *Pf*mrk identified in *Plasmodium* as the homologs to Human (Bracchi-Ricard et al., 2000; Li, Robson, Chen, Targett, & Baker, 1996; Ross-Macdonald, Graeser, Kappes, Franklin, & Williamson, 1994). Despite the structural and molecular similarities between CDKs of *Plasmodium* and human, many inhibitors of CDKs fail to act on the homologous *Plasmodium* target. However, the structural differences between the protein kinases in question may be exploited in the development of selective inhibitors to the parasite (Holton et al., 2003; Woodard et al., 2003). (*R*)-roscovitine (**1,** Fig. 4), developed by Cyclacel Pharmaceuticals, Inc., Scotland, has been assessed in clinical trials against: adenocarcinoma, advanced malignancies, B-cell malignancies, brest cancer (metatastic), corticosurrenaloma, hepatocellular carcinoma, nasopharyngeal carcinoma, non-small cell lung cancer, parotid cylindroma and thymic

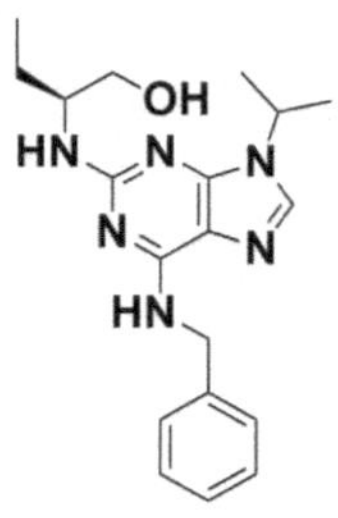
(1)

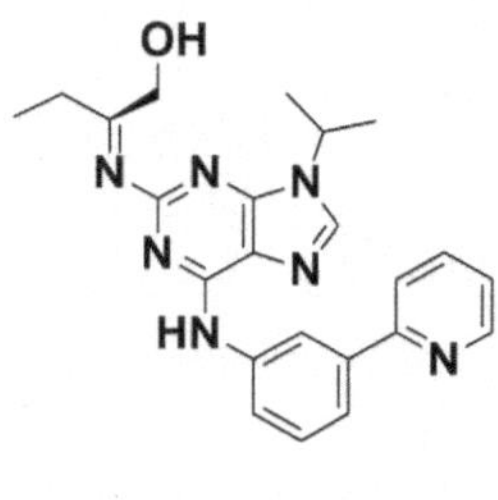
(2)

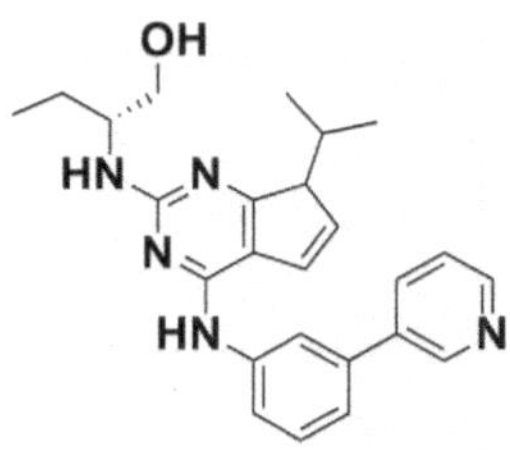
(3)

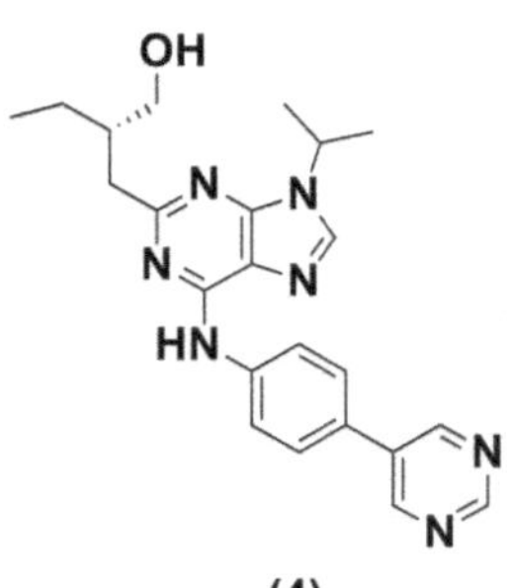
(4)

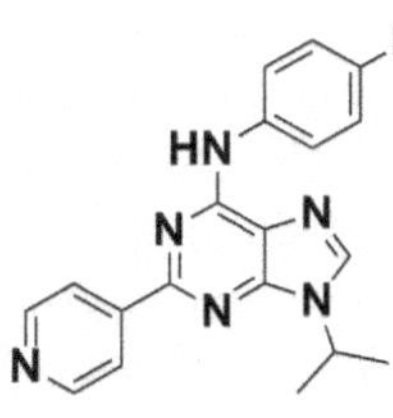
(5)

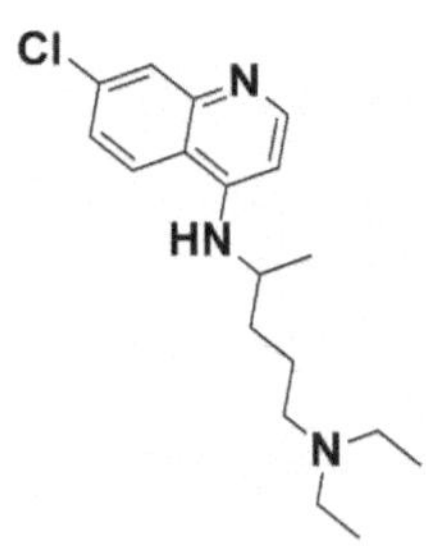
(6)

Fig. 4 Chemical structures of CDK inhibitor (**1**) ((*R*)-roscovitine) and compounds synthetized from 2,6,9-trisubstituted purines (**2–5**) and chloroquine (**6**).

carcinoma (Cicenas et al., 2015). This compound has been used as a starting point in the identification of new *Pf*CDK inhibitors. 2,6,9-trisubstituted purines structurally associated with (*R*)-roscovitine, have been synthetized and assessed in different mammalian and *P. falciparum* assays (Houzé et al., 2014). IC_{50} values were determined for mammalian kinases (CDK1, CDK2, CDk5, GSK-3αβ, CK1δ/ε, DYRK1A), and *Pf*3D7 (chloroquine sensitive African strain) and *Pf*7G8 (chloroquine resistant Brazilian strain). Although the best compounds (**2–5,** Fig. 4) synthesized showed results of inhibition of growth for both *Plasmodium* strains lower than against the antimalarial drug chloroquine (**6**), the compounds exhibited inhibition values greater than roscovitine, as presented in Table 2. When compounds screened in *P. falciparum* strains and human kinases (Table 2) were compared, it can be concluded that the human CDK inhibitors could be used as *Plasmodium* inhibitors. The study concluded that similar sequences and structures of kinases in both organisms could be a starting point to discover potent and selective inhibitors for *Plasmodium*. However, further studies need to be conducted to access selectivity of these inhibitors.

A selective inhibitor of *P. falciparum* GSK3 was developed exploring the differences of human and parasite's enzymes. The gatekeeper residue is a Leu in hsGSK3 and a Met in *Pf*GSK3 and as a consequence the interactions with the ligands were different. There is also a pocket at the bottom of ATP site unique to *Pf*GSK3 that was filled by the selective inhibitor (Masch & Kunick, 2015). Apicomplexan CDPKs were also targets for selective inhibitors (Hui, El Bakkouri, & Sibley, 2015). Although CDPKs are only present in plants and apicomplexan parasites, they have similarities with human CamKs. Selectivity could be accomplished probing the fact that some

Table 2 IC_{50} values for *P. falciparum* and *Homo sapiens* assays [111].

	***Plasmodium* assay IC_{50} (μM)**		***Homo sapiens* assay IC_{50} (μM)**	
Compounds	***Pf*3D7**	***Pf*7G8**	***Hs*CDK1**	***Hs*CDK5**
2	6.53	2.06	0.8	0.4
3	5.37	1.55	0.2	0.1
4	3.05	0.69	0.04	0.06
5	1.39	0.99	0.6	0.6
(*R*)-roscovitine, **1**	10.48	12.56	–	–
chloroquine, **6**	0.03	0.18	–	–

CDPKs have a Gly residue at the gatekeeper that is not present in any mammalian kinases. The small gatekeeper gives access to the back pocket, which was the target for type 1.5 selective kinase inhibitors (Hui et al., 2015).

6.2 Leishmaniasis

The studies of the kinomes of *L. major, L. infantum* and *L. braziliensis* identified 199, 224 and 221 protein kinases, respectively (Borba et al., 2019; Parsons et al., 2005). These numbers correspond to about 30% of human kinome and double of number present in *Plasmodium* sp. Interestingly, there are no protein kinases classified as members of the TK or TK-L group for these organisms. The rest were classified into the other six kinase groups, though some of those groups are poorly represented (CAMK and AGC) while others are over represented (CMGC and STE) (Parsons et al., 2005). The definition of the kinome of *L. major* allowed the exploration of protein kinases as targets for antileishmanial agents. Some of the targets being studied include the cyclin dependent kinases CRK1 and CRK3, the inhibition of which leads to abnormal morphology, cell cycle disruption and aberrant DNA. Casein kinase 1 was elucidated as one of the targets of the inhibitor purvalonol B (Naula, Marilyn, & Mottram, 2005) and MAPKs have been shown to be a viable target as null mutants were incapable of proliferation inside host macrophages. The elucidation of *L. infantum* and *L. braziliensis* kinomes also allowed for target prioritization of abou 30 targets and a drug repurposing of compounds that might inhibit those targets, seven prioritized compounds were tested an two showed promising ability to inhibit *L. infantum* and *L. braziliensis* amastigotes and promastigotes (Borba et al., 2019).

In 2007, the FDA approved lapatinib (7), registered by GlaxoSmithKline plc., United Kingdom, as Tykerb®, for use in the treatment of solid tumor and breast cancer (Moy, Kirkpatrick, Kar, & Goss, 2007). This inhibitor has the ability to interact with tyrosine kinases of the epidermal growth factor receptor (EGFR) family (Moy et al., 2007; Rusnak et al., 2001). Based on studies showing that lapatinib and their analogues can be potent inhibitors of growth of *Trypanosoma brucei* (Behera, Thomas, & Mensa-Wilmot, 2014; Patel et al., 2013), Woodring and co-workers proposed to evaluate the anti-parasitic activity of *T. brucei* using quinazoline (scaffold of lapatinib) and thienopyrimidine-based inhibitors (Woodring et al., 2015). In addition, they performed a screening against *Leishmania major, T. cruzi* and *P. falciparum*. The highest potency (EC50) values obtained for *T. brucei, T. cruzi, L. major*

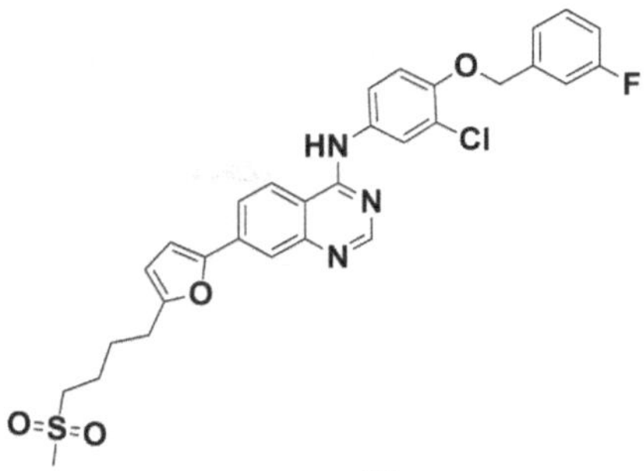

(7)

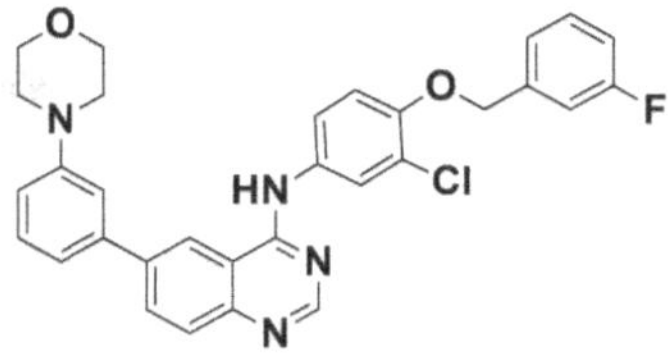

(8)

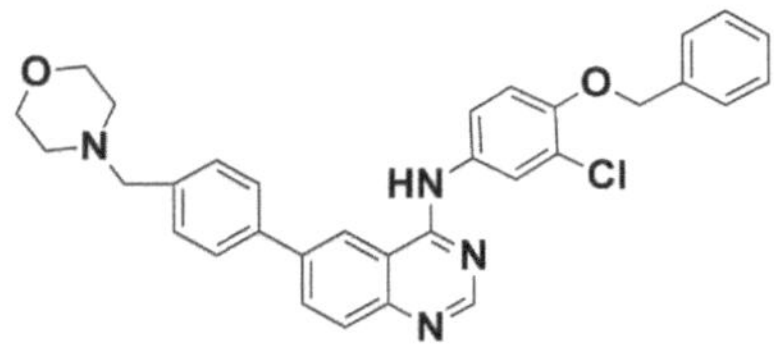

(9)

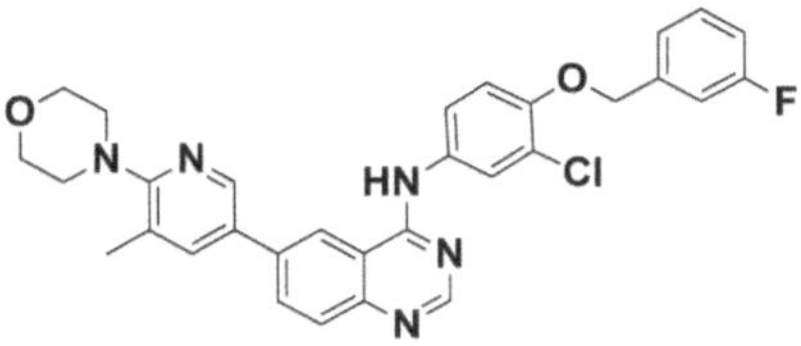

(10)

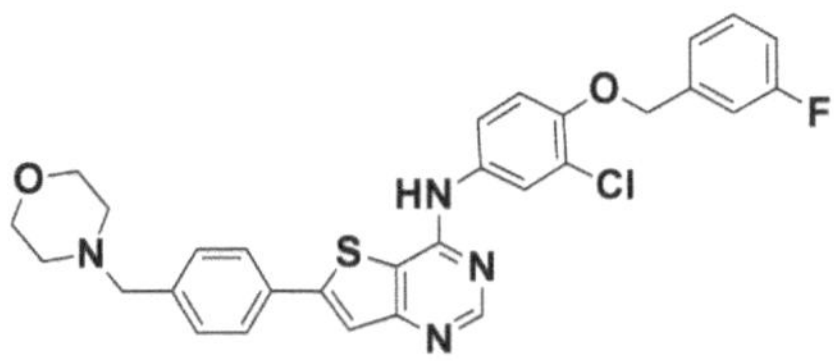

(11)

Fig. 5 Chemical structures of thienopyrimidine-based inhibitors **(7)** and their analogs matched with quinazoline scaffolds **(8,9,10).**

amastigote and *P. falciparum* were: 42 nM (compound 8), 0.60 μM (compound 9), 1.14 μM (compound 10), 0.027 μM (compound 11) (Fig. 5), showing that this is a good basis for the new range of anti-parasitic drug design.

The inhibitors of human aurora kinase hesperadin (12), VX-680 (13) and danusertib (14) (Fig. 6) were shown to be effective inhibitors of *T. brucei* (Jetton et al., 2009; Li, Umeyama, & Wang, 2009; Ochiana et al., 2013). Therefore, Patel and collaborators, guided by the existence of aurora kinase in *Leishmania* (Siman-Tov, Ivens, & Jaffe, 2001) and *Plasmodium* (Reininger, Wilkes, Bourgade, Miranda-Saavedra, & Doerig, 2011), evaluated

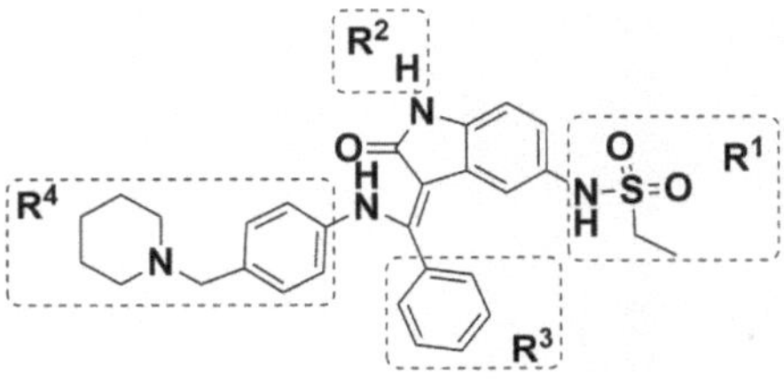

(12)

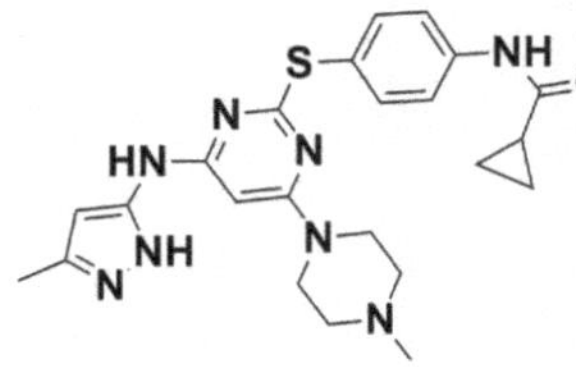

(13)

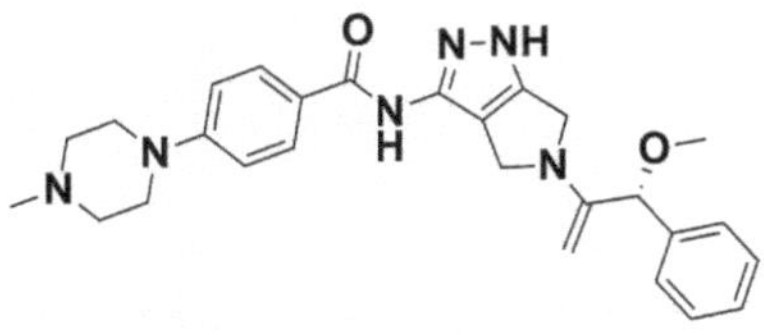

(14)

Fig. 6 Chemical structures of aurora kinase inhibitors hesperadin (**12**), VX-680 (**13**) and danusertib (**14**).

the inhibition of analogues of hesperadin in *L. major* (promastigote and amastigote forms) and D6 strain of *P. falciparum* (Patel et al., 2014). Between them, compound 12 showed the best inhibition results for all pathogens. Therefore, they synthetized analogues of 12, in order to derive structure-activity relationships (SAR). The changes in 12, were made in four regions (R1 to R4) highlighted in Fig. 6. These derivates exhibited good potencies against all the protozoan tested.

The hexokinase of *T. brucei* was the target of selective inhibitors based on sequence dissimilarity (less than 36%) with other organisms (Willson, Sanejouand, Perie, Hannaert, & Opperdoes, 2002). Selectivity could also be accomplished for inhibitors of the kinase GSK3 over CRK3 of *Leishmania*. There are differences on residues of these kinase's at the gatekeeper, P-loop and ribose-phosphate pockets. Targeting those residues could lead to LGSK3 selectivity over LCRK3 (Efstathiou et al., 2014).

6.3 Schistossomiasis

The kinome of *Shistosoma mansoni* was defined in 2011 by Andrade et al. (2011). A total of 252 ePKs were identified, corresponding to 1,9% of the proteome. There were protein kinases classified into all of the kinase groups. Only 10% of these proteins have been experimentally studied and shown to be essential for the parasite's survival. Some of the identified proteins were predicted to be inactive and 2% are still unclassified (Andrade et al., 2011). In 2015, Stroehlein and co-workers elucidated the kinome of *Shistosoma haematobium* (Stroehlein et al., 2015). They identified 269 protein kinases, of which 261 were ePKs and 8 were aPKs and representatives of all kinase groups were present. The comparison of the kinomes of *S. mansoni* and *S. haematobium* showed high overall sequence similarity (87–94%) and identity (82–92%). The transcription profile of kinase genes indicated that 214 protein kinases were transcribed in the three development stages of life of the parasite and the others were selectively transcribed in one or two of the development stages. Also, the authors could identify 216 kinases that were orthologs to kinases with lethal phenotypes in model organisms, leading to a target prioritization for future studies (Stroehlein et al., 2015). In 2018, Giuliani and co-workers built the kinome of *Shistisoma japonicum* (Giuliani et al., 2018). They identifyed 222 protein kinases that were classified into the nine kinome groups describer above. This number represents 1.7% of this organism's proteome. A target priorization was conducted and allowed for the identification 26 kinases targets and inhibitors that could be repurposed for shistosomiasis (Giuliani et al., 2018).

Giuliani et al. (2018) employed an extensive bioinformatics approach for drug repositioning to treat schistosomiasis. Their approach was based on protein domains shared by biological targets of FDA-approved drugs and *S. mansoni* proteins, relying on the principle that similar proteins can be targeted by similar ligands. Hence, all FDA-approved drugs deposited in ChEMBL database (Gaulton et al., 2017), their respective targets with or without co-cristallized ligands deposited in PDB (Berman et al., 2000), and *S. mansoni* proteome were collected. After carefully data curation process, filtering, and visual inspection, a final list of 97 FDA-approved drugs was selected for experimental testing against all lifestages of *S. mansoni* and *S. japonicum* adult worms. Further, a list of 528 potential targets in *S. mansoni* was obtained, in which 157 of them were protein kinases. Therefore, to fill the literature gap and dive deeper into schistosomal kinases, the authors also performed a thorough bioinformatics approach aiming to unveil *S. japonicum* kinome and identify potential biological targets within its kinases subset. To get this done, a diverse set of tools and databases were used: ArchSchema (Tamuri & Laskowski, 2010) and CATH (Sillitoe et al., 2015) database were queried for the multi-architecture search that guided the identification of shared domains between FDA-approved drug targets and schistosomals proteins; FunTree (Furnham et al., 2012) was used for the domains association; PDBSum (Laskowski, Jabłońska, Pravda, Vařeková, & Thornton, 2018) was used for the deeper analysis of protein-ligand interactions, which has filtered only those domains effectively involved in the binding process; WormBase Parasite (Howe et al., 2016) was largely employed to get *Schistosoma* proteomes, as well as GeneDB (Logan-Klumpler et al., 2012); InterPro was used to get the kinase consensus domains within each protein sequence composing the investigated kinomes; OrthoMCL (Li, Stoeckert, & Roos, 2003) and OrthoVenn (Wang, Coleman-Derr, Chen, & Gu, 2015; Xu et al., 2019) were used for clustering kinases, allowing their classification by comparative genomics.

Also, kinase inhibitors have been broadly studied in schistosoma drug discovery. BI 2536 (Fig. 7) has been assessed in clinical trials against advanced solid tumors (as head and neck cancer, breast adenocarcinomas, epithelial ovarian cancer, soft tissue sarcomas and melanoma), non-Hodgkin's lymphoma, non-small cell lung cancer, hormone-refractory prostate cancer and advanced pancreatic cancer. This compound became a first-in-class PlK1 (Polo-like kinase 1) inhibitor, inhibiting cells proliferation with nanomolar potency (Schöffski, 2009). The homolog protein of the target in *Schistosoma mansoni* (*Sm*PlK1) revealed 49% of sequence identity with the human protein (Long et al., 2010) and had conserved residues (Trp414, His538 and Lys540 in HsPlK1 *versus* Trp388, His512 and Lys514 *Sm*PlK1) important for the binding of phosphosubstrates (Elia et al., 2003; Long et al., 2010). In addition, the *Sm*PlK1 has a leucine residue (Leu104), equivalent to the human kinase Leu132, identified as decisive for the BI2536 interaction with *Hs*PlK1 (Kothe et al., 2007; Long et al., 2010). BI2536 was tested in a culture of adult *Schistosoma*, and induced gonad alteration of both genders with a dose of 100nM (Long et al., 2010). This phenotype highlights its potential for use in transmission blocking (Dissous & Grevelding, 2011). Thus, this inhibitor may be a starting point for designing new selective inhibitors for *S. mansoni*.

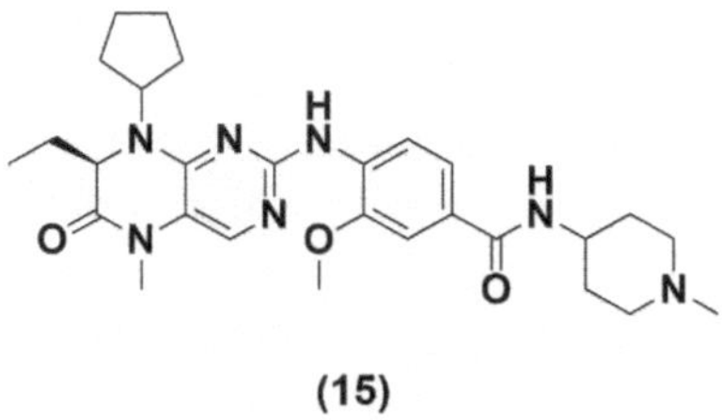

(15)

Fig. 7 Chemical structure of PlK1 inhibitor BI2536 (**15**).

7. Concluding remarks and future directions

Protein kinases can represent potential drug targets for developing new and effective drugs against NTDs. In this chapter, we show how bioinformatics and chemogenomics approaches can be used to define the kinome of causing organisms of infectious diseases and how it can be used to prioritize protein kinases as drug targets for NTDs. Moreover, key aspects of working with protein kinases and their inhibitors (single or multiple), considering the type of inhibitor, selectivity and the balance between efficacy and toxicity were discussed. Overall, we emphasize the possibility of repurposing drugs, to accelerate the drug discovery process for NTDs.

In summary, the progress in the field of kinase inhibitor drug discovery, including new tools for studing and handling chemical biology data through bioinformatics techniques and approaches to extract information hidden in the vast quantities of data generated in labs worldwide and to provide meaningful and useful biochemical insight that could be applied in drug discovery. Furthermore, the growing availability of 3D protein structures co-crystalysed with inhibitors can be useful for the identification an optimization of new inhibitors as drug candidates, and should support the rational design and development of novel drugs for NTDs.

Acknowledgments

The authors would like to thank Brazilian funding agencies, CNPq, CAPES, FAPESP and FAPEG for financial support and fellowships. FTMC and CHA are supported by FAPESP grants #2017/18611-7 and #2018/07777-4. and JVVBB is sponsored by FAPESP fellowship (#2019/21854-4). MNN and SMS are supported by a CAPES fellowship (code 001). CHA and FTMC are CNPq researcher fellows. NF is supported by the Medical Reserch Council (Grant No: MR/T000171/1). We are grateful to ChemAxon and OpenEye Inc., for providing academic license of their software for our projects. We are also grateful to DataCamp, for providing academic license for our students perform training courses.

References

Aaby, P., Fisker, A. B., Björkman, A., & Benn, C. S. (2020). WHO's rollout of malaria vaccine in Africa: Can safety questions be answered after only 24 months? *BMJ*, *368*, l6920. https://doi.org/10.1136/bmj.l6920.

Altschul, S. F., Gish, W., Miller, W., Myers, E. W., & Lipman, D. (1990). Basic local alignment search tool. *Journal of Molecular Biology*, *215*, 403–410.

Altschul, S. F., Madden, T. L., Schäffer, A. A., Zhang, J., Zhang, Z., Miller, W., et al. (1997). Gapped BLAST and PSI-BLAST: A new generation of protein database search programs. *Nucleic Acids Research*, *25*(17), 3389–3402. https://doi.org/10.1093/nar/25.17.3389.

Amata, E., Xi, H., Colmenarejo, G., Gonzalez-Diaz, R., Cordon-Obras, C., Berlanga, M., et al. (2016). Identification of "preferred" human kinase inhibitors for sleeping sickness lead discovery. Are some kinases better than others for inhibitor repurposing? *ACS Infectious Diseases*, *2*(3), 180–186. https://doi.org/10.1021/acsinfecdis.5b00136.

Andrade, L. F., Nahum, L. A., Avelar, L. G. A., Silva, L. L., Zerlotini, A., Ruiz, J. C., et al. (2011). Eukaryotic protein kinases (ePKs) of the helminth parasite *Schistosoma mansoni*. *BMC Genomics*, *12*(1), 1–19. 215. https://doi.org/10.1186/1471-2164-12-215.

Andrade, C. H., Neves, B. J., Melo-Filho, C. C., Rodrigues, J., Silva, D. C., Braga, R. C., et al. (2019). In silico chemogenomics drug repositioning strategies for neglected tropical diseases. *Current Medicinal Chemistry*, *26*, 1–22. https://doi.org/10.2174/0929867325666180309114824.

Ashburn, T. T., & Thor, K. B. (2004). Drug repositioning: Identifying and developing new uses for existing drugs. *Nature Reviews Drug Discovery*, *3*(8), 673–683. https://doi.org/10.1038/nrd1468.

Ashkenazy, H., Erez, E., Martz, E., Pupko, T., & Ben-Tal, N. (2010). ConSurf 2010: Calculating evolutionary conservation in sequence and structure of proteins and nucleic acids. *Nucleic Acids Research*, *38*, W529–W533. Web Server https://doi.org/10.1093/nar/gkq399.

Bain, J., McLauchlan, H., Elliott, M., & Cohen, P. (2003). The specificities of protein kinase inhibitors: An update. *The Biochemical Journal*, *371*(Pt. 1), 199–204. https://doi.org/10.1042/BJ20021535.

Bain, J., Plater, L., Elliott, M., Shpiro, N., Hastie, C. J., McLauchlan, H., et al. (2007). The selectivity of protein kinase inhibitors: A further update. *The Biochemical Journal*, *408*(3), 297–315. https://doi.org/10.1042/BJ20070797.

Baumert, C., & Hilgeroth, A. (2009). Recent advances in the development of multi-kinase inhibitors. *Anti-Cancer Agents in Medicinal Chemistry*, *9*(4), 415–436. https://doi.org/10.1016/j.jpowsour.2012.07.123.

Behera, R., Thomas, S. M., & Mensa-Wilmot, K. (2014). New chemical scaffolds for human african trypanosomiasis lead discovery from a screen of tyrosine kinase inhibitor drugs. *Antimicrobial Agents and Chemotherapy*, *58*(4), 2202–2210. https://doi.org/10.1128/AAC.01691-13.

Ben Salah, A., Ben Messaoud, N., Guedri, E., Zaatour, A., Ben Alaya, N., Bettaieb, J., et al. (2013). Topical Paromomycin with or without gentamicin for cutaneous Leishmaniasis. *New England Journal of Medicine*, *368*(6), 524–532. https://doi.org/10.1056/NEJMoa1202657.

Berman, J. (2015). Miltefosine, an FDA-approved drug for the 'orphan disease', leishmaniasis. *Expert Opinion on Orphan Drugs*, *3*(6), 727–735. https://doi.org/10.1517/21678707.2015.1039510.

Berman, H. M., Bhat, T. N., Bourne, P. E., Feng, Z., Gilliland, G., Weissig, H., et al. (2000). The protein data Bank and the challenge of structural genomics. *Nature Structural Biology*, 7(Suppl(november)), 957–959. https://doi.org/10.1038/80734.

Borba, J. V. B., Silva, A. C., Ramos, P. I. P., Grazzia, N., Miguel, D. C., Muratov, E. N., et al. (2019). Unveiling the kinomes of *Leishmania infantum* and *L. braziliensis* empowers the discovery of new kinase targets and antileishmanial compounds. *Computational and Structural Biotechnology Journal*, *17*, 352–361. https://doi.org/10.1016/j.csbj.2019.02.005.

Bossemeyer, D., Engh, R.a., Kinzel, V., Ponstingl, H., & Huber, R. (1993). Phosphotransferase and substrate binding mechanism of the cAMP-dependent protein kinase catalytic subunit from porcine heart as deduced from the 2.0 A structure of the complex with Mn2+ adenylyl imidodiphosphate and inhibitor peptide PKI(5-24). *The EMBO Journal*, *12*(3), 849–859.

Bracchi-Ricard, V., Barik, S., Delvecchio, C., Doerig, C., Chakrabarti, R., & Chakrabarti, D. (2000). PfPK6, a novel cyclin-dependent kinase/mitogen-activated protein kinase-related protein kinase from Plasmodium falciparum. *The Biochemical Journal*, *347*(Pt. 1), 255–263. https://doi.org/10.1042/0264-6021:3470255.

Cavalli, A., & Bolognesi, M. L. (2009). Neglected tropical diseases: Multi-target-directed ligands in the search for novel lead candidates against Trypanosoma and Leishmania. *Journal of Medicinal Chemistry*, *52*(23), 7339–7359. https://doi.org/10.1021/jm9004835.
Chen, W.-H., Strych, U., Hotez, P. J., & Bottazzi, M. E. (2020). The SARS-CoV-2 vaccine pipeline: An overview. *Current Tropical Medicine Reports*, 7, 1–4. https://doi.org/10.1007/s40475-020-00201-6.
Chen, Z., Han, L., Xu, M., Xu, Y., & Qian, X. (2013). Rationally designed multitarget anticancer agents. *Current Medicinal Chemistry*, *20*(13), 1694–1714. https://doi.org/10.217410.2174/0929867311320130009/0929867311320130011.
Cheng, H., & Force, T. (2010). Why do kinase inhibitors cause cardiotoxicity and what can be done about it? *Progress in Cardiovascular Diseases*, *53*(2), 114–120. https://doi.org/10.1016/j.pcad.2010.06.006.
Cheng, V. C. C., Lau, S. K. P., Woo, P. C. Y., & Yuen, K. Y. (2007). Severe acute respiratory syndrome coronavirus as an agent of emerging and reemerging infection. *Clinical Microbiology Reviews*, *20*(4), 660–694. https://doi.org/10.1128/CMR.00023-07.
Chenna, R., Sugawara, H., Koike, T., Lopez, R., Gibson, T. J., Higgins, D. G., et al. (2003). Multiple sequence alignment with the Clustal series of programs. *Nucleic Acids Research*, *31*(13), 3497–3500. https://doi.org/10.1093/nar/gkg500.
Cicenas, J., Kalyan, K., Sorokinas, A., Stankunas, E., Levy, J., Meskinyte, I., et al. (2015). Roscovitine in cancer and other diseases. *Annals of Translational Medicine*, *3*(10), 135. https://doi.org/10.3978/j.issn.2305-5839.2015.03.61.
Cohen, P. (2002). Protein kinases—The major drug targets of the twenty-first century? *Nature Reviews. Drug Discovery*, *1*(April), 309–315. https://doi.org/10.1038/nrd773.
Davies, S. P., Reddy, H., Caivano, M., & Cohen, P. (2000). Specificity and mechanism of action of some commonly used protein kinase inhibitors. *The Biochemical Journal*, *351*(Pt. 1), 95–105. https://doi.org/10.1517/13543776.11.3.405.
Deshmukh, K., Anamika, K., & Srinivasan, N. (2010). Evolution of domain combinations in protein kinases and its implications for functional diversity. *Progress in Biophysics and Molecular Biology*, *102*(1), 1–15. https://doi.org/10.1016/j.pbiomolbio.2009.12.009.
Dichiara, M., Marrazzo, A., Prezzavento, O., Collina, S., Rescifina, A., & Amata, E. (2017). Repurposing of human kinase inhibitors in neglected protozoan diseases. *ChemMedChem*, *12*(16), 1235–1253. https://doi.org/10.1002/cmdc.201700259.
Dissous, C., & Grevelding, C. G. (2011). Piggy-backing the concept of cancer drugs for schistosomiasis treatment: A tangible perspective? *Trends in Parasitology*, *27*(2), 59–66. https://doi.org/10.1016/j.pt.2010.09.001.
Doerig, C., Billker, O., Pratt, D., & Endicott, J. (2005). Protein kinases as targets for antimalarial intervention: Kinomics, structure-based design, transmission-blockade, and targeting host cell enzymes. *Biochimica et Biophysica Acta—Proteins & Proteomics*, *1754*(1–2), 132–150. https://doi.org/10.1016/j.bbapap.2005.08.027.
Doerig, C., & Grevelding, C. G. (2015). Targeting kinases in Plasmodium and Schistosoma: Same goals, different challenges. *Biochimica et Biophysica Acta—Proteins & Proteomics*, *1854*(10), 1637–1643. https://doi.org/10.1016/j.bbapap.2015.03.002.
Duffy, B. C., Zhu, L., Decornez, H., & Kitchen, D. B. (2012). Early phase drug discovery: Cheminformatics and computational techniques in identifying lead series. *Bioorganic and Medicinal Chemistry*, *20*(18), 5324–5342. https://doi.org/10.1016/j.bmc.2012.04.062.
Eddy, S. (1998). Profile hidden Markov models. *Bioinformatics*, *14*(9), 755–763. https://doi.org/btb114 [pii].
Edgar, R. C. (2004). MUSCLE: Multiple sequence alignment with high accuracy and high throughput. *Nucleic Acids Research*, *32*(5), 1792–1797. https://doi.org/10.1093/nar/gkh340.
Efstathiou, A., Gaboriaud-Kolar, N., Smirlis, D., Myrianthopoulos, V., Vougogiannopoulou, K., Alexandratos, A., et al. (2014). An inhibitor-driven study for enhancing the selectivity of indirubin derivatives towards leishmanial glycogen synthase Kinase-3 over leishmanial

cdc2-related protein kinase 3. *Parasites & Vectors*, 7(1), 234. https://doi.org/10.1186/1756-3305-7-234.

Ehrt, C., Brinkjost, T., & Koch, O. (2016). Impact of binding site comparisons on medicinal chemistry and rational molecular design. *Journal of Medicinal Chemistry*, *59*(9), 4121–4151. https://doi.org/10.1021/acs.jmedchem.6b00078.

Ekins, S., Mietchen, D., Coffee, M., Stratton, T., Freundlich, J., Freitas-Junior, L., et al. (2016). Open drug discovery for the Zika virus. *F1000Research*, *5*, 150. https://doi.org/10.12688/f1000research.8013.1.

Ekins, S., Mottin, M., Ramos, P. R. P. S., Sousa, B. K. P., Neves, B. J., Foil, D. H., et al. (2020). Déjà vu: Stimulating open drug discovery for SARS-CoV-2. *Drug Discovery Today*, *25*, 928–941. https://doi.org/10.1016/j.drudis.2020.03.019.

Elia, A. E. H., Rellos, P., Haire, L. F., Chao, J. W., Ivins, F. J., Hoepker, K., et al. (2003). The molecular basis for phosphodependent substrate targeting and regulation of Plks by the Polo-box domain. *Cell*, *115*(1), 83–95. https://doi.org/10.1016/S0092-8674(03)00725-6.

Fabbro, D., Cowan-Jacob, S. W., & Moebitz, H. (2015). Ten things you should know about protein kinases: IUPHAR review 14. *British Journal of Pharmacology*, *172*(11), 2675–2700. https://doi.org/10.1111/bph.13096.

Ferguson, F. M., & Gray, N. S. (2018). Kinase inhibitors: The road ahead. *Nature Reviews Drug Discovery*, *17*(5), 353–377. https://doi.org/10.1038/nrd.2018.21.

Finn, R. D., Coggill, P., Eberhardt, R. Y., Eddy, S. R., Mistry, J., Mitchell, A. L., et al. (2015). The Pfam protein families database: Towards a more sustainable future. *Nucleic Acids Research*, *44*(D1), D279–D285. https://doi.org/10.1093/nar/gkv1344.

Fu, L., Niu, B., Zhu, Z., Wu, S., & Li, W. (2012). CD-HIT: Accelerated for clustering the next-generation sequencing data. *Bioinformatics*, *28*(23), 3150–3152. https://doi.org/10.1093/bioinformatics/bts565.

Furnham, N., Sillitoe, I., Holliday, G. L., Cuff, A. L., Rahman, S. A., Laskowski, R. A., et al. (2012). FunTree: A resource for exploring the functional evolution of structurally defined enzyme superfamilies. *Nucleic Acids Research*, *40*(D1), 776–782. https://doi.org/10.1093/nar/gkr852.

Gaither, L. A. (2007). Chemogenomics approaches to novel target discovery. *Expert Review of Proteomics*, *4*(3), 411–419. https://doi.org/10.1586/14789450.4.3.411.

Garuti, L., Roberti, M., & Bottegoni, G. (2015). Multi-kinase inhibitors. *Current Medicinal Chemistry*, *22*(6), 695–712.

Gaulton, A., Hersey, A., Nowotka, M., Bento, A. P., Chambers, J., Mendez, D., et al. (2017). The ChEMBL database in 2017. *Nucleic Acids Research*, *45*(D1), D945–D954. https://doi.org/10.1093/nar/gkw1074.

Giuliani, S., Silva, A. C., Borba, J., B, V., Ramos, P. I. P., Paveley, R. A., et al. (2018). Computationally-guided drug repurposing enables the discovery of kinase targets and inhibitors as new schistosomicidal agents. *PLoS Computational Biology*, *14*(10), e1006515. https://doi.org/10.1371/journal.pcbi.1006515.

Glinca, S., & Klebe, G. (2013). Cavities tell more than sequences: Exploring functional relationships of proteases via binding pockets. *Journal of Chemical Information and Modeling*, *53*(8), 2082–2092. https://doi.org/10.1021/ci300550a.

Grant, J. A., Gallardo, M. A., & Pickup, B. T. (1996). A fast method of molecular shape comparison: A simple application of a Gaussian description of molecular shape. *Journal of Computational Chemistry*, *17*(14), 1653–1666. https://doi.org/10.1002/(SICI)1096-987X(19961115)17:14<1653::AID-JCC7>3.0.CO;2-K.

Hamill, L. C., Haslam, D., Abrahamsson, S., Hill, B., Dixon, R., Burgess, H., et al. (2019). People are neglected, not diseases: The relationship between disability and neglected tropical diseases. *Transactions of the Royal Society of Tropical Medicine and Hygiene*, *113*(12), 828–833. https://doi.org/10.1093/trstmh/trz036.

Hanks, S. K. (2003). Genomic analysis of the eukaryotic protein kinase superfamily: A perspective. *Genome Biology*, *4*(5), 111. https://doi.org/10.1186/gb-2003-4-5-111.

Hanks, S. K., & Hunter, T. (1995). Protein kinases 6. The eukaryotic protein kinase superfamily: Kinase (catalytic) domain structure and classification. *FASEB Journal : Official Publication of the Federation of American Societies for Experimental Biology*, *9*(8), 576–596. https://doi.org/7768349.

Hendlich, M., Rippmann, F., & Barnickel, G. (1997). LIGSITE: Automatic and efficient detection of potential small molecule-binding sites in proteins. *Journal of Molecular Graphics and Modelling*, *15*(6), 359–363. https://doi.org/10.1016/S1093-3263(98)00002-3.

Hernandez, H. W., Soeung, M., Zorn, K. M., Ashoura, N., Mottin, M., Andrade, C. H., et al. (2019). High throughput and computational repurposing for neglected diseases. *Pharmaceutical Research*, *36*(2), 27. https://doi.org/10.1007/s11095-018-2558-3.

Hieronymus, H., Lamb, J., Ross, K. N., Peng, X. P., Clement, C., Rodina, A., et al. (2006). Gene expression signature-based chemical genomic prediction identifies a novel class of HSP90 pathway modulators. *Cancer Cell*, *10*(4), 321–330. https://doi.org/10.1016/j.ccr.2006.09.005.

Hill, Z. B., Perera, G. K., Andrews, S. S., & Maly, D. J. (2012). Targeting diverse signaling interaction sites allows the rapid generation of bivalent kinase inhibitors. *ACS Chemical Biology*, 7(3), 487–495. https://doi.org/10.7243/2050-2273-1-6.Characterizing.

Holton, S., Merckx, A., Burgess, D., Doerig, C., Noble, M., & Endicott, J. (2003). Structures of P. falciparum PfPK5 test the CDK regulation paradigm and suggest mechanisms of small molecule inhibition. *Structure*, *11*(11), 1329–1337. https://doi.org/10.1016/j.str.2003.09.020.

Hopkins, A. L. (2008). Network pharmacology: The next paradigm in drug discovery. *Nature Chemical Biology*, *4*(11), 682–690. https://doi.org/10.1038/nchembio.118.

Houzé, S., Hoang, N.-T., Lozach, O., Bras, J., Meijer, L., Galons, H., et al. (2014). Several human cyclin-dependent kinase inhibitors, structurally related to Roscovitine, are new anti-malarial agents. *Molecules*, *19*(9), 15237–15257. https://doi.org/10.3390/molecules190915237.

Howe, K. L., Bolt, B. J., Cain, S., Chan, J., Chen, W. J., Davis, P., et al. (2016). WormBase 2016: Expanding to enable helminth genomic research. *Nucleic Acids Research*, *44*, D774–D780. https://doi.org/10.1093/nar/gkv1217.

Hui, R., El Bakkouri, M., & Sibley, L. D. (2015). Designing selective inhibitors for calcium-dependent protein kinases in apicomplexans. *Trends in Pharmacological Sciences*, *36*(7), 452–460. https://doi.org/10.1016/j.tips.2015.04.011.

Jetton, N., Rothberg, K. G., Hubbard, J. G., Wise, J., Li, Y., Ball, H. L., et al. (2009). The cell cycle as a therapeutic target against Trypanosoma brucei: Hesperadin inhibits Aurora kinase-1 and blocks mitotic progression in bloodstream forms. *Molecular Microbiology*, *72*(2), 442–458. https://doi.org/10.1111/j.1365-2958.2009.06657.x.

Johnson, S. A., & Hunter, T. (2005). Kinomics: Methods for deciphering the kinome. *Nature Methods*, *2*(1), 17–25. https://doi.org/10.1038/nmeth731.

Jurberg, A. D., & Brindley, P. J. (2015). Gene function in schistosomes: Recent advances toward a cure. *Frontiers in Genetics*, *6*(April), 144. https://doi.org/10.3389/fgene.2015.00144.

Kanev, G. K., de Graaf, C., de Esch, I. J. P., Leurs, R., Würdinger, T., Westerman, B. A., et al. (2019). The landscape of atypical and eukaryotic protein kinases. *Trends in Pharmacological Sciences*, *40*(11), 818–832. https://doi.org/10.1016/j.tips.2019.09.002.

Kannan, N., Taylor, S. S., Zhai, Y., Venter, J. C., & Manning, G. (2007). Structural and functional diversity of the microbial kinome. *PLoS Biology*, *5*(3), e17. https://doi.org/10.1371/journal.pbio.0050017.

Katara, P. (2013). Role of bioinformatics and pharmacogenomics in drug discovery and development process. *Network Modeling Analysis in Health Informatics and Bioinformatics*, *2*(4), 225–230. https://doi.org/10.1007/s13721-013-0039-5.

Katoh, K., & Standley, D. M. (2013). MAFFT multiple sequence alignment software version 7: Improvements in performance and usability. *Molecular Biology and Evolution*, *30*(4), 772–780. https://doi.org/10.1093/molbev/mst010.

Kaufmann, S. H. E. (2008). Paul Ehrlich: Founder of chemotherapy. *Nature Reviews. Drug Discovery*, 7(5), 373. https://doi.org/10.1038/nrd2582.

Kelly, A. M. (2009). Evidence-based practice: An introduction and overview. *Seminars in Roentgenology*, *44*(3), 131–139. https://doi.org/10.1053/j.ro.2009.03.010.

Klug, D. M., Gelb, M. H., & Pollastri, M. P. (2016). Repurposing strategies for tropical disease drug discovery. *Bioorganic & Medicinal Chemistry Letters*, *26*(11), 2569–2576. https://doi.org/10.1016/j.bmcl.2016.03.103.

Knighton, D., Zheng, J., Ten Eyck, L., Ashford, V., Xuong, N., Taylor, S., et al. (1991). Crystal structure of the catalytic subunit of cyclic adenosine monophosphate-dependent protein kinase. *Science*, *253*(5018), 407–414. https://doi.org/10.1126/science.1862342.

Kola, I., & Landis, J. (2004). Can the pharmaceutical industry reduce attrition rates? *Nature Reviews. Drug Discovery*, *3*(8), 711–715. https://doi.org/10.1038/nrd1470.

Konc, J., & Janežič, D. (2010). ProBiS algorithm for detection of structurally similar protein binding sites by local structural alignment. *Bioinformatics*, *26*(9), 1160–1168. https://doi.org/10.1093/bioinformatics/btq100.

Konc, J., Miller, B. T., Štular, T., Lešnik, S., Woodcock, H. L., Brooks, B. R., et al. (2015). ProBiS-CHARMMing: Web interface for prediction and optimization of ligands in protein binding sites. *Journal of Chemical Information and Modeling*, *55*(11), 2308–2314. https://doi.org/10.1021/acs.jcim.5b00534.

Kothe, M., Kohls, D., Low, S., Coli, R., Cheng, A. C., Jacques, S. L., et al. (2007). Structure of the catalytic domain of human polo-like kinase 1. *Biochemistry*, *46*(20), 5960–5971. https://doi.org/10.1021/bi602474j.

Kubinyi, H. (2006). Chemogenomics in drug discovery. In *Chemical genomics. Ernst Schering research foundation workshop* (pp. 1–19). Berlin Heidelberg: Springer. https://doi.org/10.1007/978-3-540-37635-4_1.

Kuhn, D., Weskamp, N., Hüllermeier, E., & Klebe, G. (2007). Functional classification of protein kinase binding sites using cavbase. *ChemMedChem*, *2*(10), 1432–1447. https://doi.org/10.1002/cmdc.200700075.

Kyu, H. H., Abate, D., Abate, K. H., Abay, S. M., Abbafati, C., Abbasi, N., et al. (2018). Global, regional, and national disability-adjusted life-years (DALYs) for 359 diseases and injuries and healthy life expectancy (HALE) for 195 countries and territories, 1990-2017: A systematic analysis for the global burden of disease study 2017. *The Lancet*, *392*(10159), 1859–1922. https://doi.org/10.1016/S0140-6736(18)32335-3.

Lai, C. C., Shih, T. P., Ko, W. C., Tang, H. J., & Hsueh, P. R. (2020). Severe acute respiratory syndrome coronavirus 2 (SARS-CoV-2) and coronavirus disease-2019 (COVID-19): The epidemic and the challenges. *International Journal of Antimicrobial Agents*, *55*(3), 105924. https://doi.org/10.1016/j.ijantimicag.2020.105924.

Laskowski, R. A., Jabłońska, J., Pravda, L., Vařeková, R. S., & Thornton, J. M. (2018). PDBsum: Structural summaries of PDB entries. *Protein Science*, *27*(1), 129–134. https://doi.org/10.1002/pro.3289.

Li, W., & Godzik, A. (2006). Cd-hit: A fast program for clustering and comparing large sets of protein or nucleotide sequences. *Bioinformatics*, *22*(13), 1658–1659. https://doi.org/10.1093/bioinformatics/btl158.

Li, J. L., Robson, K. J., Chen, J. L., Targett, G.a., & Baker, D.a. (1996). Pfmrk, a MO15-related protein kinase from Plasmodium falciparum. Gene cloning, sequence, stage-specific expression and chromosome localization. *European Journal of Biochemistry*, *241*(3), 805–813.

Li, L., Stoeckert, C. J., Jr., & Roos, D. S. (2003). OrthoMCL: Identification of ortholog groups for eukaryotic genomes. *Genome Research*, *13*, 2178–2189. https://doi.org/10.1101/gr.1224503.candidates.

Li, Z., Umeyama, T., & Wang, C. C. (2009). The Aurora kinase in Trypanosoma brucei plays distinctive roles in metaphase-anaphase transition and cytokinetic initiation. *PLoS Pathogens*, *5*(9), e1000575. https://doi.org/10.1371/journal.ppat.1000575.

Lima, M. N. N., Cassiano, G. C., Tomaz, K. C. P., Silva, A. C., Sousa, B. K. P., Ferreira, L. T., et al. (2019). Integrative multi-kinase approach for the identification of potent antiplasmodial hits. *Frontiers in Chemistry*, 7, 773. https://doi.org/10.3389/fchem.2019.00773.

Liu, Q., Sabnis, Y., Zhao, Z., Buhrlage, S. J., Jones, L. H., & Gray, N. S. (2013). Developing irreversible inhibitors of the protein kinase cysteinome. *Chemical Biology*, *20*(2), 146–159. https://doi.org/10.3851/IMP2701.Changes.

Liu, X., Zhu, F., Ma, X. H., Shi, Z., Yang, S. Y., Wei, Y. Q., et al. (2013). Predicting targeted polypharmacology for drug repositioning and multi- target drug discovery. *Current Medicinal Chemistry*, *20*(13), 1646–1661. https://doi.org/10.2174/0929867311320130005.

Logan-Klumpler, F. J., De Silva, N., Boehme, U., Rogers, M. B., Velarde, G., McQuillan, J. A., et al. (2012). GeneDB—An annotation database for pathogens. *Nucleic Acids Research*, *40*, 98–108. https://doi.org/10.1093/nar/gkr1032.

Long, T., Cailliau, K., Beckmann, S., Browaeys, E., Trolet, J., Grevelding, C. G., et al. (2010). Schistosoma mansoni Polo-like kinase 1: A mitotic kinase with key functions in parasite reproduction. *International Journal for Parasitology*, *40*(9), 1075–1086. https://doi.org/10.1016/j.ijpara.2010.03.002.

Loytynoja, A., & Goldman, N. (2005). An algorithm for progressive multiple alignment of sequences with insertions. *Proceedings of the National Academy of Sciences of the United States of America*, *102*(30), 10557–10562. https://doi.org/10.1073/pnas.0409137102.

Lucet, I. S., Tobin, A., Drewry, D., Wilks, A. F., & Doerig, C. (2012). *Plasmodium* kinases as targets for new-generation antimalarials. *Future Medicinal Chemistry*, *4*(18), 2295–2310. https://doi.org/10.4155/fmc.12.183.

Luscombe, N. M., Greenbaum, D., & Gerstein, M. (2001). What is bioinformatics? A proposed definition and overview of the field. *Methods of Information in Medicine*, *40*(4), 346–358. http://www.ncbi.nlm.nih.gov/pubmed/11552348.

Macarron, R., Banks, M. N., Bojanic, D., Burns, D. J., Cirovic, D.a., Garyantes, T., et al. (2011). Impact of high-throughput screening in biomedical research. *Nature Reviews. Drug Discovery*, *10*(3), 188–195. https://doi.org/10.1038/nrd3368.

Manning, G., Whyte, D. B., Martinez, R., Hunter, T., & Sudarsanam, S. (2002). The protein kinase complement of the human genome. *Science*, *298*(5600), 1912–1934. https://doi.org/10.1126/science.1075762.

Masch, A., & Kunick, C. (2015). Selective inhibitors of Plasmodium falciparum glycogen spi1;synthase-3 (PfGSK-3): New antimalarial agents? *Biochimica et Biophysica Acta—Proteins & Proteomics*, *1854*(10), 1644–1649. https://doi.org/10.1016/j.bbapap.2015.03.013.

Melnikova, I., & Golden, J. (2004). Targeting protein kinases. *Nature Reviews Drug Discovery*, *3*, 993–994. https://doi.org/10.1038/nrd1582.

Mitra, A. K., & Mawson, A. R. (2017). Neglected tropical diseases: Epidemiology and global burden. *Tropical Medicine and Infectious Disease*, *2*(3), 36. https://doi.org/10.3390/tropicalmed2030036.

Moffat, J. G., Vincent, F., Lee, J. A., Eder, J., & Prunotto, M. (2017). Opportunities and challenges in phenotypic drug discovery: An industry perspective. *Nature Reviews Drug Discovery*, *16*(8), 531–543. https://doi.org/10.1038/nrd.2017.111.

Moran, M., Chapman, N., Lisette, A.-O., Chowdhary, V., Doubell, A., Whittall, C., et al. (2012). *Neglected disease research and development : The Ebola effect.* Policy Cure Research. The G-FINDER project: https://www.policycuresresearch.org/g-finder#:~:text=The

%20G%2DFINDER%20project%20tracks,address%20priority%20global%20health%20challenges.&text=The%20data%20collected%20in%20this,%2C%20donors%2C%20researchers%20and%20industry.

Morphy, R., & Rankovic, Z. (2005). Designed multiple ligands. An emerging drug discovery paradigm. *Journal of Medicinal Chemistry, 48*(21), 6523–6543. https://doi.org/10.1021/jm058225d.

Moy, B., Kirkpatrick, P., Kar, S., & Goss, P. (2007). Lapatinib. *Nature Reviews Drug Discovery, 6*(6), 431–432. https://doi.org/10.1038/nrd2332.

Muegge, I., & Mukherjee, P. (2016). An overview of molecular fingerprint similarity search in virtual screening. *Expert Opinion on Drug Discovery, 11*(2), 137–148. https://doi.org/10.1517/17460441.2016.1117070.

Müller, S., Chaikuad, A., Gray, N. S., & Knapp, S. (2015). The ins and outs of selective kinase inhibitor development. *Nature Chemical Biology, 11*(11), 818–821. https://doi.org/10.1038/nchembio.1938. www.nature.com/naturechemicalbiology.

Naula, C., Marilyn, P., & Mottram, J. C. (2005). Protein kinases as drug targets in trypanosomes and Leishmania. *Biochimica et Biophysica Acta, 1754*(1–2), 151–159. https://doi.org/10.1016/j.micinf.2011.07.011.Innate.

Neves, B. J., Braga, R. C., Bezerra, J. C. B., Cravo, P. V. L., & Andrade, C. H. (2015). In silico repositioning-chemogenomics strategy identifies new drugs with potential activity against multiple life stages of Schistosoma mansoni. *PLoS Neglected Tropical Diseases, 9*(1), e3435. https://doi.org/10.1371/journal.pntd.0003435.

Notredame, C., Higgins, D. G., & Heringa, J. (2000). T-coffee: A novel method for fast and accurate multiple sequence alignment. *Journal of Molecular Biology, 302*(1), 205–217. https://doi.org/10.1006/jmbi.2000.4042.

Ochiana, S. O., Pandarinath, V., Wang, Z., Kapoor, R., Ondrechen, M. J., Ruben, L., et al. (2013). The human Aurora kinase inhibitor danusertib is a lead compound for anti-trypanosomal drug discovery via target repurposing. *European Journal of Medicinal Chemistry, 62*(6), 777–784. https://doi.org/10.1016/j.ejmech.2012.07.038.

Ouzounis, C. A. (2012). Rise and demise of bioinformatics? Promise and progress. *PLoS Computational Biology, 8*(4), e1002487. https://doi.org/10.1371/journal.pcbi.1002487.

Paolini, G. V., Shapland, R. H. B., van Hoorn, W. P., Mason, J. S., & Hopkins, A. L. (2006). Global mapping of pharmacological space. *Nature Biotechnology, 24*(7), 805–815. https://doi.org/10.1038/nbt1228.

Parsons, M., Worthey, E.a., Ward, P. N., & Mottram, J. C. (2005). Comparative analysis of the kinomes of three pathogenic trypanosomatids: Leishmania major, Trypanosoma brucei and Trypanosoma cruzi. *BMC Genomics, 6*, 127. https://doi.org/10.1186/1471-2164-6-127.

Patel, G., Karver, C. E., Behera, R., Guyett, P. J., Sullenberger, C., Edwards, P., et al. (2013). Kinase scaffold repurposing for neglected disease drug discovery: Discovery of an efficacious, lapatinib-derived lead compound for trypanosomiasis. *Journal of Medicinal Chemistry, 56*(10), 3820–3832. https://doi.org/10.1021/jm400349k.

Patel, G., Roncal, N. E., Lee, P. J., Leed, S. E., Erath, J., Rodriguez, A., et al. (2014). Repurposing human Aurora kinase inhibitors as leads for anti-protozoan drug discovery. *MedChemComm, 5*(5), 655–658. https://doi.org/10.1039/C4MD00045E.

Paul, M. L. S., Kaur, A., Geete, A., & Sobhia, M. E. (2014). Essential gene identification and drug target prioritization in Leishmania species. *Molecular BioSystems, 10*(5), 1184–1195. https://doi.org/10.1039/c3mb70440h.

Peters, J.-U. (2013). Polypharmacology—Foe or friend? *Journal of Medicinal Chemistry, 56*, 8955–8971. https://doi.org/10.1021/jm400856t.

Pollastri, M. P., & Campbell, R. K. (2011). Target repurposing for neglected diseases. *Future Medicinal Chemistry, 3*(10), 1307–1315. https://doi.org/10.4155/fmc.11.92.

Pushpakom, S., Iorio, F., Eyers, P. A., Escott, K. J., Hopper, S., Wells, A., et al. (2019). Drug repurposing: Progress, challenges and recommendations. *Nature Reviews Drug Discovery*, *18*(1), 41–58. https://doi.org/10.1038/nrd.2018.168.

Reininger, L., Wilkes, J. M., Bourgade, H., Miranda-Saavedra, D., & Doerig, C. (2011). An essential Aurora-related kinase transiently associates with spindle pole bodies during Plasmodium falciparum erythrocytic schizogony. *Molecular Microbiology*, *79*(1), 205–221. https://doi.org/10.1111/j.1365-2958.2010.07442.x.

Robertson, S. A., & Renslo, A. R. (2011). Drug discovery for neglected tropical diseases at the Sandler center. *Future Medicinal Chemistry*, *3*(10), 1279–1288. https://doi.org/10.4155/fmc.11.85.

Rognan, D. (2007). Chemogenomic approaches to rational drug design. *British Journal of Pharmacology*, *152*(1), 38–52. https://doi.org/10.1038/sj.bjp.0707307.

Ross-Macdonald, P. B., Graeser, R., Kappes, B., Franklin, R., & Williamson, D. H. (1994). Isolation and expression of a gene specifying a cdc2-like protein kinase from the human malaria parasite Plasmodium falciparum. *European Journal of Biochemistry*, *220*(3), 693–701.

Rotella, D. P. (2012). Recent results in protein kinase inhibition for tropical diseases. *Bioorganic & Medicinal Chemistry Letters*, *22*(22), 6788–6793. https://doi.org/10.1016/j.bmcl.2012.09.044.

Rusnak, D. W., Lackey, K., Affleck, K., Wood, E. R., Alligood, K., Rhodes, N., et al. (2001). The effects of the novel, reversible epidermal growth factor receptor/ErbB-2 tyrosine kinase inhibitor, GW2016, on the growth of human normal and tumor-derived cell lines in vitro and in vivo. *Molecular Cancer Therapeutics*, *1*(2), 85–94.

Schöffski, P. (2009). Polo-like kinase (PLK) inhibitors in preclinical and early clinical development in oncology. *The Oncologist*, *14*(6), 559–570. https://doi.org/10.1634/theoncologist.2009-0010.

Schuster-Böckler, B., Schultz, J., & Rahmann, S. (2004). HMM Logos for visualization of protein families. *BMC Bioinformatics*, *5*, 7. https://doi.org/10.1186/1471-2105-5-7.

Shah, N. P., Nicoll, J. M., Nagar, B., Gorre, M. E., Paquette, R. L., Kuriyan, J., et al. (2002). Multiple BCR-ABL kinase domain mutations confer polyclonal resistance to the tyrosine kinase inhibitor imatinib (STI571) in chronic phase and blast crisis chronic myeloid leukemia. *Cancer Cell*, *2*(2), 117–125. https://doi.org/10.1016/S1535-6108(02)00096-X.

Shchemelinin, I., Šefc, L., & Nečas, E. (2006). Protein kinases, their function and implication in cancer and other diseases. *Folia Biologica*, *52*(3), 81–101. https://doi.org/CAN0010.

Sibley, C. H. (2013). Tracking artemisinin resistance in Plasmodium falciparum. *The Lancet Infectious Diseases*, *13*(12), 999–1000. https://doi.org/10.1016/S1473-3099(13)70260-3.

Sillitoe, I., Lewis, T. E., Cuff, A., Das, S., Ashford, P., Dawson, N. L., et al. (2015). CATH: Comprehensive structural and functional annotations for genome sequences. *Nucleic Acids Research*, *43*(D1), D376–D381. https://doi.org/10.1093/nar/gku947.

Silva-Moraes, V., Shollenberger, L. M., Siqueira, L. M. V., Castro-Borges, W., Harn, D. A., Grenfell, R. F. Q. E., et al. (2019). Diagnosis of Schistosoma mansoni infections: What are the choices in Brazilian low-endemic areas? *Memorias Do Instituto Oswaldo Cruz*, *114*, e180478. https://doi.org/10.1590/0074-02760180478.

Siman-Tov, M. M., Ivens, A. C., & Jaffe, C. L. (2001). Identification and cloning of Lmairk, a member of the Aurora/Ipl1p protein kinase family, from the human protozoan parasite Leishmania. *Biochimica et Biophysica Acta*, *1519*(3), 241–245. http://www.ncbi.nlm.nih.gov/pubmed/11418192.

Sleno, L., & Emili, A. (2008). Proteomic methods for drug target discovery. *Current Opinion in Chemical Biology*, *12*(1), 46–54. https://doi.org/10.1016/j.cbpa.2008.01.022.

Stenberg, K. A. E., Riikonen, P. T., & Vihinen, M. (2000). KinMutBase, a database of human disease-causing protein kinase mutations. *Nucleic Acids Research*, *28*(1), 369–371. https://doi.org/10.1093/nar/27.1.362.

Stroehlein, A. J., Young, N. D., Jex, A. R., Sternberg, P. W., Tan, P., Boag, P. R., et al. (2015). Defining the Schistosoma haematobium kinome enables the prediction of essential kinases as anti-schistosome drug targets. *Scientific Reports*, *5*(August), 17759. https://doi.org/10.1038/srep17759.

Sundar, S., Chakravarty, J., Agarwal, D., Rai, M., & Murray, H. W. (2010). Single-dose liposomal amphotericin B for visceral leishmaniasis in India. *New England Journal of Medicine*, *362*(6), 504–512. https://doi.org/10.1056/NEJMoa0903627.

Tamuri, A. U., & Laskowski, R. A. (2010). ArchSchema: A tool for interactive graphing of related Pfam domain architectures. *Bioinformatics*, *26*(9), 1260–1261. https://doi.org/10.1093/bioinformatics/btq119.

Taylor, C. M., Martin, J., Rao, R. U., Powell, K., Abubucker, S., & Mitreva, M. (2013). Using existing drugs as leads for broad spectrum anthelmintics targeting protein kinases. *PLoS Pathogens*, *9*(2), e1003149. https://doi.org/10.1371/journal.ppat.1003149.

Traxler, P., & Furet, P. (1999). Strategies toward the design of novel and selective protein tyrosine kinase inhibitors. *Pharmacology & Therapeutics*, *82*(98), 195–206. https://doi.org/10.1016/S0163-7258(98)00044-8.

Ubersax, J. A., & Ferrell, J. E., Jr. (2007). Mechanisms of specificity in protein phosphorylation. *Nature Reviews Molecular Cell Biology*, *8*(7), 530–541. https://doi.org/10.1038/nrm2203.

Uniting to Combat Neglected Tropical Diseases. (2012). *London declaration on neglected tropical diseases*. http://unitingtocombatntds.org/resource/london-declaration.

Viayna, E., Sola, I., Di Pietro, O., & Muñoz-Torrero, D. (2013). Human disease and drug pharmacology, complex as real life. *Current Medicinal Chemistry*, *20*(13), 1623–1634. https://doi.org/10.2174/0929867311320130002.

Vieth, M., Sutherland, J. J., Robertson, D. H., Campbell, R. M., Vieth, M., Sutherland, J. J., et al. (2005). Kinomics: Characterizing the therapeutically validated kinase space. *Drug Discovery Today*, *10*(12), 839–846. https://doi.org/10.1016/S1359-6446(05)03477-X.

Wang, Y., Coleman-Derr, D., Chen, G., & Gu, Y. Q. (2015). OrthoVenn: A web server for genome wide comparison and annotation of orthologous clusters across multiple species. *Nucleic Acids Research*, *43*(W1), W78–W84. https://doi.org/10.1093/nar/gkv487.

Ward, P., Equinet, L., Packer, J., & Doerig, C. (2004). Protein kinases of the human malaria parasite Plasmodium falciparum: The kinome of a divergent eukaryote. *BMC Genomics*, *5*, 79. https://doi.org/10.1186/1471-2164-5-79.

Willson, M., Sanejouand, Y. H., Perie, J., Hannaert, V., & Opperdoes, F. (2002). Sequencing, modeling, and selective inhibition of Trypanosoma brucei hexokinase. *Chemistry and Biology*, *9*(7), 839–847. https://doi.org/10.1016/S1074-5521(02)00169-2.

Woodard, C. L., Li, Z., Kathcart, A. K., Terrell, J., Gerena, L., Lopez-Sanchez, M., et al. (2003). Oxindole-based compounds are selective inhibitors of Plasmodium falciparum cyclin dependent protein kinases. *Journal of Medicinal Chemistry*, *46*(18), 3877–3882. https://doi.org/10.1021/jm0300983.

Woodring, J. L., Patel, G., Erath, J., Behera, R., Lee, P. J., Leed, S. E., et al. (2015). Evaluation of aromatic 6-substituted thienopyrimidines as scaffolds against parasites that cause trypanosomiasis, leishmaniasis, and malaria. *MedChemComm*, *6*(2), 339–346. https://doi.org/10.1039/C4MD00441H.

World Health Organization. (2004). *Intensified control of neglected diseases*. Report of an international workshop (p. 57). WHO Communicable Diseases. December 2003 http://apps.who.int/iris/bitstream/10665/68529/1/WHO_CDS_CPE_CEE_2004.45.pdf.

World Health Organization. (2007). *Report of the first global partners' meeting on neglected tropical diseases: A turning point*. Geneva: World Health Organization. WHO/CDS/NTD/2007.4.

World Health Organization. (2012). *Accelerating work to overcome the global impact of neglected tropical diseases: A roadmap for implementation*. World Health Organization.

World Health Organization. (2019a). Global updete on implementation of preventive chemotherapy against neglected tropical diseases. *Weekly Epidemiological Record*, *94*(38), 425–438.

World Health Organization. (2019b). WHO. In *WHO Regional Office for Africa*.

World Health Organization. (2020). *Neglected tropical diseases*. https://www.who.int/neglected_diseases/diseases/en/.

Wu, P., Nielsen, T. E., & Clausen, M. H. (2015). FDA-approved small-molecule kinase inhibitors. *Trends in Pharmacological Sciences*, *36*(7), 422–439. https://doi.org/10.1016/j.tips.2015.04.005.

Xu, L., Dong, Z., Fang, L., Luo, Y., Wei, Z., Guo, H., et al. (2019). OrthoVenn2: A web server for whole-genome comparison and annotation of orthologous clusters across multiple species. *Nucleic Acids Research*, *47*(W1), W52–W58. https://doi.org/10.1093/nar/gkz333.

You, H., McManus, D. P., & Gobert, G. N. (2015). Current and prospective chemotherapy options for schistosomiasis. *Expert Opinion on Orphan Drugs*, *3*(2), 195–205. https://doi.org/10.1517/21678707.2015.998998.

Zheng, J., Knighton, D. R., ten Eyck, L. F., Karlsson, R., Xuong, N., Taylor, S. S., et al. (1993). Crystal structure of the catalytic subunit of cAMP-dependent protein kinase complexed with MgATP and peptide inhibitor. *Biochemistry*, *32*(9), 2154–2161.

Zimmermann, G. R., Lehár, J., & Keith, C. T. (2007). Multi-target therapeutics: When the whole is greater than the sum of the parts. *Drug Discovery Today*, *12*(1–2), 34–42. https://doi.org/10.1016/j.drudis.2006.11.008.

Zuccotto, F., Ardini, E., Casale, E., & Angiolini, M. (2010). Through the "gatekeeper door": Exploiting the active kinase conformation. *Journal of Medicinal Chemistry*, *53*(7), 2681–2694. https://doi.org/10.1021/jm901443h.

CHAPTER SEVEN

Targeting malaria protein kinases

G.C. Cassiano[a,b,*], T.A. Tavella[b], M.N. Nascimento[c], D.A. Rodrigues[b], P.V.L. Cravo[a], Carolina Horta Andrade[b,c], and Fabio Trindade Maranhão Costa[b,*]

[a]Global Health and Tropical Medicine (GHTM), Instituto de Higiene e Medicina Tropical, Universidade Nova de Lisboa, Lisbon, Portugal
[b]Laboratory of Tropical Diseases—Prof. Luiz Jacintho da Silva, Department of Genetics, Evolution and Bioagents, University of Campinas, Campinas, SP, Brazil
[c]LabMol—Laboratory for Molecular Modeling and Drug Design, Faculty of Pharmacy, Federal University of Goiás, Goiânia, GO, Brazil
*Corresponding authors: e-mail address: gcapatti@hotmail.com; fabiotmc72@gmail.com

Contents

Abstract

Malaria is one of the most impacting public health problems in tropical and subtropical areas of the globe, with approximately 200 million cases worldwide annually. In the absence of an effective vaccine, rapid treatment is vital for effective malaria control. However, parasite resistance to currently available drugs underscores the urgent need for identifying new antimalarial therapies with new mechanisms of action. Among potential drug targets for developing new antimalarial candidates, protein kinases are attractive. These enzymes catalyze the phosphorylation of several proteins, thereby regulating a variety of cellular processes and playing crucial roles in the development of all stages of the malaria parasite life cycle. Moreover, the large phylogenetic distance between *Plasmodium* species and its human host is reflected in marked differences

Advances in Protein Chemistry and Structural Biology, Volume 124
ISSN 1876-1623
https://doi.org/10.1016/bs.apcsb.2020.10.004

in structure and function of malaria protein kinases between the homologs of both species, indicating that selectivity can be attained. In this review, we describe the functions of the different types of *Plasmodium* kinases and highlight the main recent advances in the discovery of kinase inhibitors as potential new antimalarial drug candidates.

1. Introduction

Malaria is a parasitic disease characterized by high intermittent fever, caused by protozoa of the genus *Plasmodium*, and transmitted to humans through the bite of female *Anopheles* mosquitoes. Among the different *Plasmodium* species, five are known etiologic agents of human malaria: *P. falciparum*, *P. vivax*, *P. malariae*, *P. ovale* and *P. knowlesi*; the first two being responsible for the majority of cases. Almost all cases of complications and deaths related to malaria are caused by *P. falciparum*, the species that bears the highest burden in Africa, especially in children under 5-years-old (WHO, 2019). *P. vivax* is the most prevalent species outside Africa where around 2.5 billion people worldwide live in areas where *P. vivax* malaria is transmitted, with a large proportion living in countries in Southeast Asia, Eastern Mediterranean and Latin America. This causes a significant burden on the socioeconomic development of these regions (Anstey, Douglas, Poespoprodjo, & Price, 2012). The scenario becomes even more complex taking into account the worrying increase in clinical complications associated with *P. vivax* infections that have been reported, such as acute renal failure, anemia, metabolic acidosis, and death (Dos-Santos et al., 2014; Lacerda et al., 2012). Currently, the World Health Organization (WHO) estimates that there will be between 1400 and 14,900 deaths associated with *P. vivax* outside Africa, and so the "benign" status that has traditionally been associated with infections related to *P. vivax* needs to be reconsidered (Global Malaria Progamme, 2015).

At present, approximately 228 million cases of malaria worldwide result in 405,000 deaths yearly. Nevertheless, there has been significant progress toward control, as the incidence of malaria was reduced by 37% between 2000 and 2015, with a 48% reduction in the number of deaths during the same period (WHO, 2019). Achieving these results was only possible with the improvement of the main control strategies that are based on vector control, such as the use of mosquito nets impregnated with insecticide and indoor residual spraying, added to rapid diagnosis and early treatment of patients with the highly efficacious Artemisinin-Based Combination therapies (ACTs). Apart from reducing the severity of infection, rapid treatment prevents the appearance of gametocytes, the infecting form for anophelines in the peripheral circulation, thereby interrupting transmission (Dunyo et al., 2006). However, among the greatest challenges in combating malaria is the recurrent evolution and spread of parasite resistance to antimalarial drugs. Currently, this is even more alarming due to early reports of *P. falciparum* resistance to ACTs (manifested as delayed clearance of parasitemia after treatment) in countries in the Greater Mekong subregion (Cambodia, Laos, Myanmar, Thailand and Vietnam) (Sibley, 2013; WHO, 2019). Thus, the development of new antimalarial drugs is extremely urgent, especially ones that have new mechanisms of action that may overcome cross-resistance. Genome sequencing and respective *in silico* or experimental protein characterization of the main species that cause human malaria, including *P. falciparum* and *P. vivax*, has facilitated the search for new target-based drug discovery strategies. This type of strategy has the advantage of using knowledge about specific targets and their mechanism of action, leading to the potential identification of new chemical entities with novel modes of action. Protein kinases are a class of proteins that have been largely studied as potential targets for new antimalarial drugs. These enzymes mediate almost all cellular processes and are essential in the development of all stages of *Plasmodium* spp. (Solyakov et al., 2011; Zhang, Chavchich, & Waters, 2012), rendering them as promising targets for the development of prophylactic, curative, and transmission-blocking drugs. In this review, we discuss the potential of protein kinases as drug targets and summarize efforts to design kinase inhibitors that could be used for the development of novel antimalarials.

2. Antimalarial drug resistance

To date, drug resistance has already been documented for three of the five species that cause malaria in humans: *P. falciparum*, *P. vivax* and *P. malariae* (Sinha, Medhi, & Sehgal, 2014). This recurring problem has been a threat to disease control policies for decades. In the early 1950s, chloroquine, an inexpensive low-toxicity parasiticide with a rapid action was the best treatment against all species of human malaria. However, by the end of that decade, chloroquine resistance in *P. falciparum* evolved both in Thailand and South America (Wernsdorfer & Payne, 1991). In the following years and by the mid-1980s resistance had spread to vast areas of the African continent, contributing significantly to the increase in the number of cases and deaths during that period (Carter & Mendis, 2002). At present, chloroquine remains effective against *P. falciparum* only in some areas of Central America (Morton et al., 2016). Furthermore, *P. falciparum* resistance against several other antimalarial drugs, such as pyrimethamine (Abdul-Ghani, Farag, & Allam, 2013), mefloquine (Mockenhaupt, 1995) and amodiaquine (Holmgren et al., 2006), evolved even faster, compared to chloroquine. For example, sulfadoxine-pyrimethamine resistance was observed in the same year as its introduction (Talisuna, Bloland, & D'Alessandro, 2004). Hence, in order to minimize the emergence of resistant parasites, the WHO recommended the use of ACTs since 2006 (Lin, Juliano, & Wongsrichanalai, 2010). However, as early as 2008, the efficacy of ACTs against *P. falciparum* in the Greater Mekong region started to decline, as reflected by delayed parasite clearance and sporadic treatment failure, jeopardizing chemotherapy as one of the pillars of malaria control (WHO, 2019). If resistance to ACTs spreads globally, new malaria epidemics are likely to incur in striking increases in morbidity and mortality.

The first cases of *P. vivax* resistance against chloroquine were reported only in 1989 in Papua New Guinea, nearly 30 years after the first cases of resistance observed for *P. falciparum* (Rieckmann, Davis, & Hutton, 1989). This delay in the detection of resistance of *P. vivax* is probably in part due to the lack of robust methodologies to detect and monitor the sensitivity of *P. vivax* to antimalarial drugs. Moreover, the fact that mature gametocytes are produced before the onset of clinical symptoms allows the transmission of the parasite before exposure to drugs, thus reducing the levels of selective pressure and consequently the development of resistance to antimalarial drugs (Popovici & Ménard, 2015). At present, according to a survey by the WHO, although chloroquine is the first-line treatment against *P. vivax* infections in most endemic areas, evidence of decreased efficacy of the drug has already been detected in 21 countries (Global Malaria Progamme, 2015). According to a meta-analysis of the last 5 years, clear signs of *P. vivax* resistance to chloroquine occurred in Thailand, Cambodia, Ethiopia and South America (Price et al., 2014). In Manaus, Brazil, a failure rate of 10.1% in 166 patients with uncomplicated *P. vivax* malaria treated using chloroquine has been reported (de Santana Filho et al., 2007).

The vast majority of endemic countries recommend a combination of chloroquine and primaquine to treat *P. vivax* infections. This is because only primaquine is able to eliminate the hypnozoites of this species from the liver. In addition, as primaquine is also active against the blood stages of the parasite, it is possible that *P. vivax* resistance to chloroquine will spread more slowly (or be more difficult to detect), as primaquine might eliminate chloroquine-resistant strains (Baird & Hoffman, 2004). However, the use of primaquine is constrained by its oxidizing nature, which can induce severe hemolytic anemia, especially in patients with glucose 6-phosphate dehydrogenase (G6PD) enzyme deficiency, a common condition in malaria-endemic areas (Baird, 2015). In addition, patients often do not adhere to the recommended treatment, a regimen for up to 14 days, which can favor the selection of resistant strains (Baird & Hoffman, 2004). A new drug named Tafenoquine, administered in a single dose, has been approved by Food and Drug Administration and has emerged as a possible alternative to primaquine. Nevertheless, treatment of G6PD deficient patients is still not recommended (Price & Nosten, 2014). Hence, there is an urgent need of chemical compounds that differ from the antimalarial therapies currently employed.

3. Protein kinases

Based on the pressing need to discover new antimalarial drugs, large scale phenotypic assays (High-Throughput Screening—HTS) against the intraerythrocytic stages of *P. falciparum* were performed, both by pharmaceutical companies (GSK, Novartis) (Gamo et al., 2010; Plouffe et al., 2008) and by academic institutions (St. Jude Children's Research Hospital) (Guiguemde et al., 2010). Consequently, out of nearly 4 million compounds screened, about 20,000 hits were identified and deposited in the open access databases, such as ChEMBL Neglected Tropical Disease (ChEMBL-NTD) (Bento et al., 2014). Spitzmüller & Mestres used bioinformatics tools and statistical tests to predict and prioritize proteins of *P. falciparum* as the most likely targets for the active compounds deposited in the ChEMBL-NTD (Spitzmüller & Mestres, 2013), generating a list of 39 high-priority targetable proteins. Of these, 20 (51%) belong to the protein kinase class, consistent with the observed major representation of kinase inhibitors as phenotypically active compounds against *P. falciparum* in the GlaxoSmithKline (GSK) compound library (Gamo et al., 2010).

Predictably, the above results were unsurprising, as it has been widely reported that protein kinases are involved in almost all cellular processes such as proliferation, differentiation, motility, cell survival and death (Manning, Whyte, Martinez, Hunter, & Sudarsanam, 2002). However, the activity of protein kinases is ubiquitous; regardless of their substrate, each protein kinase catalyzes the transfer of a phosphate group from adenosine triphosphate (ATP) to an amino acid residue (Ser/Thr/Tyr) of a substrate protein. Thus, due to this common nature, most kinases have structural homology of the ATP binding site, and therefore the selectivity of competitive inhibitors in this region is not easily achieved. Nevertheless, in 2001, imatinib, a small molecule kinase inhibitor was approved by the FDA for use in humans, revolutionizing the treatment of chronic myeloid leukemia (Schiffer, 2007), thus breaking the uncertainty about the feasibility of using selective kinase inhibitors. Since then, several other small molecule kinase inhibitors have been approved for therapeutic use, predominantly against cancer but also against other diseases such as rheumatoid arthritis and pulmonary fibrosis (Roskoski, 2015). Thus, a large number of known or potential kinase inhibitors are part of the small molecule libraries of large pharmaceutical companies, and are available for antimalarial drug screening. Importantly, the large phylogenetic distance between the Apicomplexa and human kinome renders kinases as attractive targets for the development of new antimalarial drugs.

4. *Plasmodium* kinome

Following sequencing of the *P. falciparum* genome, it was possible to analyze the entire panoply of the parasite's kinases, collectively known as the kinome. These studies revealed that the genome of *P. falciparum* has 85–99 kinases (the variation in the number of kinases is due to the stringency criteria used in bioinformatics analysis) (Anamika, Srinivasan, & Krupa, 2004; Talevich, Tobin, Kannan, & Doerig, 2012; Ward, Equinet, Packer, & Doerig, 2004). In addition, there are at least three atypical protein kinases (aPKs) of the RIO family and three members the phosphatidylinositol kinases (PIPK) family (Trivedi & Nag, 2012; Vaid, Ranjan, Smythe, Hoppe, & Sharma, 2010; Ward et al., 2004). This number of kinases is relatively small compared to other free-living alveolates, which probably reflects the reduction in the overall size of the genome associated with these parasites (Doerig & Grevelding, 2015).

Of all the kinases of *P. falciparum*, at least 65 are clearly related to families of eukaryotic protein kinases (ePKs) (Ward et al., 2004). According to the nomenclature of Manning et al. based on the similarity of the sequences in the catalytic domains, ePKs are divided into nine major groups: AGC [named after the protein kinase A, G, and C families (PKA, PKG, PKC)], Ca^{2+}/calmodulin-dependent protein kinase (CaMK), CMGC (containing CDK, MAPK, GSK3 and CLK kinases), tyrosine kinases (TK), tyrosine-kinase like group of kinases (TKL), sterile kinases (STE), casein kinase I (CKI), receptor guanylate cyclases (RGC), and others (Manning et al., 2002). The most widely represented group in *Plasmodium* is CMGC with 26 members in *P. falciparum*, while the AGC group with 5–6 members is relatively small compared to the other eukaryotes (Lucet, Tobin, Drewry, Wilks, & Doerig, 2012; Talevich et al., 2012). While most groups are represented in the genome of *Plasmodium* spp., there is complete absence of members of the TK and RGC groups, similar to what is observed in other Apicomplexa (Talevich et al., 2012). Another notable absence in the kinome of *P. falciparum* relates to members of the STE group. Members of this group include a range of kinases that form a signaling cascade that culminates in phosphorylation in the activation loop of the mitogen-activated protein kinase (MAPK) pathway, thus playing an essential role in cell regulation and differentiation related to extracellular stimulation (Boutros, Chevet, & Metrakos, 2008). However, even in the absence of STE kinases, *P. falciparum*, alike other Apicomplexa has two members of the MAPK family (Pfmap-1 and Pfmap-2), suggesting that these kinases are regulated differently from other eukaryotes. Indeed, phosphorylation of at least one (Pfmap-2) appears to involve the participation of another NIMA-related kinase family, Pfnek-1, which has an activation motif similar to the STE family (Dorin et al., 2001). It is worth noting that although some species of *Plasmodium*, such as *P. vivax* and *P. knowlesi* have one member of the STE group, the regulatory potential of these kinases has yet to be elucidated. Also, another example involving the particular regulatory kinases of malaria parasites comes from studies involving phosphoproteomics that identified 70 residues of phosphorylated tyrosine in schizonts of *P. falciparum*, despite the fact that the parasites do not have TK group members (Collins, Wright, Jones, Rayner, & Choudhary, 2014). Possible reasons for tyrosine phosphorylation events include signaling pathways mediated by host tyrosine kinases or dual-specificity kinases in the parasite (Pease et al., 2013). Tyrosine autophosphorylation has been observed in the activation loops of dual-specificity PfGSK3 and PfCLK3 (Solyakov et al., 2011; Treeck, Sanders, Elias, & Boothroyd, 2011). Although many kinases of *Plasmodium* are grouped in the ePK families, a large number are considered orphans (when they are not members of established ePK families) or semi-orphans (when they belong to a particular ePK family but they do not have orthologs within the kinase family of mammals) (Doerig & Grevelding, 2015). This is the case of calcium-dependent protein kinases (CDPKs), a kinase family of plants and some alveolates that are not found in metazoans, represented by 5–7 members in *Plasmodium* species (Billker et al., 2004; Talevich et al., 2012). Moreover, certain kinases share characteristics of more than one ePK family. As mentioned above, Pfnek-1 has homology to the NIMA-kinase family, but also bears a mitogen-activated protein kinase (MAPKK)-like activation motif. Other examples of "hybrid" kinases include the *P. falciparum* protein kinase 7 (PfPK7) which shows similarities to both the AGC and STE groups, and *P. falciparum* cdc2-related protein kinase 4 (Pfcrk-4), which relates to the CDC and MAPKK kinase families (Ward et al., 2004).

The most notable orphan kinases are those belonging to the FIKK family, characterized by presenting an amino acid sequence of Phe-Ile-Lys-Lys in subdomain II of the ePK catalytic domain (Ward et al., 2004). This family is unique to Apicomplexa, and is mainly responsible for the differences in the kinome of *P. falciparum* compared to the other *Plasmodium* species; while most species, including *P. vivax*, have only one member, the FIKK family is expanded to 20 members in *P. falciparum* (Talevich et al., 2012). Some members of this family are transported to the membrane of infected erythrocytes, and appear to be implicated in changes in the conformation and rigidity of the erythrocyte (Brandt & Bailey, 2013). For instance, erythrocytes infected with knockout parasites for the *fikk4.2* kinase-coding gene had a reduced number of knobs, important structures for cytoadherence of *P. falciparum*, which is responsible for much of the pathogenesis associated to this species (Kats et al., 2014). Furthermore, many of the FIKK genes are refractory to deletion, emphasizing the important role of these kinases in the survival of the parasite (Lucet et al., 2012). This suggests that this family of kinases may be a good therapeutic target. However, the fact that this family is expanded only in *P. falciparum*, and to a lesser extent in *P. reichenowi* limits their importance as drug targets of other *Plasmodium* species that cause human malaria.

Although the characterization of *Plasmodium* kinases is still in its infancy, reverse genetic studies using specific kinase inhibitors have shown that, as expected, this group of proteins acts on all stages of the parasite life cycle (Alam et al., 2015; Collins et al., 2013; Deshmukh et al., 2015; Dorin-Semblat, Sicard, Doerig, Ranford-Cartwright, & Doerig, 2008; Tewari et al., 2010). Solyakov et al. showed that intraerythrocytic development is prevented by knocking out approximately half of the 65 kinases of *P. falciparum* grouped within the ePK families (Solyakov et al., 2011). Even kinases that appear to be redundant for asexual proliferation are required for the sexual stages of the parasite (Tewari et al., 2010). Furthermore, structural changes in *Plasmodium* kinases have been identified, which that may result in new functions, such as insertions, deletions and substitutions of amino acids in, or adjacent to, the catalytic domain regions

(Talevich, Mirza, & Kannan, 2011). Thus, the conservation of the functional domains of eukaryotic kinases allows the initial identification of compounds that have human kinases as targets, and which are bioactive against malaria parasites. At the same time, structural and regulatory characteristics of parasite kinases enable hit-to-lead processes for the development of new antimalarial drugs that are specific to *Plasmodium* protein kinases with low host toxicity. With this in mind, the discovery of antiparasitic kinase inhibitors that may serve as lead scaffolds, have been investigated extensively and have been previously reviewed elsewhere (Cabrera et al., 2018; Doerig et al., 2010; Masch & Kunick, 2015; Rotella, 2012; Zhang et al., 2012). In this chapter, we will make a case as for why kinases are likely targets for the development of new antimalarial drugs and present the latest available results involving the identification and optimization of new kinase inhibitors (Table 1).

Table 1 Selected examples of kinase inhibitors with antimalarial activity and potency against plasmodial kinases found in the literature.

Compound	Potential target	Kinase inhibitory assay [IC_{50} (μM)]	Antimalarial activity[a] [IC_{50} (μM)]	References
(**1**)	PfPKA	–	2.5	Syin et al. (2001)
(**4**)	PfPKG	0.00016	0.002	Baker et al. (2017)
(**8**)	PfPKB	0.20	0.25	Vaid, Thomas, and Sharma (2008)
(**10**)	PfPK5	0.06	2.0	Graeser, Wernli, Franklin, and Kappes (1996)
(**32**)	PfMRK	1.3	4.6	Li, Robson, Chen, Targett, and Baker (1996)
(**35**)	PfCDPK1	0.008	0.21	Green et al. (2016)
(**46**)	PfCDPK4	0.04	–	Ojo et al. (2012)
(**48**)	PfCDPK5	–	0.018	Vaidya et al. (2014)
(**55**)	PfPK7	4.6	1.0	Merckx et al. (2008)
(**62**)	PfNEK-1	1.1	3.0	Laurent et al. (2006)
(**73**)	PfGSK-3	5.5	0.48	Fugel et al. (2013)

[a] *In vitro* assays with *P. falciparum*.

4.1 AGC kinase group

Cyclic nucleotides, such as cyclic adenosine- and guanosine-monophosphate (cAMP and cGMP, respectively) are molecules involved in cell communication that appear to participate in various stages of the development of the parasite. For example, several studies indicate that cAMP triggers the release of ligands located in the micronemes that subsequently participate in both the motility and invasion capacity of sporozoites (Kebaier & Vanderberg, 2010; Ono et al., 2008) and in the development and invasion of merozoites into erythrocytes (Dawn et al., 2014; Syin et al., 2001). Moreover, it has been reported that cAMP induces gametogenesis (Read & Mikkelsen, 1991; Tracer & Gill, 1989) and facilitates the transmission of gametocytes to the vectors by regulating the deformability of erythrocytes infected with gametocytes (Ramdani et al., 2015). The only kinase known to be dependent on cAMP in *P. falciparum* is PfPKA (Talevich et al., 2012). PfPKA is composed of a catalytic subunit (PfPKAc) and a regulatory subunit (PfPKAr) (Wurtz, Chapus, Desplans, & Parzy, 2011). The homology between the PKA in humans and in *P. falciparum* suggests that PfPKA is regulated in a manner similar to that in humans. In this case, the regulatory subunit binds to and inhibits the catalytic subunit. However, the binding of cAMP to the regulatory subunit alters its affinity, releasing the active catalytic subunit, which can then phosphorylate several substrates (Sudo, Kato, Kobayashi, Tohya, & Akashi, 2008; Wurtz et al., 2011). A phosphoproteomics study indicated that 425 proteins in the schizont stage of *P. falciparum* share an extended PKA phosphorylation motif, thereby reinforcing the suggestion that the PfPKA has a great influence on parasite biology (Lasonder et al., 2012). An important example involves the phosphorylation of a particular residue, S610, in the cytoplasmic tail of the AMA-1 protein, which is essential for merozoite invasion of erythrocytes (Leykauf et al., 2010; Wilde et al., 2019).

However, despite the essential role of PfPKA in the development of *Plasmodium*, as well as the significant differences in the sequence and structure between the PKAs of the parasite and of humans, surprisingly, few studies have considered PfPKA as a potential drug target. A previous study evaluated the *in vitro* potential of a potent PKA inhibitor in mammals, H89 (**1**), in respect to the growth of *P. falciparum* blood stages and found a half maximal inhibitory concentration (IC_{50}) of less than 3 μM (Syin et al., 2001). However, subsequent studies using enzyme activity assays of the PfPKAc recombinant protein showed that H89 weakly inhibited the parasite's kinase (Sudo et al., 2008). This suggests that the inhibitory potential of H89 is more likely to occur due to the inhibition of the PKA of the host's erythrocytes, or an alternative target, since H89 may inhibit at least eight other human protein kinases (Lochner & Moolman, 2006). Flaherty et al. investigated the antiplasmodial activity of a stapled peptide (STAD-2), which disrupts the interaction between the PKA regulatory subunit and human A-Kinase Anchoring Proteins (AKAPs), which have a vital role in the compartmentalization of PKA to specific subcellular locations. The STAD-2 compound is selectively permeable to infected erythrocytes and showed good antiplasmodial activity ($IC_{50} \approx 1\,\mu M$). Nevertheless, although STAD-2 is located inside the parasite, a study using specific antibodies against PfPKAr did not find evidence of any association between STAD-2 and the PKA of the parasite (Flaherty et al., 2015).

Another kinase of the same group, PfPKG, has been widely studied as a potential target toward the elimination of asexual blood forms, or to break parasite transmission by acting against the sexual forms. Increases in cGMP concentration seem to be important in the transition of the different forms during the life cycle of the parasite, and yet, only PKG has been identified as the main effector molecule of cGMP in *Plasmodium* (Deng & Baker, 2002). The essential role of PKG in the development of *Plasmodium* is evident, as the PKG gene is refractory to deletion in both *P. falciparum* and *P. berghei* (Solyakov et al., 2011; Tewari et al., 2010). The role of PKG during the life cycle of malaria parasites was determined mainly by exploiting kinase inhibitors that were previously associated to PKG in two coccidia, *Toxoplasma gondii* and *Eimeria tenella* (Donald et al., 2006; Gurnett et al., 2002). Two structurally distinct compounds, trisubstituted pyrrole 4-[2-(4-fluorophenyl)-5-(1-methylpiperidine-4-yl)-1*H*-pyrrol-3-yl] pyridine (**2**) and the imidazopyridine compound 4-[7-[(dimethylamino) methyl]-2-(4-fluorophenyl) imidazo [1,2-a] pyridin-3-yl] pyrimidin-2-amine (**3**) were highly selective for the PKGs of Apicomplexa. These inhibitors competitively block the ATP binding site of the parasite due to the presence of a threonine residue at the gatekeeper position that is relatively smaller than the methionine or leucine found in the same position of the PKG isoforms in humans, which access of the inhibitors to the hydrophobic pocket adjacent to the ATP binding site (Donald et al., 2006). Compound (**2**) was first tested over *P. falciparum* by Diaz et al. proving to be a potent PfPKG inhibitor (IC_{50} of 8.53 nM) acting against the asexual blood forms of the parasite and inhibiting the *in vitro* growth of both chloroquine-sensitive (NF54) and chloroquine-resistant (Dd2) strains (IC_{50} of 0.49 and 1.3 μM, respectively) (Diaz et al., 2006). Both compounds (**2**) and (**3**) cause visible morphologic changes in schizonts, triggering the accumulation of mature segmented schizonts that do not rupture and therefore fail to release merozoites (Taylor et al., 2010). Further evidence suggests that at least part of the mechanism of action of PfPKG inhibitors occurs due to blockage of the release of *P. falciparum* subtilisin-like protease 1 (PfSUB1) by exonemes into the parasitophorous vacuole. This prevents the processing of certain proteins, such as PfMSP-1 and PfSERA5, involved in parasite egress and also the release of *P. falciparum* apical membrane antigen 1 (PfAMA-1) of micronemes to the surface of the merozoites (Collins et al., 2013).

Although *in vitro* results against *P. falciparum* demonstrated that PfPKG inhibitors might be promising in the treatment of malaria, the *in vivo* activity of compound (**2**) is limited in models of *P. berghei* infection using C57BL/6 mice (Diaz et al., 2006). The administration of 50 mg/kg of compound (**2**) twice daily, starting 2 h after initial infection and ending on the eighth day, increased survival time, but no changes were observed in the clearance of parasitemia and all mice succumbed by day 25. In order to check whether the same inhibitors show inhibitory effects on PKG of *P. berghei* and *P. falciparum*, Tewari et al. developed a *P. berghei* transgenic strain expressing PfPKG. Although they found that development of the parasite in the mosquito vector was compromised, the asexual blood stages of these transgenic parasites occurred normally, and no difference was found in the development of both transgenic and wild *P. berghei* when treated with 25 mg/kg of Compound (**2**) for 4 days (both strains showed a 40–50% reduction in parasitemia) (Tewari, Patzewitz, Poulin, Stewart, & Baker, 2014). Interestingly, when administered prior to infection, a single dose of 50 mg/kg of compound (**2**) was able to inhibit sporozoite invasion of hepatocytes, thereby preventing blood stage infections in mice infected with *P. berghei* or *P. yoelii* (Panchal & Bhanot, 2010). However, a mutant strain of sporozoites, which does not express the PKG, remained susceptible to treatment with (**2**), suggesting that the compound has additional targets apart from PKG (Panchal & Bhanot, 2010).

Further optimization of compound (**3**) resulted in a series of potent imidazopyridine inhibitors of *Plasmodium* PKG, with the best compound of this series ML10 (**4**) showing an IC_{50} of 160 pM against recombinant PfPKG. ML10 was also able to clear *in vivo* infection using a *P. falciparum* humanized model when administered twice a day, but lower efficacy was observed in a single dose regimen, probably due to rapid compound clearance and the limited narrow window of activity of PKG during the asexual blood cycle (Baker et al., 2017).

PfPKG inhibitors also affect the development of sexual stages of the parasite. Gametocytes treated with compound (**2**) retained a crescent shape without rounding up. Furthermore, exflagellation was also inhibited in a dose-dependent manner. Moreover, PfPKG regulates ookinete gliding (Brochet et al., 2014; Moon et al., 2009), demonstrating a possible role for PfPKG inhibitors in blocking the transmission of malaria. In fact, ML10 effectively inhibited oocyst production in mosquitoes after artificial feeding, with blood containing gametocytes treated with 41 nM of ML10 (Baker et al., 2017). The selectivity of compounds (**2**), (**3**), and ML10 for PfPKG, has been verified by experiments using parasites expressing PfPKG resistant to the action of these inhibitors. Genetically modified parasites (PfPKGT618Q) expressing glutamine, which is a bulkier amino acid at the gatekeeper position, prevents access of the inhibitors making the kinase about 3000 times more resistant in assays using the recombinant enzyme (Baker et al., 2017; McRobert et al., 2008). Thus, both asexual blood forms and sexual forms of PfPKGT618Q parasites develop normally when treated with compounds (**2**), (**3**), and ML10 confirming that PfPKG is the target of these inhibitors.

Recently, using chemoproteomic and gene-editing tools, Vanaerschot et al. identified PKG as the target of a trisubstituted imidazole (**5**) (Vanaerschot et al., 2020). In agreement with the role of PKG in the malaria parasite life cycle, this compound inhibited sporozoite invasion into hepatocytes, being active against asexual blood-stage schizonts and also male gamete exflagellation. Interestingly, that study also suggested that PKG inhibitors are also less prone to the evolution of resistance (Vanaerschot et al., 2020).

PfPKB shows high similarity to human PKB (~70%), particularly within the catalytic domain. However, the N-terminal region of PfPKB is significantly different. While HsPKB has regulatory domains (PH) at the N-terminal that interact with PDK1 for its activation, the PfPKB enzyme lacks a PH domain homolog (Kumar, Vaid, Syin, & Sharma, 2004). Furthermore, no PDK1 ortholog has been annotated in *P. falciparum*, although one is predicted in *P. yoelii*. In fact, PfPKB is activated by autophosphorylation, which is triggered by binding of calmodulin (CaM) to its N-terminal region, which in turn is dependent on phospholipase C-mediated calcium release (Vaid & Sharma, 2006; Vaid et al., 2008). The role of PfPKB in *P. falciparum* intraerythrocytic development was initially evaluated using the two specific PKC inhibitors Go 6983 (**6**) and Go 6976 (**7**), as PfPKB also shares 65% sequence homology with the catalytic domain of PKC. While Go 6976 was unable to inhibit PfPKB activity, Go 6983 inhibited this kinase with an IC_{50} of ~1 μM. In *P. falciparum* cultures treated with Go 6983 (2 μM), parasitemia was reduced after 40–44 h, when parasites reached the late schizont stage, which matches the expression time of PfPKB (primarily in the schizont/merozoite stages) (Kumar et al., 2004). Further evidence using a specific inhibitor of mammalian PKB/AKT, A443654 (**8**), which inhibited the activity of PfPKB (IC_{50} = ~200 nM), suggested that PfPKB is important in merozoite invasion. Probably, PfPKB acts in the erythrocyte invasion process, phosphorylating PfGAP45, which is part of the inner membrane complex and facilitates anchoring and effective functioning of actin-myosin motor (Vaid et al., 2008) (Fig. 1).

4.2 Cyclin-dependent kinases family

CDK proteins are important regulators of the cell cycle and gene expression, and their activation requires the binding of regulatory subunits, cyclins (Malumbres, 2014). Despite important differences, seven protein kinases

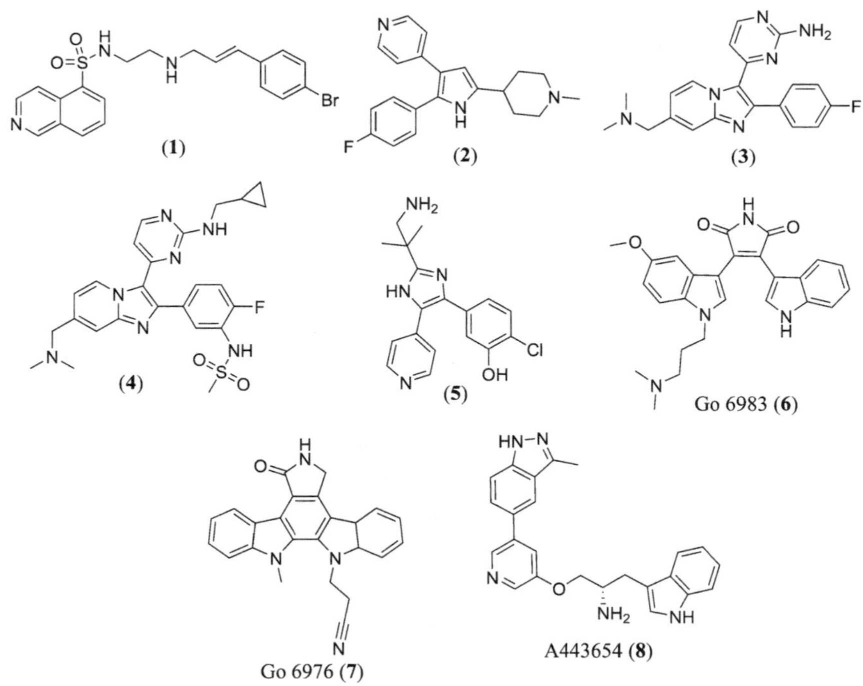

Fig. 1 AGC kinase inhibitors.

related to human CDKs have been identified in the genome of *P. falciparum* so far: PfPK5, PfPK6, PfMRK, PfCRK-1, PfCRK-3, PfCRK-4 and PfCRK-5 (Talevich et al., 2012). PfPK5 was the first CDK to be characterized in *Plasmodium*, being initially isolated in 1994 (Ross-Macdonald, Graeser, Kappes, Franklin, & Williamson, 1994) and subsequently crystallized in 2003 (Holton et al., 2003). The similarity between amino acid sequences between PfPK5 and human CDK1 and CDK2 is approximately 60%, with even more resemblance between the residues that form the ATP binding pocket (82% and 92%, respectively). Furthermore, PfPK5 contains a PSTTIRE sequence that corresponds to the PSTAIRE cyclin-binding motif of human CDKs (Ross-Macdonald et al., 1994). The activation mechanism of PfPK5 also appears to be similar to its human counterpart, as both are regulated by the phosphorylation of a conserved threonine residue located within the T loop (Graeser et al., 1996). Furthermore, PfPK5 is activated *in vitro* by human cyclins and at least four cyclins have been predicted in the *P. falciparum* genome, three of which can activate PfPK5 *in vitro* (Pfcyc-1, -3 and -4) (Merckx et al., 2003). However, unlike the CDKs of other eukaryotes, PfPK5, as well as PfPK6 and PfMRK, are able to perform autophosphorylation in the presence of cyclins (Graeser et al., 1996).

Based on the similarity between human and the parasite's CDKs, initial drug screens were performed to evaluate the antimalarial activity against well-characterized human CDK inhibitors. Flavopiridol (**9**) and olomoucine (**10**) were shown to inhibit PfPK5 *in vitro* with IC_{50} values of 0.06 and 0.15 μM, respectively. Both drugs significantly reduced the incorporation of [^{3}H]hypoxanthine by DNA, thus demonstrating an inhibition of DNA synthesis. These results, coupled with the facts that the addition of aphidicolin, a DNA synthesis blocker, increased PfPK5 activity, and that PfPK5 is mainly expressed in the schizont stage, suggest that this kinase regulates the S phase in the cell cycle of malaria parasites (Graeser et al., 1996). It was demonstrated that PfPK5 is able to phosphorylate *P. falciparum* origin recognition complex subunit 1 (PfOrc1), a regulator of chromosomal replication (Deshmukh et al., 2015). Several other CDK inhibitors, such as indirubin-3′-monoxime (**11**), purvalanol A (**12**), purvalanol B (**13**), hymenialdisine (**14**), and NU6102 (**15**), showed activity against PfPK5/cyclin, which in turn was insensitive to larger compounds such as indirubin-5-sulfonate (**16**) and staurosporine (**17**) (Holton et al., 2003; Le Roch et al., 2000). Interestingly, despite the similarities between PfPK5 and human CDK 1/2, different IC_{50} values are observed between them, thereby suggesting the existence of remarkable differences (Harmse et al., 2001), encouraging the search for selective inhibitors.

The antimalarial potential of some predicted anti-kinase compounds was evaluated *in vitro* against asexual blood stages of *P. falciparum*. Olomoucine (**10**), roscovitine (**18**) and isopentenyladenine (**19**) exhibited similar activity, with IC_{50} values of 8.45, 5.35 and 8.34 μM, respectively (Harmse et al., 2001). In that same study, a series of purvalanol compounds was tested and was shown to display a wide range of inhibition. Although several compounds displayed submicromolar IC_{50} values, all were less efficacious than all known drugs with proven efficacy against plasmodia, thereby indicating that none of these compounds has immediate potential as an antimalarial candidate. However, a number of specific structural features of these purine derivatives, which enhance the activity against the parasite, were identified. Among them, we highlight the size and spatial orientation of the chiral carbon substituents of the hydroxyethyl amino group of the purine ring, and the enantiomer orientation (R or S) (Harmse et al., 2001).

Computational tools have also assisted in the identification of PfPK5 as a potential target for antimalarial inhibitors. Derbyshire et al. evaluated 1358 small molecules that are biased for targeting an array of human kinase families during infection in the liver stage of malaria (Derbyshire et al., 2014). Of these, 31 compounds inhibit infection by *P. berghei* sporozoites of both hepatic cell lines, HepG2 and Huh-7 ($IC_{50} < 10$ μM). Of the 31 compounds, most were identified as human CDK inhibitors and later evaluations of these compounds against the asexual blood forms of *P. falciparum* showed they had similar inhibitory potential against the two different stages of *Plasmodium*, liver and blood, suggesting that they do not exhibit stage selectivity. As all selected hits against CDKs exhibit activity against hCDK2, the crystallographic structure of PfPK5 was used to identify possible interactions with two CDK inhibitors, SNS-032 (**20**) and CDK 1/2 (**21**). Molecular docking suggests that these compounds bind to PfPK5. Further biochemical studies indicated that SNS-032 binds to PfPK5 (Kd value of 7.0 μM) and to CDK 1/2 (Kd value of 0.32 μM). Interestingly, *in vivo*, the low-toxicity SNS-032 compound, which is currently being evaluated in clinical trials for cancer patients, reduces the parasite load of hepatic forms of *P. berghei* in mice when administered before infection (Derbyshire et al., 2014).

Compound CDK 1/2 (**21**) was explored for selectivity toward PfPK5. CDK 1/2 and structural analogs showed to be more selective to HsCDK2, underscoring the challenge in the design of selective PfPK5 inhibitors (Eubanks et al., 2018). The authors used a virtual screening strategy using PfPK5 ATP-binding site and found that compounds with the 4-methylumbelliferone scaffold showed improved selectivity for PfPK5. Compound XA (**22**) and compound XB (**23**) showed a Kd of 5 and 3.8 μM for PfPK5, respectively, but only moderate antiplasmodial activity was observed (Eubanks et al., 2018).

In phenotypic assays with *P. falciparum*, the pyrrolopyrazine derivative (**24**) presented an IC_{50} value of 200 nM (Reker et al., 2014). Best chemical optimization showed increased growth inhibition when the quinoxaline core was replaced by a dicyanopyrazine. Of the 32 derivatives synthesized and tested, (±)-dicyanopyrrolopyrazine (**25**) exhibited strong antiplasmodial activity against both the chloroquine-sensitive (NF54) and chloroquine-resistant (K1) strains, with IC_{50} values of 53 and 40 nM, respectively. In addition, this compound also acts against the liver forms of *P. berghei* ($IC_{50} = 0.3$). Subsequent protein sequence-, advanced ligand-, and pocket-based computational studies suggest inhibitor binding (±)-dicyanopyrrolopyrazine (**25**) to the active site of PfPK5 (Reker et al., 2014). However, this compound showed poor selectivity against a panel of human kinases.

PfMRK is another well-characterized CDK of *Plasmodium*, sharing 46% identity with the human CDK7 (Li et al., 1996). Despite being mainly expressed during the sexual reproduction stages of the parasite, PfMRK is also expressed during the trophozoite and schizont stages. Similar to PfPK5, PfMRK is activated *in vitro* by the presence of human cyclin H and Pfcyc-1, its parasite ortholog (Le Roch et al., 2000), and shows its maximum activity after binding to the effector, PfMAT1 (Chen et al., 2006; Jirage et al., 2010). Although the function of PfMRK during the parasite's life cycle is incompletely understood, evidence suggests that it acts during DNA replication. Additional to this kinase being located near the nucleus of the parasite together with its regulator, Pfcyc-1, it has been shown that two regulators of DNA synthesis, PfRFC-5 and PfMCM6 are *in vitro* substrates of PfMRK (Jirage et al., 2010). Furthermore, the PfMRK-PfMAT1 complex phosphorylates the carboxy-terminal domain (CTD) of RNA polymerase II (Li et al., 2001).

Prioritizing research for PfMRK inhibitors is justified by the unique characteristics of this kinase, which favors the development of specific compounds (Keenan, Geyer, Welsh, Prigge, & Waters, 2005). Olomoucine (**10**) and roscovitine (**18**), which inhibit PfPK5, failed to inhibit PfMRK, apparently due to residues in the ATP binding site that block the binding of these compounds to PfMRK (Waters, Woodard, & Prigge, 2000). An initial search for PfMRK inhibitors evaluated a series of quinolines, but none displayed PfMRK IC_{50} activity below 10 μM *in vitro* (Xiao, Waters, Woodard, Li, & Li, 2001). A series of 18 compounds containing an oxindole core motif was consequently assessed in respect to Pfmrk, with IC_{50} values ranging from 1.4 to 57 μM. Molecular docking of a selection of the best inhibitors of this series (**26**) indicated that the oxindole ring is anchored at the active site by two hydrogen bonds to the main chains of the Glu-92 and Met-94 residues. Furthermore, additional interactions in the hydrophobic active site involving both Leu-16 and Tyr-96 residues are important in the binding of oxindoles to PfMRK, and probably contribute to the specificity of these compounds, since differences in these positions are observed both for PfPK5 and for human CDKs (Woodard et al., 2003).

In the absence of structural information for PfMRK, the three dimensional Quantitative Structure-Activity Relationship (QSAR) pharmacophore was developed based on one a set of structurally distinct kinase inhibitors in order to predict characteristics of compounds that may help to identify new PfMRK inhibitors (Bhattacharjee et al., 2004). The best pharmacophore model consists of a spatial arrangement of four functional groups: two hydrogen bond acceptors, one hydrophobic group and one planar aromatic ring. This pharmacophore was then used to search for other PfMRK inhibitors, and 16 compounds were selected with predictive IC_{50} values below 25 μM. Of these, three compounds evaluated experimentally in kinase assays with PfMRK had IC_{50} values below 5 μM: 5-chloro-4-nitrothiophene-2-sulfonic acid (4-trifluoromethylphenyl)-amide (**27**) (IC_{50} = 2.5 μM); 4-aza-9-chloroindolo[2,1-b]-quinazoline-6,12-dione (**28**) (IC_{50} = 3.5 μM) and; 4-aza-8-fluoroindolo[2,1-b]-quinazoline-6,12-dione (**29**) (IC_{50} = 4.7 μM). Overall, there was good correlation between predictive and experimental values of the evaluated compounds, with the exception of four compounds that share a nitrophenyl group, which is not compatible with the inhibition of PfMRK (Bhattacharjee et al., 2004).

Isoquinoline sulfonamides may also have variable inhibition against PfMRK with greater activity appearing to depend on the presence of less bulky substituents at ring position 5 and the number of hydrogen bonds formed with residues within the active site of PfMRK. Thus, compound (**30**), which was the only one of this series to form three hydrogen bonds in molecular modeling analysis was also the only shown to be potent against PfMRK (IC_{50} = 0.7 μM). The other isoquinoline sulfonamides were weakly active (Woodard et al., 2007). However, other sulfonamides containing thiophene and benzene moieties instead of quinoline (**31** and **32**) displayed improved *in vitro* activity against PfMRK and limited toxicity to mammalian cells (Caridha, Kathcart, Jirage, & Waters, 2010).

Nonetheless, it is noteworthy that although PfMRK can be is inhibited by a variety of chemical scaffolds, including quinolinones, oxindoles, and sulfonamides, there is often no correlation between PfMRK inhibition and antimalarial activity against *P. falciparum*, which may result from poor bioavailability or the existence of additional alternative targets (Caridha et al., 2010; Woodard et al., 2003). However, certain chalcone derivatives (1,3-diaryl-2-propanone) (**33** and **34**) selectively inhibit PfMRK and display *in vitro* activity against *P. falciparum* (Geyer et al., 2009).

Although the active site of PfPK6 is structurally similar to CDKs, this kinase has characteristics that differentiate it from other CDKs. Sequence analysis of PfPK6 showed that it shares 33% identity with to human p38 MAPK. Furthermore, the primary structure of PfPK6 does not contain any sequence related to the PSTAIRE-like cyclin-binding motif. In fact, the recombinant PfPK6 exhibit robust kinase activity in the absence of cyclin (Bracchi-Ricard et al., 2000). Immunofluorescence assays reveled that PfPK6 is manly expressed during trophozoite and schizont stages and that it is located in the both cytoplasm and nucleus. Two inhibitors with selectivity to CDK-related kinases, roscovitine (**18**) and olomoucine (**10**), inhibit PfPK6 activity with IC_{50} of 25 and 130 μM respectively, comparable to those of MAPKs rather than of CDKs (Bracchi-Ricard et al., 2000). Molecular models of PfPK6 complexed with these inhibitors suggested that the presence of two tyrosine residues (Tyr-13 and -96) in the access of the ATP-binding pocket reduces the volume for inhibitors binding in the PfPK6, and is responsible for the high IC_{50} values for PfPK6 compared to other CDKs (Manhani et al., 2005).

Four other CDK-related kinases have also been identified (PfCRK-1, PfCRK-3, PfCRK-4, and PfCRK-5) but have been less studied. Although PfCRK-1 is mainly expressed in gametocytes, one previous study using *P. berghei* indicated that this kinase is essential to intraerythrocytic asexual multiplication (Rangarajan et al., 2006). A common CDK inhibitor, olomoucine (**10**) at 1.2 mM reduced the phosphorylation of histones incubated with the PfCRK-1 recombinant protein, suggesting that olomoucine inhibits PfCRK-1 (Iwanaga et al., 2013). PfCRK-3 is also crucial for asexual proliferation since disruption of the respective gene was unviable, and further evidence point to a pivotal role of PfCRK-3 in the regulation of transcription (Halbert et al., 2010). However, to date, no small molecule inhibitor of PfCRK-3 has been reported. PfCRK-4 was found to be a key cell-cycle regulator of DNA replication throughout schizogony since the reduction of PfCRK-4 led to blocking nuclear division and DNA replication. This enzyme is also required for ookinete formation and early oocyst development. The modulation of this target can lead to efficacy in the blood stage and in the transmission stage, an interesting strategy for development of new drugs (Ganter et al., 2017). PfCRK-5 is a critical regulator of atypical mitosis in the gametogenesis and is essential for parasite transmission (Balestra et al., 2020).

CDKs play an important role in the artemisinin-induced dormancy, which is related with up-regulation of PfCRK-1 and PfCRK-4 to halt cell cycle in the S phase. In addition, the down-regulation of other CDKs is important to prevent DNA replication and consequent arrest at the G_1 phase (Gray et al., 2016). The use of CDK inhibitors targeting PfCRK-1 can have a profound effect in dormant parasites since it appears to be important in controlling and suppressing entry into S phase, having a major impact in the survival of dormant parasites (Gray et al., 2016) (Fig. 2).

4.3 Calcium-dependent protein kinase group

Calcium controls several essential pathways in apicomplexan parasites such as protein secretion, motility, and parasite development (Lourido & Moreno, 2015; Nagamune, Moreno, Chini, & Sibley, 2008); these mechanisms are tightly regulated by a serine-threonine kinase family, known as calcium-dependent protein kinases (CDPKs) that is present in plants, ciliates and protozoan, but not in metazoans (Billker, Lourido, & Sibley, 2009). The canonical structure of CDPK contains a calcium-binding domain (CAD) with four EF hands that function as a calcium sensor. When CDPKs are activated by calcium, the whole kinase undergoes radical conformational changes and this new rearrangement in the kinase active form make residues essential for interactions with ATP available for substrate recognition (Wernimont et al., 2010). Seven kinases comprise the CDPK family in *Plasmodium* species.

P. falciparum CDPK1 was the first CDPK to be described (Zhao, Kappes, & Franklin, 1993), and probably the most characterized kinase from the CDPK family so far. CDPK1 is expressed in all blood stages of *P. falciparum*, being associated with the erythrocyte membrane in ring and trophozoite stages and with the plasma membrane in merozoites. There is evidence that this kinase plays a role in microneme discharge during

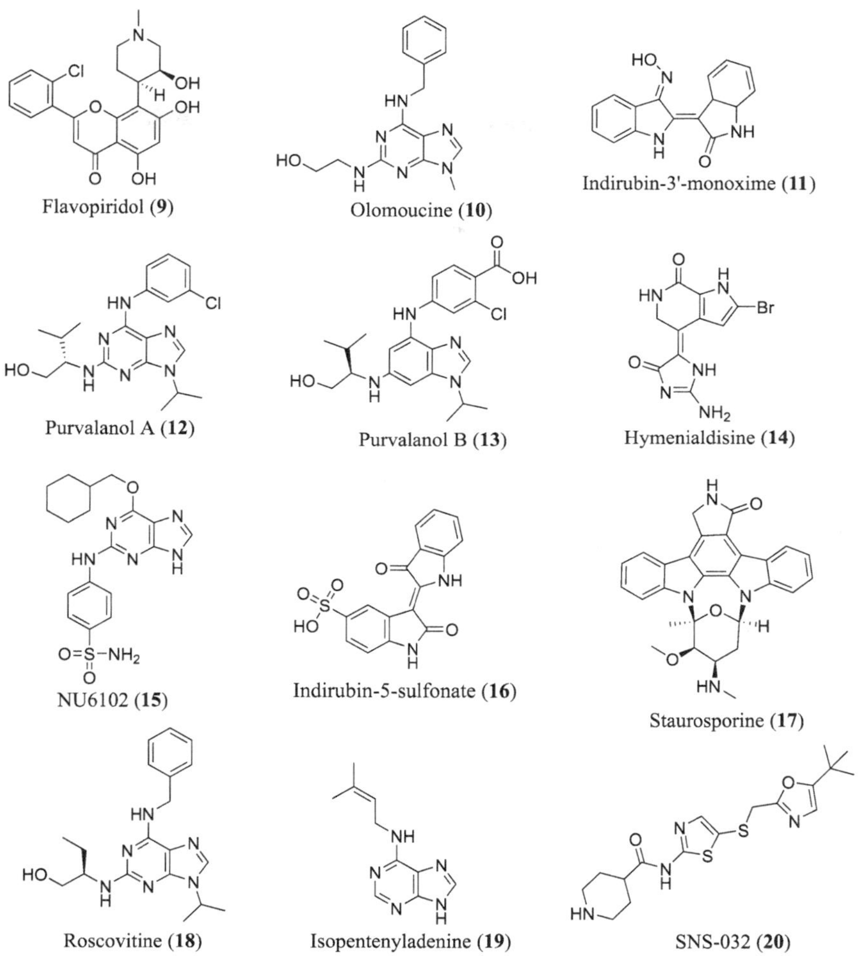

Fig. 2 CDKs kinase inhibitors.

erythrocyte invasion (Bansal et al., 2013) and in parasite motility (Green et al., 2008). Unsuccessful attempts to knockout *P. falciparum cdpk1* suggested that this gene could be essential for parasite growth and development (Solyakov et al., 2011). Studies in *P. berghei* showed that this kinase is expressed not only in blood stages, but during all life cycle stages of the parasite, including gametocytes, zygotes and oocysts. Gene knockdown in *P. berghei* demonstrated that CDPK1 participates in regulating post-translational modifications and translational activation of a subset of mRNAs in ookinete development and gliding (Sebastian et al., 2012).

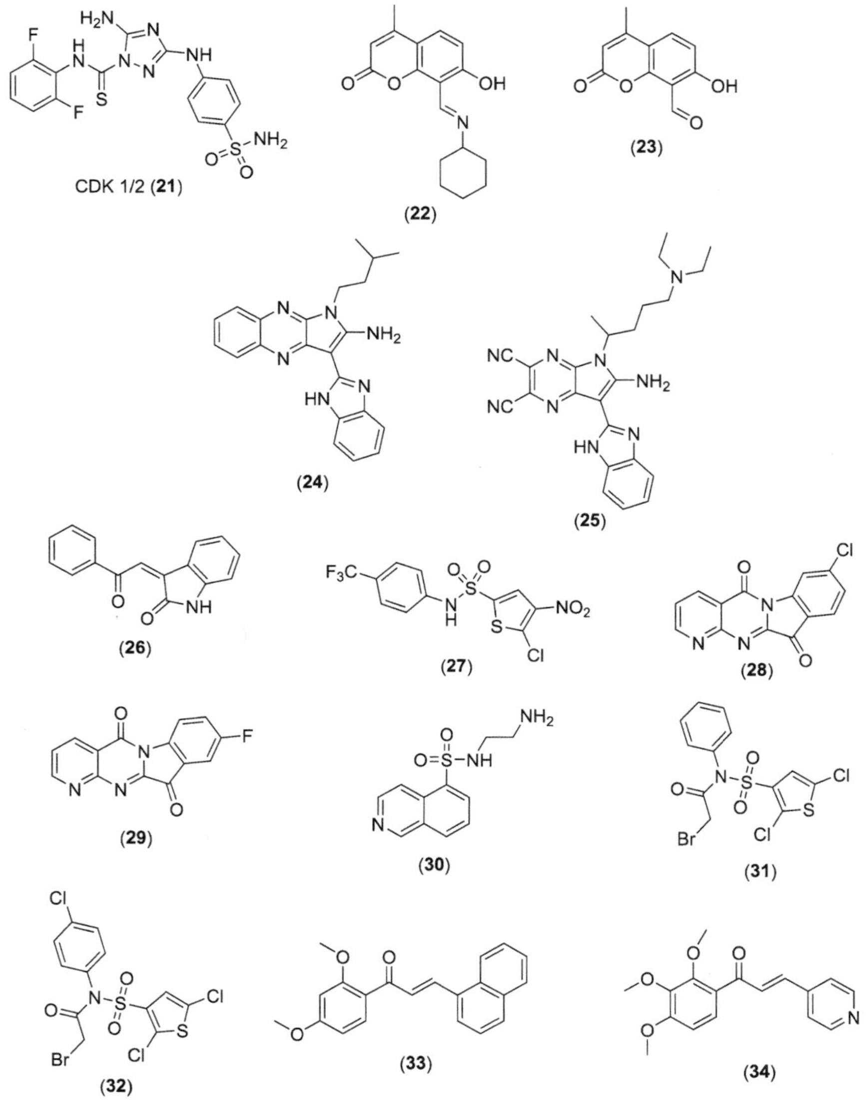

Fig.2—Cont'd

Unlike *P. falciparum*, the successful disruption of *P. berghei cdpk1* by homologous recombination showed that CDPK1 is not essential for erythrocytic invasion or intraerythrocytic development, suggesting that it is essential only for mosquito sexual stages (Jebiwott, Govindaswamy, Mbugua, & Bhanot, 2013). This is reinforced by studies in *T. gondii* CDPK3, a homolog of

CDPK1 in *P. falciparum*, in which inhibition does not bring about major defects in either motility or microneme secretion (Garrison et al., 2012).

The lack of direct CDPK homologs in animals makes these kinases potential targets for inhibitors to treat diseases caused by apicomplexan organisms. There have thus been intensive efforts for designing CDPK1 inhibitors. Imidazopyridazine compounds are known to be pronounced ATP-competitive inhibitors of CDPK1, being the first shown to display potent *in vitro* activity and favorable ADMET properties (**35**–**38**) (Chapman et al., 2013; Large et al., 2013). Despite displaying activity in the nanomolar range in both kinase and parasite in several studies, no efficacy was observed when imidazopyridazine compounds were used against malaria *in vivo* (Ansell et al., 2014; Chapman et al., 2014). A recent study divided imidazopyridazine compounds into two classes: class 1 compounds inhibit parasite growth at late schizogony, while class 2 compounds inhibit trophozoite growth (Green et al., 2016).

Although it was initially thought that class 1 compounds targeted CDPK1, it was demonstrated that they probably act against PKG, as decreased inhibition when the gatekeeper residue of PKG was substituted was observed (Green et al., 2016). Class 2 compounds were shown to inhibit HSP90, a chaperone with an ATP binding site in its N-terminal domain. Reducing the gatekeeper size of the ATP binding domain in CDPK1 allows inhibition by bumped kinase inhibitors. However, when parasites expressing this gatekeeper mutation were incubated with class 2 compounds, no change in growth sensitivity was observed (Green et al., 2016). These findings suggest that imidazopyridazines do not specifically target CDPK1, and their activity against *Plasmodium* occurs primarily through inhibition of PKG or HSP90 and not of CDPK1, as previously assumed.

Recently, a virtual screening campaign identified 18 small molecules with propensity to bind to PfCDPK1. Of these, two compounds (**39**, **40**) showed limited inhibitory potency against PfCDPK1, with IC_{50} values of 33.8 and 42.6 μM, respectively. However, there was only a moderate correlation between enzyme inhibition and antiplasmodial activity (Jain, Gupta, Munde, Pati, & Singh, 2020).

CDPK3 is the third CDPK-family protein, which was first reported to play a role in sexual stages of the parasite (Li, Baker, & Cox, 2000). Reverse genetic studies showed that *cdpk3* knockout parasites had no detectable defect in asexual erythrocytic development in mice (Siden-Kiamos et al., 2006). Further work showed that CDPK3 is predominantly found in late stage of ookinete formation, being undetectable in blood stages parasites

(Ishino, Orito, Chinzei, & Yuda, 2006). This pattern is similar to that observed for microneme proteins CTRP and WARP, suggesting a role in the invasion of the mosquito midgut by ookinetes (Dessens et al., 1999; Yuda, Yano, Tsuboi, Torii, & Chinzei, 2001). PbCDPK3(−) parasites displayed normal exflagellation, ookinete development, CTRP distribution and normal morphology, but when mosquitos were fed on mice infected with mutant parasites, a considerable decrease of oocyst number in the mosquito midgut was observed (Ishino et al., 2006; Li et al., 2000). The same work reported that ookinetes with *cdpk3* disruption were unable to invade midgut cells and presented abnormal movement, with frequent flexing and bending. Staurosporine (**17**) (a broad-spectrum kinase inhibitor) and another two calmodulin kinase inhibitors, W-7 (**41**) and KN-93 (**42**) were tested in *P. berghei* wild-type parasites and the phenotype of ookinete migration was similar to that observed in CDPK3(−) parasites in all cases. ML-7 (**43**) (a myosin light chain kinase inhibitor) showed no effect over ookinete migration (Ishino et al., 2006). Thus, CDPK3 seems to be necessary for ookinete invasive motility, being a potential target for transmission-blocking drugs.

Plasmodium CDPK4 was first characterized in 2004. It is suggested that this calcium-dependent kinase is involved in the molecular switch that translates the xanthurenic acid-induced calcium signal into a cellular response, playing a role in the male gametocyte cell cycle, proposed to be essential for sexual development and consequently, mosquito transmission of *P. berghei* (Billker et al., 2004). A model for PfCDPK4 activation was suggested, which is dependent on junction domain regulation to reach autophosphorylation and activation (Ranjan, Ahmed, Gourinath, & Sharma, 2009). Additional studies demonstrated that the kinase's phosphorylation is activated not only by an increase in Ca^{2+} concentration but is also sensitive to pH and a decrease in temperature, both factors being associated with gametogenesis (Kato et al., 2009). Calcium activates a myristoylated isoform of CDPK4 that is required to complete cytokinesis during microgametogenesis (Fang et al., 2017). These studies indicated that CDPK4 is essential for mosquito transmission of malaria parasites by regulating gametogenesis in response to xanthurenic acid-induced Ca^{2+} signals, being an attractive target for blocking malaria transmission.

The first attempt to find *Plasmodium* CDPK4 inhibitors was performed by Cavagnino et al. using calmodulin antagonists (Cavagnino, Rossi, & Rizzi, 2011). As a result, it was shown that the phenothiazines (TFP, thioredazine and chlorpromazine) (**44**, **45** and **46**), exerted a significant inhibition on PfHisCDPK4, with residual enzyme activity decreasing to

1.5%, 7.0% and 21.2% respectively at 1 mM concentration. Testing a chemical Tg/CpCDPK1 library, which is significantly similar to PfCDPK4, various compounds showed potency against PfCDPK4, with BKI-1 (**47**) (IC_{50} of 4 nM) being the most potent. Compound BKI-1291 (**48**) had an IC_{50} of 10 nM against PfCDPK4 enzymatic activity and an EC_{50} of 0.047 μM for blocking *P. falciparum* exflagellation, comparable to that of BKI-1. Induced mutation in the PfCDPK4 gatekeeper (rPfCDPK4 S147M), that has a methionine instead of a serine at the gatekeeper position, blocked sensitivity ($IC_{50} > 3000$ nM) to all of the BKIs tested, showing the importance of the gatekeeper size in kinase activation (Andreadaki et al., 2015; Ojo et al., 2012, 2014). BKI compounds have high selectivity and low toxicity, being able to block exflagellation, oocyst formation and reducing sporozoite numbers at nanomolar ranges (Ojo et al., 2012, 2014). Due to low toxicity in mice, oral bioavailability and relative ease of generation from simple starting materials (Murphy et al., 2010), BKI compounds seems to fit important criteria for drug delivery in challenging malaria-endemic regions.

The *P. falciparum cdpk5* gene is transcribed in mature blood-stage schizonts and in invasive merozoites, suggesting a role in egress and/or erythrocyte invasion. In fact, *cdpk5* knockout parasites arrested as mature schizonts, evidencing an essential role of PfCDPK5 in parasite proliferation and egress from the erythrocyte (Dvorin et al., 2010). Pyrazoleamides are potent antiplasmodial agents, displaying activity against *Plasmodium* strains at nanomolar ranges. Considering biological and pharmaceutical profiles, PA21A092 (**49**) is a compound with the required characteristics of a promising a pre-clinical drug candidate. In addition, PA21A092 also inhibited both male and female *P. falciparum* gametocytes with an EC_{50} of 39 and 74 nM respectively. Importantly, this compound showed potency against *P. falciparum* and *P. vivax* multi-drug resistant field isolates with EC_{50}'s of 18 and 10 nM, respectively (Vaidya et al., 2014). In a bid to understanding the mode of action of pyrazoleamides, *P. falciparum* resistant to C2-1 was selected after continuous exposure with compound under sub lethal doses, resulting in the generation of three resistant strains. Whole genome sequencing of these resistant strains revealing SNPs in five genes, including CDPK5 (T392A) (Vaidya et al., 2014) (Fig. 3).

4.4 Mitogen-activated protein kinases family

Mitogen-activated protein kinases (MAPK) have been isolated from a broad range of species, comprising serine-threonine protein kinases that are known to play a central role in eukaryotic responses to a variety of *stimuli*.

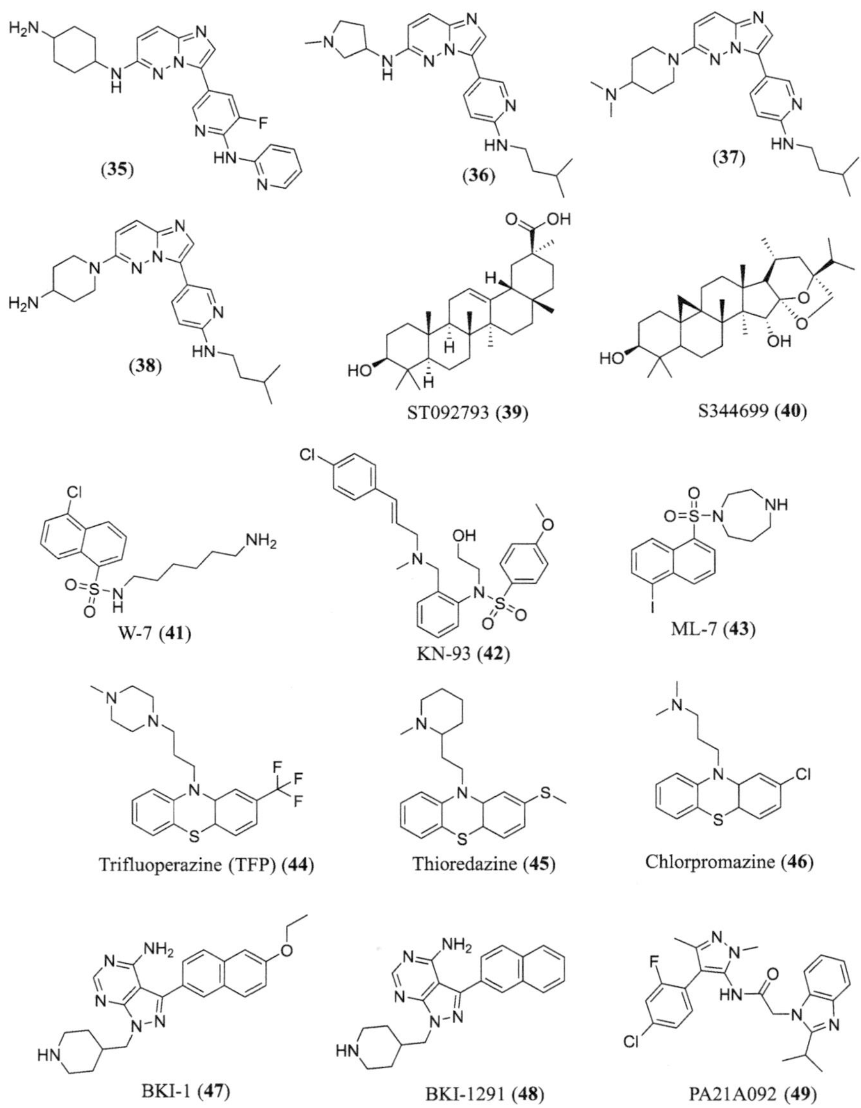

Fig. 3 Calcium-dependent protein kinases inhibitors.

The canonical MAPK activation motif is formed by the threonine-X-tyrosine (T-X-Y) site between the VII and VIII protein domains. The typical MAP kinase signaling cascade is composed of three hierarchical protein kinases: MAPKK, MAPK and MAP. Following a sequential activation pathway, MAPKK phosphorylates and activates MAPK, that when activated, phosphorylates and activates MAPK, that once activated regulates gene expression and several cellular functions through activation of transcription factors and other protein kinases (Widmann, Gibson, Jarpe, & Johnson, 1999). However, to date, no homologs for MAPKKs were identified in the *P. falciparum* kinome, indicating that parasites from the *Plasmodium* genus do not undergo the typical MAPK activation cascade (Dorin et al., 2004). In fact, the regulation at least one MAPK of *P. falciparum* (PfMAP-2) appears to involve two other kinases related to the NIMA family (PfNEK-1 and PfNEK-3) (Dvorin et al., 2010; Trivedi & Nag, 2012). When PfMAP-1 (but not PfMAP-1) was co-incubated with either PfNEK-1 or PfNEK-3, there was a synergistic increase in kinase activity (Dorin et al., 2001; Low, Lye, & Sim, 2007).

The PfMAP-1 kinase was first isolated using PCR with degenerate oligodeoxyribonucleotides designed to hybridize to conserved regions of cdc2-related kinases (Doerig et al., 1996). The second MAP kinase, named PfMAP-2, was isolated from *P. falciparum* also using degenerate PCR targeting two conserved regions of the MEKs (MAP/ERK) family (Dorin et al., 1999). PfMAP-1 comprises the classical activation motif threonine-X-tyrosine (TXY), and is able to auto-phosphorylate the tyrosine and threonine residues, alike other MAP kinases. *P. berghei map1* knockout showed no apparent defects during liver stage infection *in vitro* (Tewari et al., 2010). Additionally, no difference in *P. falciparum* blood stage development was found in *pfmap1* knockout parasite compared with wild-type controls, suggesting that this kinase may not be essential for parasite development (Dorin-Semblat et al., 2007).

Despite high homology with classical MAPKs, PfMAP-2 present differences when compared to canonical MAPKs. Instead of the conserved TXY at the activation motif like other MAPKs, PfMAP-2 has a TSH motif at the same place and the regulation of PfMAP-2 by PfNEK-3 seems to involve phosphorylation of the threonine residue 290 motif within the activation loop. It also contains insertions in its catalytic domain, culminating in distinct features from usual MAPKs (Dorin et al., 1999). PfMAP-2 is expressed mostly in gametocytes, although it is present at low levels in all blood stages. Studies in *P. berghei* have shown that the PbMAP-2 kinase is essential for exflagellation *in vitro* as well as for completion of the parasite's sexual cycle. Despite *pbmap2* knockout parasites undergoing differentiation into male gametocytes, these mutants are unable to progress from mitosis III to cytokinesis, to initiate axoneme motility, and to condense chromatin, suggesting a kinase role in these events (Rangarajan et al., 2005; Tewari, Dorin, Moon, Doerig, & Billker, 2005).

Protozoan MAPKs are more closely related to each other when compared to human MAPKs, therefore being considered promising drug targets. Five human p38 MAPK inhibitors were tested against *P. falciparum* strains *in vitro*. Of these, the pyridinylimidazole RWJ67657 (**50**) and the pyrrolobenzimidazole RWJ68198 (**51**) showed considerable activity against both chloroquine-resistant (W2) and chloroquine-sensitive (HB3) strains of *P. falciparum*, having an impact in trophozoite size and stage differentiation (Brumlik et al., 2011). However, the trisubstituted pyrroles from pyrrolobenzimidazole and pyridinylimidazole-based compounds were shown target parasite PKG from *Eimeria* species and not the p38 MAPK homolog (Biftu et al., 2005). Similarity modeling of the catalytic site from parasite PKG using the crystallized mammalian p38α MAPK predicted that these trisubstituted compounds exploit a hydrophobic binding pocket that overlaps the ATP binding site in a similar way to known p38 MAPK inhibitors. Pyridinylimidazole p38 MAPK inhibitors are smaller than ATP, therefore, they do not fully occupy the ATP binding site, leaving two large hydrophobic pockets on either side of the pyridine ring (Tong et al., 1997). Designing new pyridinylimidazole or similar pharmacophores that properly fill the ATP binding pocket can be a tool for developing specific pathogen kinase drugs. Modeling studies of the PfMAP-2 kinase pinpointed cubebin (**52**), a lignan of tetrahydrofuran-2-ol substituted class, as a predicted strong inhibitor. Kinase docking analysis demonstrated that cubebin presented favorable free binding energy predicted values (−9.5 Kcal/mol) and a good inhibition constant (Ki = 0.1 μM). Still, laboratory experiments are required to validate the computational modeling results (Gupta et al., 2016) (Fig. 4).

4.5 PfPK7

Although two MAP kinases have been identified in the *P. falciparum* kinome, to date, no typical MEK (which is responsible for the phosphorylation of

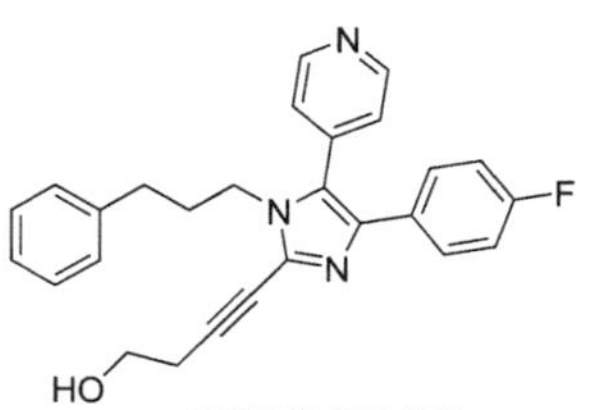

RWJ67657 (**50**)

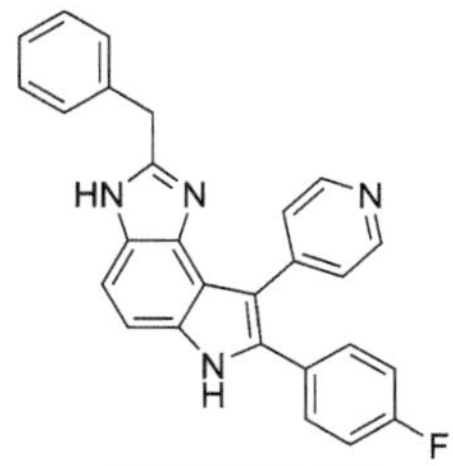

RWJ68198 (**51**)

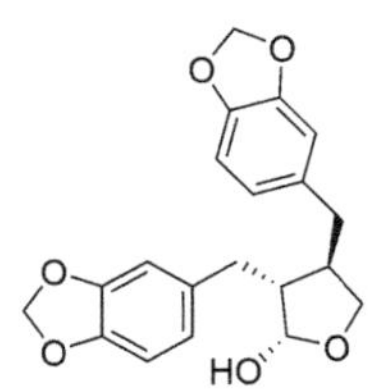

Cubebin (**52**)

Fig. 4 Mitogen-activated kinases inhibitors.

MAP kinases) homolog was found. In an attempt to discover MEK enzymes in *P. falciparum*, alignments of MEK sequences from different organisms showed that PfPK7 has high homology with the MEK3/6 family. However, their N-terminal region displays high similarity with the fungal PKA. Besides this divergence between PfPK7 with members of the MEK family, PfPK7 does not demonstrate the ability to phosphorylate *P. falciparum* MAK kinases (PfMAP-1 and PfMAP-2), suggesting that PfPK7 is not a MEK kinase (Dorin et al., 2004; Merckx et al., 2008).

Whereas PfPK7 shares characteristics of different kinase families, its classification into an appropriate ePK group is challenging, as many common characteristic functions of some groups are not present in this kinase. For instance, the glycine triad motif (GXGXXG) responsible for ATP binding in many eukaryotes is substituted by an NQGKFN sequence in PfPK7. Additionally, kinase regulation by phosphorylation of one or two residues in the typical motif (SXXXS/T present in animal and yeast, or S/TXXXXXS/T in plants) does not occur in PfPK7 (Dorin et al., 2004). However, several important amino acids for kinase function have been identified, such as (i): Lys55, that interacts with ATP; (ii): Asp172 and Asn177, that are catalytic residues acting as base acceptors; (iii): Asp190 within the DFG motif, that binds Mg^{2+} or Mn^{2+} ions; (iv): Glu215, that contributes for structural stability of the C-terminal lobe through the formation of a salt bridge with Arg of subdomain IX; (v): Asp229, that promotes stability of the catalytic loop of subdomain VI and; (vi): Asp172, that aids in controlling phosphate transfer (Dorin et al., 2004).

Efforts have been made to find potent PfPK7 inhibitors through HTS, and a class of imidazopyridazines was signaled as promising PfPK7 inhibitors, such as compound (**53**) (IC_{50} of 11.6 μM) (Bouloc et al., 2008). Co-crystallization of PfPK7 with an inhibitor and screening of a library of 568 small molecules, also resulted in the identification of compounds with the same imidazopyridazine scaffold (Merckx et al., 2008). Based on Structure-Activity Relationship (SAR) analyses of (**53**), 11 molecules were designed through optimization of aryl and amine substituents, resulting in compound (**54**) which displayed improved potency (IC_{50} of 0.28 μM) (Bouloc et al., 2008). It has also been proposed that modifications in the 4-cyanophenyl motif could help to optimize imidazopyridazines, since modifications in this position could assist for salt bridge formation involving Lys55, but these modified compounds had lower efficacy than (**54**). Additionally, changes in the amino alcohol group led to further optimization of imidazopyridazines, and other compounds (such as **55** and **56**), which displayed activity against the *P. falciparum* 3D7 and K1 strains. However, cytotoxicity and selectivity were under optimal. Cell-active compound screening of 14,000 compounds against PfPK7, also identified two others imidazopyridazines hits (**57** and **58**), with IC_{50} of 0.44 and 0.42 μM, respectively (Crowther et al., 2016). In addition, predictive 2D and 3D QSAR models based on imidazopyridazine scaffold were developed, which may aid guiding rational design of novel inhibitors (Sahu & Kohli, 2012).

Further work based on a pyrazolopyrimidine scaffold, generated two compounds (**59** and **60**) that inhibited PfPK7 in micromolar ranges (IC_{50} 10–20 μM). Docking studies with (**61**), demonstrated that the molecule forms a hydrogen bond with Met120 and Glu118 of PfPK7 (Klein, Dinér, Dorin-Semblat, Doerig, & Grøtli, 2009). Collectively and as described above, PfPK7 is likely to play a crucial role in the asexual parasite cycle as well as in oocyst formation, flagging this kinase as an attractive multi-stage target. Still, more efforts need to be made toward clarification of its selectivity and cytotoxicity parameters (Fig. 5).

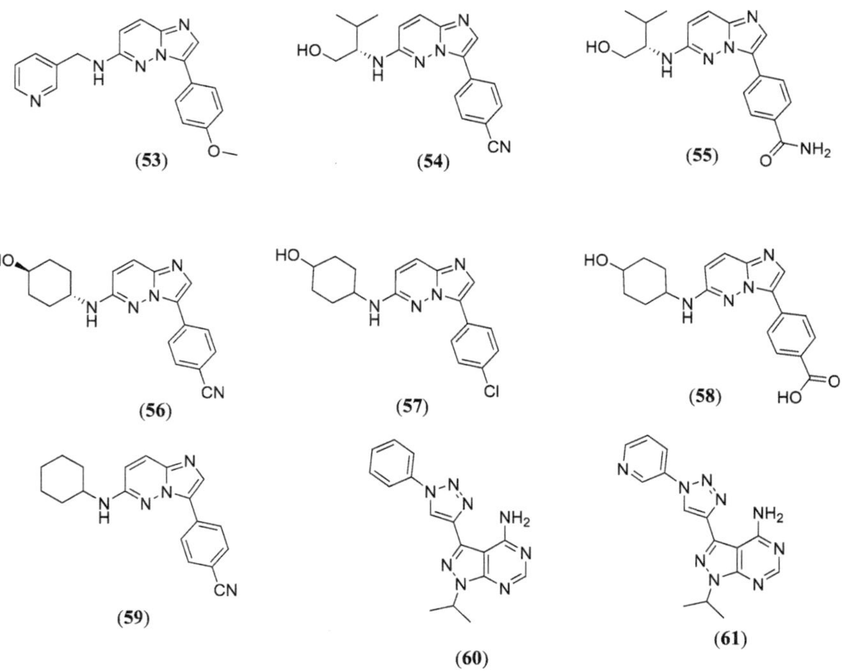

Fig. 5 PfPK7 inhibitors.

4.6 NIMA/NEK kinases family

The *P. falciparum* kinome includes a cluster of four protein kinases related to the NIMA/Nek family (PfNEK-1 to -4). The predominant expression of three of these four kinases (PfNEK-2, -3 and -4) in gametocytes suggests a probable role of this family in the sexual development of the parasite (Le Roch, 2003). This is further corroborated by studies that have demonstrated that these three kinases are considered as redundant for the erythrocyte asexual cycle (Reininger et al., 2005, 2009; Solyakov et al., 2011).

Although it has been shown that schizont-stage parasites expressing PfNEK-4 are more likely to differentiate into gametocytes, this kinase is not required for gametogenesis in both *P. falciparum* and *P. berghei* (Reininger, Garcia, Tomlins, Müller, & Doerig, 2012). However, experiments using *P. berghei* have indicated that PbNEK-4 is essential in the control of zygote to ookinete differentiation, by regulating DNA replication in the zygote, prior to meiosis (Reininger et al., 2005). Interestingly, PbNEK-4 is expressed only in female gametes and *pbnek4*-male gametes were able to fertilize wild-type female gametes that developed into ookinetes (Khan et al., 2005). A similar phenotype to that caused by the loss of PbNEK-4 was observed in *pbnek2*-female gametes, which, added to the fact that the PbNEK-2 and PbNEK-4 do not activate each other *in vitro*, suggests that they have an important function in ookinete development (Reininger et al., 2009), and may be thus flagged as promising targets for transmission-blocking intervention.

PfNEK-1 is expressed in both asexual and gametocyte stages, being the only essential kinase of the NIMA family for the erythrocytic stage of the life cycle. Although the precise role of this protein kinase in the development of the parasite is not yet known, it has been suggested that, as in other eukaryotes, *Plasmodium* NEK-1 acts in microtubule organization during mitosis (Carvalho, Doerig, & Reininger, 2013). Biochemical characterization and recombinant production of this kinase has allowed inhibition assays to screen for potential drugs. A medium-throughput screen using a panel of 10,480 small molecules revealed 23 compounds with an IC_{50} below $10\,\mu M$ (Dorin-Semblat et al., 2011). Additionally, natural compounds (xestoquinone and halenaquinone) (**62** and **63**) isolated from a Vanuatu marine sponge *Xestospongia* sp. were firstly evaluated as potential inhibitors of PfNEK-1. Xestoquinone exhibited an IC_{50} of $1.1\,\mu M$, whereas halenaquinone inhibited PfNEK-1 with an IC_{50} near to $3\,\mu M$, showing a very good correlation between PfNEK-1 and antiplasmodial activity. In addition, treatment with 5 mg/kg for 4 days in mice infected with *P. berghei* NK65 reduced the mortality rates by 45% (Laurent et al., 2006). What's more, three alisiaquinones (A, C and alisiaquinol) (**64**, **65** and **66**), which are related to xestoquinone, were isolated from a New Caledonian deep water sponge and tested against PfNEK-1. The IC_{50} on PfNEK-1 for alisiaquinone A and alisiaquinol were comparable to that observed with xestoquinone (~1 μM) whereas alisiaquinone C exhibited poor inhibition. However, alisiaquinone C, which differs from alisiaquinone A by a taurine substituent linked to the quinone, showed better *in vitro* activity on *P. falciparum*, suggesting that this compound has different targets apart from PfNEK-1 (Desoubzdanne et al., 2008) (Fig. 6).

4.7 Aurora kinases family

Aurora kinases belong to a family of highly conserved serine/threonine kinases, important for mitotic events and cytokinesis. While only one member was originally identified in *Saccharomyces cerevisiae*, mammalian cells contain three members, named Aurora kinase-A, -B, and -C, whose roles include centrosome maturation, creation of mitotic spindle and alignment of chromosomes, resulting in equal chromosome segregation into daughter cells (D'Assoro, Haddad, & Galanis, 2016). Three Aurora-related serine/threonine kinases have been identified in the *P. falciparum* kinome (PfARK-1, -2, and -3) (Reininger, Wilkes, Bourgade, Miranda-Saavedra, & Doerig, 2011). PfARK-2 clusters within the "classic" Aurora kinases due to the existence of a similar signature motif of Aurora kinases in the potential activation loop,

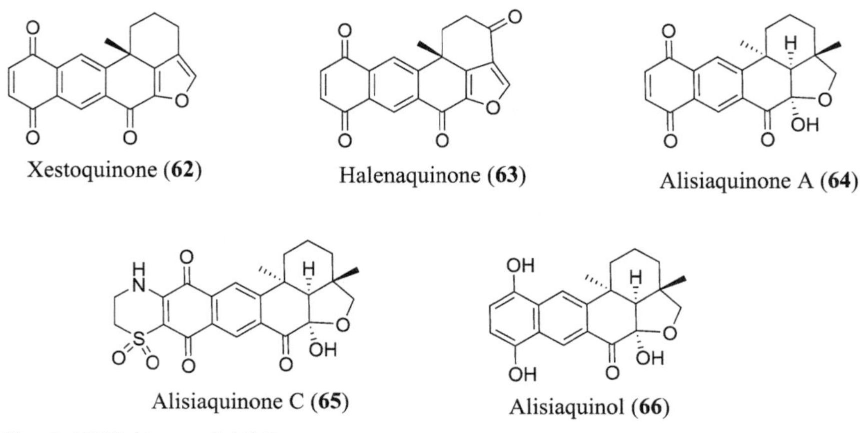

Fig. 6 NIMA-kinases inhibitors.

(DFGWSxxxxxxxxRxTxCGTxDYLPPE), while both PfARK-1 and PfARK-3 do not hold such a signature motif. Imaging of PfARK-1 tagged with a C-terminal green fluorescent protein (GFP) showed that it is located at the spindle pole body embedded in the nuclear membrane during early stages of nuclear division of erythrocytic schizogony, suggesting that PfARK-1 acts in the mitotic spindle formation of *Plasmodium* species (Reininger et al., 2011).

Unsuccessful attempts to disrupt the *pfark1*, *pfark2* or *pfark3 loci* indicate that these Aurora-related kinases are essential for *P. falciparum* asexual erythrocytic development (Reininger et al., 2011; Solyakov et al., 2011). However, to date, no inhibition assay with these kinases was carried out to try to identify possible malaria inhibitors. Nevertheless, screening of around 2 million compounds from the GSK library identified both the PfARK-1 and PfARK as potential targets (Gamo et al., 2010). Based on the importance of these Aurora kinases in the biology of *P. falciparum* development, Patel et al. repurposed three inhibitors of human Aurora kinases as antimalarial agents, which interestingly, had been previously shown to also be effective against *Trypanosoma cruzi* (Patel et al., 2014). One of these, hesperadin (**67**), displayed good *in vitro* activity against the D6 *P. falciparum* strain, with an IC_{50} of 0.01 μM, even though cytotoxicity was detected at concentrations of >0.2 μM. Analogs of hesperidin with variations of the R4 substituent showed reduction in cytotoxicity, but this was also met with a decrease in antimalarial activity (Patel et al., 2014). Optimization of this chemotype may hold promise for the development of novel antimalarials (Fig. 7).

4.8 Glycogen synthase kinase-3 family

Plasmodium glycogen synthase kinase-3 (GSK-3) is a serine/threonine kinase classified as ePK, with maximal expression in the early trophozoite stage. While present at low levels in the parasite, this kinase tends to be exported

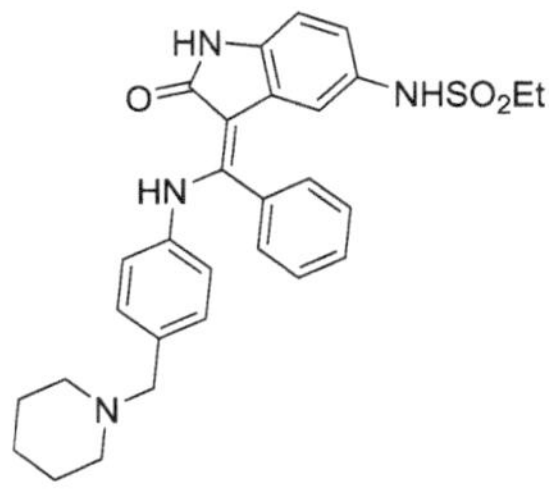

Hesperadin (**67**)

Fig. 7 Aurora kinase inhibitor.

quickly to the cytoplasm of the host red blood cell. This suggests that PfGSK-3 may be involved in metabolism and cell cycle, as well in the differentiation of ring stages to schizonts and to gametocytes. Because PfGSK-3 is rapidly transported to the cytoplasm of the erythrocyte, it is believed that it may also act in the host cell (Droucheau et al., 2004).

When compared to both human homologs HsGSK-3α and HsGSK-3β, PfGSK-3 displays a high state of amino acid conservation (62.4/58.9% similarity with human GSK-3α/GSK-3β, respectively). The similarity between the human and *Plasmodium* kinases is evidenced by the presence of similar residues in the active site (Lys85, Glu97, Asp181, Asp200 in humans; and Lys108, Glu20, Asp206, Asp225 in the parasite), in the phosphate binding site (Arg96, Arg180, Lys205, Val214 in humans; and Arg119, Arg205, Lys230, Val239 in parasite), and in the residues believed to be involved in protein activation (Arg220, Arg223, Tyr216 in humans; and Arg245, Arg248, Tyr241 in the parasite). One particular characteristic of PfGSK-3 is an N-terminal extension with an extra 23 amino acids. The function of this extension is still unclear, but it is believed that this domain may be related to excretion of PfGSK-3 into the RBC cytoplasm and/or the Maurer's clefts (Droucheau et al., 2004).

PfGSK3 differs from its human homolog mainly in the ATP-binding pocket. This important difference can aid in the design of selective inhibitors. Such alterations in the ATP pocket may cause modifications in the binding sites, due to volume changes caused by a difference in the gatekeeper residue Leu132 for Met145 in the parasite, as well as the inability to form the common salt bridge between Arg141 and Glu137 in humans, and the presence of a residue of Lys154 and Gln150 in *P. falciparum*, that creates a steric clash (Osolodkin, Zakharevich, Palyulin, Danilenko, & Zefirov, 2011). Another important difference may be related to regulation of GSK-3, since while there is corresponding Tyr 216 activation in *Plasmodium* species (Tyr229 in *P. falciparum* and Tyr 281 in *P. vivax*), there shouldn't be an inhibitory Ser9, because this residue is absent in malaria parasites (Droucheau et al., 2004).

Manzamines are cytotoxic marine alkaloids extracted from sponges, characterized by a 5-,6-,6-,8-,13-membered heterocyclic ring system attached to a β-carboline moiety (Hu, Hamann, Hill, & Kelly, 2003), that display significant biological activities against diverse microbes (Ang et al., 2000; Radwan, Hanora, Khalifa, & Abou-el-ela, 2012; Rao et al., 2003) and neuronal inflammatory diseases (Yousaf et al., 2004). Manzamine A (**68**) has been shown to be a promising antimalarial agent, with IC_{50} value of 0.0045 and 0.008 μg/mL against the *P. falciparum* D6 and W3 strains, respectively (Rao et al., 2003). Since (**68**) has been shown to inhibit HsGSK-3β (Hamann et al., 2007), there is a good probability that it will also be active against the *Plasmodium* homolog GSK-3.

Paullones are 7,12-dihydroindolo[3,2-d][1]benzazepin-6(5H)-ones that showed potential inhibition of HsCDKs and HsGSK-3 (Tolle & Kunick, 2011). Paullone (**69**), kenpaullone (**70**), alsterpaullone (**71**) and gwennpaullone (**72**) were tested against both PfGSK-3 and HsGSK-3β, with IC_{50} values of 18.0, 8.0, 0.32, and 1.5 μM against *P. falciparum*, respectively and 0.62, 0.023, 0.004, and 0.04 μM, for the mammalian homolog, respectively. Thus, paullones are potential inhibitors of PfGSK-3, albeit the lack of selectivity presents a serious challenge for their use in humans. Nevertheless, an HTS identified 3,6-diamino-4-(2-halophenyl)-2-benzoylthieno[2,3-*b*]pyridine-5-carbonitriles (**73**) as a common structure that is able to inhibit PfGSK-3 under submicromolar concentrations, while inhibiting HsGSK-3β at micromolar concentrations (Fugel et al., 2013). Based on the work describing the compounds with the 3,6-diamino-4-(2-halophenyl)-2-benzoylthieno[2,3-*b*]pyridine-5-carbonitriles scaffold, QSAR models were developed to help the design and prediction of novel PfGSK-3 inhibitors since there are no crystallographic structures available (Balasaheb Aher & Roy, 2015).

In order to improve the aqueous solubility and antiplasmodial activity, molecular modifications were performed in the 4-phenylthieno[2,3-*b*]pyridine core, in which the use of alkylamino side chains at position 4 led to the improvement of antiplasmodial activity and aqueous solubility of selective PfGSK-3 inhibitors (Masch et al., 2019). Compound **74** showed a promising profile for selective inhibition of PfGSK-3, with a PfGSK-3 IC_{50} = 0.72 μM and HsGSK-3 IC_{50} = 40.2 μM.

The use of a Biginelli multicomponent reaction and subsequent one-pot substitution/cyclization led to the synthesis of several dihydropyrimidinones and pyrrolo[3,4-d]-pyrimidinodiones, which were evaluated against the chloroquine-resistant *P. falciparum* W2 strain. Compound **75** showed a *P. falciparum* W2 IC_{50} of 1.76 μM and a selectivity index of 12.5. Although compound **75** has not been tested against the PfGSK-3 enzyme, an *in silico* target fishing approach identified it as a likely target for **75** (Rogerio et al., 2018).

The fact the PfGSK-3 is an essential kinase for the parasite (Derbyshire et al., 2014; Solyakov et al., 2011), makes it a promising target for the

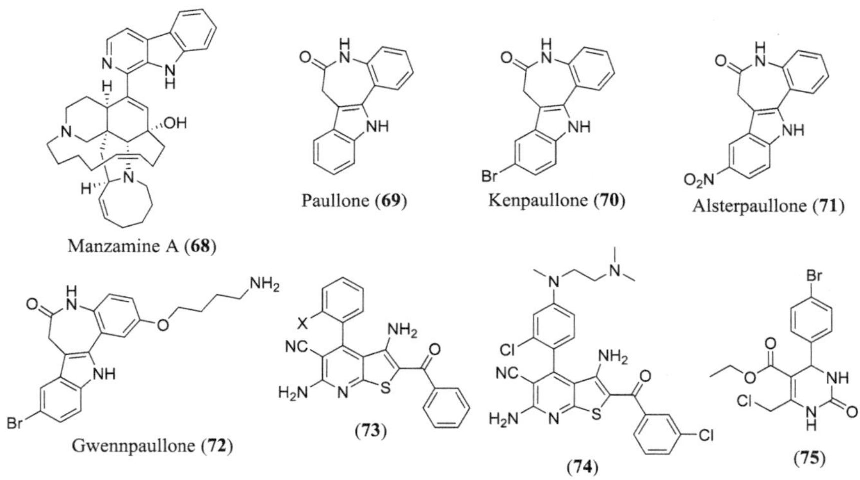

Fig. 8 Glycogen synthase kinase-3 inhibitors.

development of antimalarial drugs. However, due to the high similarity between human and parasite GSK-3, discovery of a selective inhibitor becomes somewhat challenging. Interestingly, the GSK-3 of *Plasmodium* species appears to be more sensitive to small inhibitors, because of the occurrence of valine residues (Val110 and Val135, human sequence) rather than isoleucine in the parasite kinase, which reduces its active site volume (Droucheau et al., 2004; Osolodkin et al., 2011). Thus, efforts should be concentrated in these and other differences toward the design of potent and selective inhibitors of *Plasmodium* GSK-3 (Fig. 8).

5. Conclusions and perspectives

Decades of research in human kinases led to development of several drugs approved by the FDA (predominantly anticancer agents), lending weight to the hypothesis that kinases are attractive drug targets. A large number of compound libraries were developed to inhibit human kinases, and phenotypic screening efforts have been demonstrated that these kinase inhibitors can also act as new antimalarial drugs. Unsurprisingly, several studies have shown that malarial protein kinases are involved in a large number of cellular pathways, being essential for the development of all stages of parasite life cycle. However, protein kinases also are essential to cellular process in humans, and thus the identification of selective *Plasmodium* kinase inhibitors is a major challenge. Nonetheless, there is encouraging recent evidence to suggest that selectivity can be achieved, as some inhibitors have demonstrated significant differential activity between parasite and human kinases. Moreover, the large phylogenetic distance between malaria parasites and its human host as reflected in marked differences in protein structures, that can be exploited to develop selective inhibitors. For example, the malaria kinome contains kinases with small gatekeeper residues (such as PKG and some of the CDPKs), which can be targeted by bulky analogs of ATP mimetics (Hui, El Bakkouri, & Sibley, 2015).

New advances in the field of kinase inhibitor discovery may assist toward the rational design and development of novel drugs for malaria. Currently, many assays using parasite recombinant protein kinases have been described. Additionally, chemical genetic approaches, phosphoproteomic analysis, molecular docking and the availability of protein structure data will be useful for the identification of new inhibitors, which can be used for further hit-to-lead optimization to increase selectivity, for example. In summary, although still in its infancy, great strides have been made to increase knowledge about the *Plasmodium* kinome, and kinase inhibitors may be a new hope for the development of novel antimalarials.

Acknowledgments

This work was supported by Fundação de Amparo à Pesquisa do Estado de São Paulo (FAPESP, Grant 2017/18611-7), Conselho Nacional de Desenvolvimento Científico e Tecnológico (CNPq) and a joint grant of FCT of Portugal and FAPESP of Brazil, within the scope of the project "Identification of new antimalarial treatments through a target-centered "drug repositioning" approach" (ref PTDC/SAU-PAR/28459/2017). GCC and TAT were sponsored by a FAPESP fellowship (Grants 2015/20774-6 and 2019/27626-3, respectively). MNN was supported by a CAPES fellowship. CHA and FTMC are CNPq researcher fellows.

References

Abdul-Ghani, R., Farag, H. F., & Allam, A. F. (2013). Sulfadoxine-pyrimethamine resistance in *Plasmodium falciparum*: A zoomed image at the molecular level within a geographic context. *Acta Tropica, 125*(2), 163–190. https://doi.org/10.1016/j.actatropica.2012.10.013.

Alam, M. M., Solyakov, L., Bottrill, A. R., Flueck, C., Siddiqui, F. A., Singh, S., et al. (2015). Phosphoproteomics reveals malaria parasite Protein Kinase G as a signalling hub regulating egress and invasion. *Nature Communications, 6*, 7285. https://doi.org/10.1038/ncomms8285.

Anamika, Srinivasan, N., & Krupa, A. (2004). A genomic perspective of protein kinases in *Plasmodium falciparum*. *Proteins: Structure, Function, and Bioinformatics, 58*(1), 180–189. https://doi.org/10.1002/prot.20278.

Andreadaki, M., Mollenkopf, H.-J., Nika, F., Brady, D., Tewari, R., Matuschewski, K., et al. (2015). Global expression profiling reveals shared and distinct transcript signatures in arrested act2(−) and CDPK4(−) Plasmodium berghei gametocytes. *Molecular and Biochemical Parasitology, 201*(2), 100–107. https://doi.org/10.1016/j.molbiopara.2015.07.001.

Ang, K. K. H., Holmes, M. J., Higa, T., Hamann, M. T., Kara, U. A. K., & Hemother, A. N. A. G. C. (2000). In vivo antimalarial activity of the beta-carboline alkaloid manzamine A. *Antimicrobial Agents and Chemotherapy*, *44*(6), 1645–1649.

Ansell, K. H., Jones, H. M., Whalley, D., Hearn, A., Taylor, D. L., Patin, E. C., et al. (2014). Biochemical and antiparasitic properties of inhibitors of the *Plasmodium falciparum* calcium-dependent protein kinase PfCDPK1. *Antimicrobial Agents and Chemotherapy*, *58*(10), 6032–6043. https://doi.org/10.1128/AAC.02959-14.

Baird, K. (2015). Origins and implications of neglect of G6PD deficiency and primaquine toxicity in *Plasmodium vivax* malaria. *Pathogens and Global Health*, *109*(3), 93–106. https://doi.org/10.1179/2047773215y.0000000016.

Anstey, N. M., Douglas, N. M., Poespoprodjo, J. R., & Price, R. N. (2012). *Plasmodium vivax*: Clinical spectrum, risk factors and pathogenesis. In S. I. Hay, R. Price, & J. K. Baird (Eds.), *Vol. 80. Advances in parasitology* Elsevier. https://doi.org/10.1016/B978-0-12-397900-1.00003-7.

Baird, J. K., & Hoffman, S. L. (2004). Primaquine therapy for malaria. *Clinical Infectious Diseases*, *39*(9), 1336–1345. https://doi.org/10.1086/424663.

Baker, D. A., Stewart, L. B., Large, J. M., Bowyer, P. W., Ansell, K. H., Jiménez-Díaz, M. B., et al. (2017). A potent series targeting the malarial cGMP-dependent protein kinase clears infection and blocks transmission. *Nature Communications*, *8*(1), 1–9. https://doi.org/10.1038/s41467-017-00572-x.

Balasaheb Aher, R., & Roy, K. (2015). First report on exploring classification and regression based QSAR modelling of *Plasmodium falciparum* glycogen synthase kinase (Pf GSK-3) inhibitors. *SAR and QSAR in Environmental Research*, *26*(11), 959–976. https://doi.org/10.1080/1062936X.2015.1104518.

Balestra, A. C., Zeeshan, M., Rea, E., Pasquarello, C., Brusini, L., Mourier, T., et al. (2020). A divergent cyclin/cyclin-dependent kinase complex controls the atypical replication of a malaria parasite during gametogony and transmission. *eLife*, *9*, 1–25. https://doi.org/10.7554/eLife.56474.

Bansal, A., Singh, S., More, K. R., Hans, D., Nangalia, K., Yogavel, M., et al. (2013). Characterization of *Plasmodium falciparum* calcium-dependent protein kinase 1 (PFCDPK1) and its role in microneme secretion during erythrocyte invasion. *Journal of Biological Chemistry*, *288*(3), 1590–1602. https://doi.org/10.1074/jbc.M112.411934.

Bento, A. P., Gaulton, A., Hersey, A., Bellis, L. J., Chambers, J., Davies, M., et al. (2014). The ChEMBL bioactivity database: An update. *Nucleic Acids Research*, *42*(Database issue), D1083–D1090. https://doi.org/10.1093/nar/gkt1031.

Bhattacharjee, A. K., Geyer, J. A., Woodard, C. L., Kathcart, A. K., Nichols, D. A., Prigge, S. T., et al. (2004). A three-dimensional in silico pharmacophore model for inhibition of *Plasmodium falciparum* cyclin-dependent kinases and discovery of different classes of novel Pfmrk specific inhibitors. *Journal of Medicinal Chemistry*, *47*(22), 5418–5426. https://doi.org/10.1021/jm040108f.

Biftu, T., Feng, D., Ponpipom, M., Girotra, N., Liang, G., Qian, X., et al. (2005). Synthesis and SAR of 2,3-diarylpyrrole inhibitors of parasite cGMP-dependent protein kinase as novel anticoccidial agents. *Bioorganic & Medicinal Chemistry Letters*, *15*(13), 3296–3301. https://doi.org/10.1016/j.bmcl.2005.04.060.

Billker, O., Dechamps, S., Tewari, R., Wenig, G., Franke-Fayard, B., & Brinkmann, V. (2004). Calcium and a calcium-dependent protein kinase regulate gamete formation and mosquito transmission in a malaria parasite. *Cell*, *117*(4), 503–514. https://doi.org/10.1016/S0092-8674(04)00449-0.

Billker, O., Lourido, S., & Sibley, L. D. (2009). Calcium-dependent signaling and kinases in apicomplexan parasites. *Cell Host & Microbe*, *5*(6), 612–622. https://doi.org/10.1016/j.chom.2009.05.017.

Bouloc, N., Large, J. M., Smiljanic, E., Whalley, D., Ansell, K. H., Edlin, C. D., et al. (2008). Synthesis and in vitro evaluation of imidazopyridazines as novel inhibitors of the malarial kinase PfPK7. *Bioorganic and Medicinal Chemistry Letters*, *18*(19), 5294–5298. https://doi.org/10.1016/j.bmcl.2008.08.043.

Boutros, T., Chevet, E., & Metrakos, P. (2008). Mitogen-activated protein (MAP) kinase/MAP kinase phosphatase regulation: Roles in cell growth, death, and cancer. *Pharmacological Reviews*, *60*(3), 261–310. https://doi.org/10.1124/pr.107.00106.

Bracchi-Ricard, V., Barik, S., Delvecchio, C., Doerig, C., Chakrabarti, R., & Chakrabarti, D. (2000). PfPK6, a novel cyclin-dependent kinase/mitogen-activated protein kinase-related protein kinase from *Plasmodium falciparum*. *Biochemical Journal*, *347*(1), 255. https://doi.org/10.1042/0264-6021:3470255.

Brandt, G. S., & Bailey, S. (2013). Dematin, a human erythrocyte cytoskeletal protein, is a substrate for a recombinant FIKK kinase from *Plasmodium falciparum*. *Molecular and Biochemical Parasitology*, *191*(1), 20–23. https://doi.org/10.1016/j.molbiopara.2013.08.003.

Brochet, M., Collins, M. O., Smith, T. K., Thompson, E., Sebastian, S., Volkmann, K., et al. (2014). Phosphoinositide metabolism links cGMP-dependent protein kinase G to essential Ca2 + signals at key decision points in the life cycle of malaria parasites. *PLoS Biology*, *12*(3), e1001806. https://doi.org/10.1371/journal.pbio.1001806.

Brumlik, M. J., Nkhoma, S., Kious, M. J., Thompson, G. R., Patterson, T. F., Siekierka, J. J., et al. (2011). Human p38 mitogen-activated protein kinase inhibitor drugs inhibit *Plasmodium falciparum* replication. *Experimental Parasitology*, *128*(2), 170–175. https://doi.org/10.1016/j.exppara.2011.02.016.

Cabrera, D. G., Horatscheck, A., Wilson, C. R., Basarab, G., Eyermann, C. J., & Chibale, K. (2018). Plasmodial kinase inhibitors: License to cure? *Journal of Medicinal Chemistry*, *61*(18), 8061–8077. https://doi.org/10.1021/acs.jmedchem.8b00329.

Caridha, D., Kathcart, A. K., Jirage, D., & Waters, N. C. (2010). Activity of substituted thiophene sulfonamides against malarial and mammalian cyclin dependent protein kinases. *Bioorganic and Medicinal Chemistry Letters*, *20*(13), 3863–3867. https://doi.org/10.1016/j.bmcl.2010.05.039.

Carter, R., & Mendis, K. N. (2002). Evolutionary and historical aspects of the burden of malaria evolutionary and historical aspects of the burden of malaria. *Clinical Microbiology Reviews*, *15*(4), 564–594. https://doi.org/10.1128/CMR.15.4.564.

Carvalho, T. G., Doerig, C., & Reininger, L. (2013). Nima- and Aurora-related kinases of malaria parasites. *Biochimica et Biophysica Acta—Proteins and Proteomics*, *1834*(7), 1336–1345. https://doi.org/10.1016/j.bbapap.2013.02.022.

Cavagnino, A., Rossi, F., & Rizzi, M. (2011). The potent antiplasmodial calmodulin-antagonist trifluoperazine inhibits *Plasmodium falciparum* calcium-dependent protein kinase 4. *Protein and Peptide Letters*, *18*(12), 1273–1279. https://doi.org/10.2174/092986611797642742.

Chapman, T. M., Osborne, S. A., Bouloc, N., Large, J. M., Wallace, C., Birchall, K., et al. (2013). Substituted imidazopyridazines are potent and selective inhibitors of *Plasmodium falciparum* calcium-dependent protein kinase 1 (PfCDPK1). *Bioorganic & Medicinal Chemistry Letters*, *23*(10), 3064–3069. https://doi.org/10.1016/j.bmcl.2013.03.017.

Chapman, T. M., Osborne, S. A., Wallace, C., Birchall, K., Bouloc, N., Jones, H. M., et al. (2014). Optimization of an imidazopyridazine series of inhibitors of *Plasmodium falciparum* calcium-dependent protein kinase 1 (PfCDPK1). *Journal of Medicinal Chemistry*, *57*(8), 3570–3587. https://doi.org/10.1021/jm500342d.

Chen, Y., Jirage, D., Caridha, D., Kathcart, A. K., Cortes, E. A., Dennull, R. A., et al. (2006). Identification of an effector protein and gain-of-function mutants that activate Pfmrk, a malarial cyclin-dependent protein kinase. *Molecular and Biochemical Parasitology*, *149*(1), 48–57. https://doi.org/10.1016/j.molbiopara.2006.04.004.

Collins, C. R., Hackett, F., Strath, M., Penzo, M., Withers-Martinez, C., Baker, D. A., et al. (2013). Malaria parasite cGMP-dependent protein kinase regulates blood stage merozoite secretory organelle discharge and egress. *PLoS Pathogens, 9*(5), e1003344. https://doi.org/10.1371/journal.ppat.1003344.

Collins, M. O., Wright, J. C., Jones, M., Rayner, J. C., & Choudhary, J. S. (2014). Confident and sensitive phosphoproteomics using combinations of collision induced dissociation and electron transfer dissociation. *Journal of Proteomics, 103*, 1–14. https://doi.org/10.1016/j.jprot.2014.03.010.

Crowther, G. J., Hillesland, H. K., Keyloun, K. R., Reid, M. C., Lafuente-Monasterio, M. J., Ghidelli-Disse, S., et al. (2016). Biochemical screening of five protein kinases from *Plasmodium falciparum* against 14,000 cell-active compounds. *PLoS One, 11*(3), e0149996. https://doi.org/10.1371/journal.pone.0149996.

D'Assoro, A. B., Haddad, T., & Galanis, E. (2016). Aurora-A kinase as a promising therapeutic target in cancer. *Frontiers in Oncology, 5*(January), 1–8. https://doi.org/10.3389/fonc.2015.00295.

Dawn, A., Singh, S., More, K. R., Siddiqui, F. A., Pachikara, N., Ramdani, G., et al. (2014). The central role of cAMP in regulating *Plasmodium falciparum* merozoite invasion of human erythrocytes. *PLoS Pathogens, 10*(12), e1004520. https://doi.org/10.1371/journal.ppat.1004520.

de Santana Filho, F. S., Arcanjo, A. R., Chehuan, Y. M., Costa, M. R., Martinez-Espinosa, F. E., Vieira, J. L., et al. (2007). Chloroquine-resistant *Plasmodium vivax*, Brazilian Amazon. *Emerging Infectious Diseases, 13*(7), 1125–1126. https://doi.org/10.3201/eid1307.061386.

Deng, W., & Baker, D. A. (2002). A novel cyclic GMP-dependent protein kinase is expressed in the ring stage of the *Plasmodium falciparum* life cycle. *Molecular Microbiology, 44*(5), 1141–1151. https://doi.org/10.1046/j.1365-2958.2002.02948.x.

Derbyshire, E. R., Zuzarte-Luís, V., Magalhães, A. D., Kato, N., Sanschagrin, P. C., Wang, J., et al. (2014). Chemical interrogation of the malaria kinome. *Chembiochem: A European Journal of Chemical Biology, 15*(13), 1920–1930. https://doi.org/10.1002/cbic.201400025.

Deshmukh, A. S., Agarwal, M., Mehra, P., Gupta, A., Gupta, N., Doerig, C. D., et al. (2015). Regulation of *Plasmodium falciparum* origin recognition complex subunit 1 (PfORC1) function through phosphorylation mediated by CDK-like kinase PK5. *Molecular Microbiology, 98*(1), 17–33. https://doi.org/10.1111/mmi.13099.

Desoubzdanne, D., Marcourt, L., Raux, R., Chevalley, S., Dorin, D., Doerig, C., et al. (2008). Alisiaquinones and alisiaquinol, dual inhibitors of *Plasmodium falciparum* enzyme targets from a New Caledonian deep water sponge. *Journal of Natural Products, 71*(7), 1189–1192. https://doi.org/10.1021/np8000909.

Dessens, J. T., Beetsma, A. L., Dimopoulos, G., Wengelnik, K., Crisanti, A., Kafatos, F. C., et al. (1999). CTRP is essential for mosquito infection by malaria ookinetes. *EMBO Journal, 18*(22), 6221–6227. https://doi.org/10.1093/emboj/18.22.6221.

Diaz, C. A., Allocco, J., Powles, M. A., Yeung, L., Donald, R. G. K., Anderson, J. W., et al. (2006). Characterization of Plasmodium falciparum cGMP-dependent protein kinase (PfPKG): Antiparasitic activity of a PKG inhibitor. *Molecular and Biochemical Parasitology, 146*(1), 78–88. https://doi.org/10.1016/j.molbiopara.2005.10.020.

Doerig, C., Abdi, A., Bland, N., Eschenlauer, S., Dorin-Semblat, D., Fennell, C., et al. (2010). Malaria: Targeting parasite and host cell kinomes. *Biochimica et Biophysica Acta—Proteins and Proteomics, 1804*(3), 604–612. https://doi.org/10.1016/j.bbapap.2009.10.009.

Doerig, C., & Grevelding, C. G. (2015). Targeting kinases in Plasmodium and Schistosoma: Same goals, different challenges. *Biochimica et Biophysica Acta (BBA)—Proteins and Proteomics, 1854*(10), 1637–1643. https://doi.org/10.1016/j.bbapap.2015.03.002.

Doerig, C. M., Parzy, D., Langsley, G., Horrocks, P., Carter, R., & Doerig, C. D. (1996). A MAP kinase homologue from the human malaria parasite, *Plasmodium falciparum*. *Gene*, *177*(1–2), 1–6. http://www.ncbi.nlm.nih.gov/pubmed/8921836.

Donald, R. G. K., Zhong, T., Wiersma, H., Nare, B., Yao, D., Lee, A., et al. (2006). Anticoccidial kinase inhibitors: Identification of protein kinase targets secondary to cGMP-dependent protein kinase. *Molecular and Biochemical Parasitology*, *149*(1), 86–98. https://doi.org/10.1016/j.molbiopara.2006.05.003.

Dorin, D., Alano, P., Boccaccio, I., Cicéron, L., Doerig, C., Sulpice, R., et al. (1999). An atypical mitogen-activated protein kinase (MAPK) homologue expressed in gametocytes of the human malaria parasite *Plasmodium falciparum*. Identification of a MAPK signature. *The Journal of Biological Chemistry*, *274*(42), 29912–29920. http://www.ncbi.nlm.nih.gov/pubmed/10514473.

Dorin, D., Le Roch, K., Sallicandro, P., Alano, P., Parzy, D., Poullet, P., et al. (2001). Pfnek-1, a NIMA-related kinase from the human malaria parasite *Plasmodium falciparum*. *European Journal of Biochemistry*, *268*(9), 2600–2608. https://doi.org/10.1046/j.1432-1327.2001.02151.x.

Dorin, D., Semblat, J., Poullet, P., Alano, P., Goldring, J. P. D., Whittle, C., et al. (2004). PfPK7, an atypical MEK-related protein kinase, reflects the absence of classical three-component MAPK pathways in the human malaria parasite *Plasmodium falciparum*. *Molecular Microbiology*, *55*(1), 184–186. https://doi.org/10.1111/j.1365-2958.2004.04393.x.

Dorin-Semblat, D., Quashie, N., Halbert, J., Sicard, A., Doerig, C., Peat, E., et al. (2007). Functional characterization of both MAP kinases of the human malaria parasite *Plasmodium falciparum* by reverse genetics. *Molecular Microbiology*, *65*(5), 1170–1180. https://doi.org/10.1111/j.1365-2958.2007.05859.x.

Dorin-Semblat, D., Schmitt, S., Semblat, J. P., Sicard, A., Reininger, L., Goldring, D., et al. (2011). *Plasmodium falciparum* NIMA-related kinase Pfnek-1: Sex specificity and assessment of essentiality for the erythrocytic asexual cycle. *Microbiology*, *157*(10), 2785–2794. https://doi.org/10.1099/mic.0.049023-0.

Dorin-Semblat, D., Sicard, A., Doerig, C., Ranford-Cartwright, L., & Doerig, C. (2008). Disruption of the PfPK7 gene impairs schizogony and sporogony in the human malaria parasite *Plasmodium falciparum*. *Eukaryotic Cell*, 7(2), 279–285. https://doi.org/10.1128/EC.00245-07.

Dos-Santos, J. C. K., Angerami, R. N., Castiñeiras, C. M. S., Lopes, S. C. P., Albrecht, L., Garcia, M. T., et al. (2014). Imported malaria in a non-endemic area: The experience of the university of Campinas hospital in the Brazilian Southeast. *Malaria Journal*, *13*, 280. https://doi.org/10.1186/1475-2875-13-280.

Droucheau, E., Primot, A., Thomas, V., Mattei, D., Knockaert, M., Richardson, C., et al. (2004). Plasmodium falciparum glycogen synthase kinase-3: Molecular model, expression, intracellular localisation and selective inhibitors. *Biochimica et Biophysica Acta (BBA)—Proteins and Proteomics*, *1697*(1–2), 181–196. https://doi.org/10.1016/j.bbapap.2003.11.023.

Dunyo, S., Milligan, P., Edwards, T., Sutherland, C., Targett, G., & Pinder, M. (2006). Gametocytaemia after drug treatment of asymptomatic *Plasmodium falciparum*. *PLoS Clinical Trials*, *1*(4), e20. https://doi.org/10.1371/journal.pctr.0010020.

Dvorin, J. D., Martyn, D. C., Patel, S. D., Grimley, J. S., Collins, C. R., Hopp, C. S., et al. (2010). A plant-like kinase in *Plasmodium falciparum* regulates parasite egress from erythrocytes. *Science*, *328*(5980), 910–912. https://doi.org/10.1126/science.1188191.

Eubanks, A. L., Perkins, M. M., Sylvester, K., Ganley, J. G., Posfai, D., Sanschargrin, P. C., et al. (2018). In silico screening and evaluation of *Plasmodium falciparum* protein kinase 5 (PK5) inhibitors. *ChemMedChem*, *13*(23), 2479–2483. https://doi.org/10.1002/cmdc.201800625.

Fang, H., Klages, N., Baechler, B., Hillner, E., Yu, L., Pardo, M., et al. (2017). Multiple short windows of calcium-dependent protein kinase 4 activity coordinate distinct cell cycle events during Plasmodium gametogenesis. *eLife*, *6*, 1–23. https://doi.org/10.7554/eLife.26524.

Flaherty, B. R., Wang, Y., Trope, E. C., Ho, T. G., Muralidharan, V., Kennedy, E. J., et al. (2015). The stapled AKAP disruptor peptide STAD-2 displays antimalarial activity through a PKA-independent mechanism. *PLoS One*, *10*(5), e0129239. https://doi.org/10.1371/journal.pone.0129239.

Fugel, W., Oberholzer, A. E., Gschloessl, B., Dzikowski, R., Pressburger, N., Preu, L., et al. (2013). 3,6-Diamino-4-(2-halophenyl)-2-benzoylthieno[2,3-b]pyridine-5-carbonitriles are selective inhibitors of *Plasmodium falciparum* glycogen synthase kinase-3. *Journal of Medicinal Chemistry*, *56*(1), 264–275. https://doi.org/10.1021/jm301575n.

Gamo, F.-J., Sanz, L. M., Vidal, J., de Cozar, C., Alvarez, E., Lavandera, J.-L., et al. (2010). Thousands of chemical starting points for antimalarial lead identification. *Nature*, *465*(7296), 305–310. https://doi.org/10.1038/nature09107.

Ganter, M., Goldberg, J. M., Dvorin, J. D., Paulo, J. A., King, J. G., Tripathi, A. K., et al. (2017). *Plasmodium falciparum* CRK4 directs continuous rounds of DNA replication during schizogony. *Nature Microbiology*, *2*(5), 17017. https://doi.org/10.1038/nmicrobiol.2017.17.

Garrison, E., Treeck, M., Ehret, E., Butz, H., Garbuz, T., Oswald, B. P., et al. (2012). A forward genetic screen reveals that calcium-dependent protein kinase 3 regulates egress in toxoplasma. *PLoS Pathogens*, *8*(11), e1003049. https://doi.org/10.1371/journal.ppat.1003049.

Geyer, J. A., Keenan, S. M., Woodard, C. L., Thompson, P. A., Gerena, L., Nichols, D. A., et al. (2009). Selective inhibition of Pfmrk, a *Plasmodium falciparum* CDK, by antimalarial 1,3-diaryl-2-propenones. *Bioorganic and Medicinal Chemistry Letters*, *19*(7), 1982–1985. https://doi.org/10.1016/j.bmcl.2009.02.042.

Global Malaria Progamme. (2015). *Control and elimination of plamodium vivax malaria—A technical brief* (pp. 1–64). WHO.

Graeser, R., Wernli, B., Franklin, R. M., & Kappes, B. (1996). *Plasmodium falciparum* protein kinase 5 and the malarial nuclear division cycles. *Molecular and Biochemical Parasitology*, *82*(1), 37–49. https://doi.org/10.1016/0166-6851(96)02716-8.

Gray, K. A., Gresty, K. J., Chen, N., Zhang, V., Gutteridge, C. E., Peatey, C. L., et al. (2016). Correlation between cyclin dependent kinases and artemisinin-induced dormancy in *Plasmodium falciparum* in vitro. *PLoS One*, *11*(6), 1–14. https://doi.org/10.1371/journal.pone.0157906.

Green, J. L., Moon, R. W., Whalley, D., Bowyer, P. W., Wallace, C., Rochani, A., et al. (2016). Imidazopyridazine inhibitors of *Plasmodium falciparum* calcium-dependent protein kinase 1 also target cyclic GMP-dependent protein kinase and heat shock protein 90 to kill the parasite at different stages of intracellular development. *Antimicrobial Agents and Chemotherapy*, *60*(3), 1464–1475. https://doi.org/10.1128/AAC.01748-15.

Green, J. L., Rees-Channer, R. R., Howell, S. A., Martin, S. R., Knuepfer, E., Taylor, H. M., et al. (2008). The motor complex of *Plasmodium falciparum*: Phosphorylation by a calcium-dependent protein kinase. *Journal of Biological Chemistry*, *283*(45), 30980–30989. https://doi.org/10.1074/jbc.M803129200.

Guiguemde, W. A., Shelat, A. A., Bouck, D., Duffy, S., Crowther, G. J., Davis, P. H., et al. (2010). Chemical genetics of *Plasmodium falciparum*. *Nature*, *465*(7296), 311–315. https://doi.org/10.1038/nature09099.

Gupta, C. L., Akhtar, S., Kumar, N., Ali, J., Pathak, N., & Bajpai, P. (2016). In silico elucidation and inhibition studies of selected phytoligands against mitogen-activated protein kinases of protozoan parasites. *Interdisciplinary Sciences: Computational Life Sciences*, *8*(1), 41–52. https://doi.org/10.1007/s12539-015-0269-6.

Gurnett, A. M., Liberator, P. A., Dulski, P. M., Salowe, S. P., Donald, R. G. K., Anderson, J. W., et al. (2002). Purification and molecular characterization of cGMP-dependent protein kinase from Apicomplexan parasites. A novel chemotherapeutic target. *Journal of Biological Chemistry*, *277*(18), 15913–15922. https://doi.org/10.1074/jbc.M108393200.

Halbert, J., Ayong, L., Equinet, L., le Roch, K., Hardy, M., Goldring, D., et al. (2010). A *Plasmodium falciparum* transcriptional cyclin-dependent kinase-related kinase with a crucial role in parasite proliferation associates with histone deacetylase activity. *Eukaryotic Cell*, *9*(6), 952–959. https://doi.org/10.1128/EC.00005-10.

Hamann, M., Alonso, D., Martín-Aparicio, E., Fuertes, A., Pérez-Puerto, M. J., Castro, A., et al. (2007). Glycogen synthase kinase-3 (GSK-3) inhibitory activity and structure–activity relationship (SAR) studies of the manzamine alkaloids. Potential for Alzheimer's disease. *Journal of Natural Products*, *70*(9), 1397–1405. https://doi.org/10.1021/np060092r.

Harmse, L., van Zyl, R., Gray, N., Schultz, P., Leclerc, S., Meijer, L., et al. (2001). Structure-activity relationships and inhibitory effects of various purine derivatives on the in vitro growth of *Plasmodium falciparum*. *Biochemical Pharmacology*, *62*(3), 341–348. http://www.ncbi.nlm.nih.gov/pubmed/11434907.

Holmgren, G., Gil, J. P., Ferreira, P. M., Veiga, M. I., Obonyo, C. O., & Björkman, A. (2006). Amodiaquine resistant *Plasmodium falciparum* malaria in vivo is associated with selection of pfcrt 76T and pfmdr1 86Y. *Infection, Genetics and Evolution*, *6*(4), 309–314. https://doi.org/10.1016/j.meegid.2005.09.001.

Holton, S., Merckx, A., Burgess, D., Doerig, C., Noble, M., & Endicott, J. (2003). Structures of *P. falciparum* PfPK5 test the CDK regulation paradigm and suggest mechanisms of small molecule inhibition. *Structure*, *11*(11), 1329–1337. https://doi.org/10.1016/j.str.2003.09.020.

Hu, J.-F., Hamann, M. T., Hill, R., & Kelly, M. (2003). The manzamine alkaloids. *The Alkaloids: Chemistry and Biology*, *60*(C), 207–285. https://doi.org/10.1016/S0099-9598(03)60004-0.

Hui, R., El Bakkouri, M., & Sibley, L. D. (2015). Designing selective inhibitors for calcium-dependent protein kinases in apicomplexans. *Trends in Pharmacological Sciences*, *36*(7), 452–460. https://doi.org/10.1016/j.tips.2015.04.011.

Ishino, T., Orito, Y., Chinzei, Y., & Yuda, M. (2006). A calcium-dependent protein kinase regulates Plasmodium ookinete access to the midgut epithelial cell. *Molecular Microbiology*, *59*(4), 1175–1184. https://doi.org/10.1111/j.1365-2958.2005.05014.x.

Iwanaga, T., Sugi, T., Kobayashi, K., Takemae, H., Gong, H., Ishiwa, A., et al. (2013). Characterization of *Plasmodium falciparum* cdc2-related kinase and the effects of a CDK inhibitor on the parasites in erythrocytic schizogony. *Parasitology International*, *62*(5), 423–430. https://doi.org/10.1016/j.parint.2013.05.003.

Jain, R., Gupta, S., Munde, M., Pati, S., & Singh, S. (2020). Development of novel anti-malarial from structurally diverse library of molecules, targeting plant-like CDPK1, a multistage growth regulator of *P. falciparum*. *Biochemical Journal*, *477*(10), 1951–1970. https://doi.org/10.1042/BCJ20200045.

Jebiwott, S., Govindaswamy, K., Mbugua, A., & Bhanot, P. (2013). Plasmodium berghei calcium dependent protein kinase 1 is not required for host cell invasion. *PLoS One*, *8*(11), e79171. https://doi.org/10.1371/journal.pone.0079171.

Jirage, D., Chen, Y., Caridha, D., O'Neil, M. T., Eyase, F., Witola, W. H., et al. (2010). The malarial CDK Pfmrk and its effector PfMAT1 phosphorylate DNA replication proteins and co-localize in the nucleus. *Molecular and Biochemical Parasitology*, *172*(1), 9–18. https://doi.org/10.1016/j.molbiopara.2010.03.009.

Kato, K., Sudo, A., Kobayashi, K., Sugi, T., Tohya, Y., & Akashi, H. (2009). Characterization of *Plasmodium falciparum* calcium-dependent protein kinase 4. *Parasitology International*, *58*(4), 394–400. https://doi.org/10.1016/j.parint.2009.08.001.

Kats, L. M., Fernandez, K. M., Glenister, F. K., Herrmann, S., Buckingham, D. W., Siddiqui, G., et al. (2014). An exported kinase (FIKK4.2) that mediates virulence-associated changes in *Plasmodium falciparum*-infected red blood cells. *International Journal for Parasitology*, *44*(5), 319–328. https://doi.org/10.1016/j.ijpara.2014.01.003.

Kebaier, C., & Vanderberg, J. P. (2010). Initiation of Plasmodium sporozoite motility by albumin is associated with induction of intracellular signalling. *International Journal for Parasitology*, *40*(1), 25–33. https://doi.org/10.1016/j.ijpara.2009.06.011.

Keenan, S. M., Geyer, J.a., Welsh, W. J., Prigge, S. T., & Waters, N. C. (2005). Rational inhibitor design and iterative screening in the identification of selective plasmodial cyclin dependent kinase inhibitors. *Combinatorial Chemistry & High Throughput Screening*, *8*(1), 27–38. https://doi.org/10.2174/1386207053328183.

Khan, S. M., Franke-Fayard, B., Mair, G. R., Lasonder, E., Janse, C. J., Mann, M., et al. (2005). Proteome analysis of separated male and female gametocytes reveals novel sex-specific Plasmodium biology. *Cell*, *121*(5), 675–687. https://doi.org/10.1016/j.cell.2005.03.027.

Klein, M., Dinér, P., Dorin-Semblat, D., Doerig, C., & Grøtli, M. (2009). Synthesis of 3-(1,2,3-triazol-1-yl)- and 3-(1,2,3-triazol-4-yl)-substituted pyrazolo[3,4-d]pyrimidin-4-amines via click chemistry: Potential inhibitors of the *Plasmodium falciparum* PfPK7 protein kinase. *Organic & Biomolecular Chemistry*, 7(17), 3421–3429. https://doi.org/10.1039/b906482f.

Kumar, A., Vaid, A., Syin, C., & Sharma, P. (2004). PfPKB, a novel protein kinase B-like enzyme from *Plasmodium falciparum*. I. Identification, characterization, and possible role in parasite development. *Journal of Biological Chemistry*, *279*(23), 24255–24264. https://doi.org/10.1074/jbc.M312855200.

Lacerda, M. V., Mourão, M. P., Alexandre, M. A., Siqueira, A. M., Magalhães, B. M., Martinez-Espinosa, F. E., et al. (2012). Understanding the clinical spectrum of complicated *Plasmodium vivax* malaria: A systematic review on the contributions of the Brazilian literature. *Malaria Journal*, *11*, 12. https://doi.org/10.1186/1475-2875-11-12.

Large, J. M., Osborne, S. A., Smiljanic-Hurley, E., Ansell, K. H., Jones, H. M., Taylor, D. L., et al. (2013). Imidazopyridazines as potent inhibitors of *Plasmodium falciparum* calcium-dependent protein kinase 1 (PfCDPK1): Preparation and evaluation of pyrazole linked analogues. *Bioorganic and Medicinal Chemistry Letters*, *23*(21), 6019–6024. https://doi.org/10.1016/j.bmcl.2013.08.010.

Lasonder, E., Green, J. L., Camarda, G., Talabani, H., Holder, A. A., Langsley, G., et al. (2012). The *Plasmodium falciparum* schizont phosphoproteome reveals extensive phosphatidylinositol and cAMP-protein kinase A signaling. *Journal of Proteome Research*, *11*(11), 5323–5337. https://doi.org/10.1021/pr300557m.

Laurent, D., Jullian, V., Parenty, A., Knibiehler, M., Dorin, D., Schmitt, S., et al. (2006). Antimalarial potential of xestoquinone, a protein kinase inhibitor isolated from a Vanuatu marine sponge Xestospongia sp. *Bioorganic & Medicinal Chemistry*, *14*(13), 4477–4482. https://doi.org/10.1016/j.bmc.2006.02.026.

Le Roch, K. G. (2003). Discovery of gene function by expression profiling of the malaria parasite life cycle. *Science*, *301*(5639), 1503–1508. https://doi.org/10.1126/science.1087025.

Le Roch, K., Sestier, C., Dorin, D., Waters, N., Kappes, B., Chakrabarti, D., et al. (2000). Activation of a *Plasmodium falciparum* cdc2-related kinase by heterologous p25 and cyclin H. Functional characterization of a *P. falciparum* cyclin homologue. *Journal of Biological Chemistry*, *275*(12), 8952–8958. https://doi.org/10.1074/jbc.275.12.8952.

Leykauf, K., Treeck, M., Gilson, P. R., Nebl, T., Braulke, T., Cowman, A. F., et al. (2010). Protein kinase a dependent phosphorylation of apical membrane antigen 1 plays an important role in erythrocyte invasion by the malaria parasite. *PLoS Pathogens*, *6*(6), e1000941. https://doi.org/10.1371/journal.ppat.1000941.

Li, J. L., Baker, D. A., & Cox, L. S. (2000). Sexual stage-specific expression of a third calcium-dependent protein kinase from *Plasmodium falciparum*. *Biochimica et Biophysica Acta—Gene Structure and Expression*, *1491*(1–3), 341–349. https://doi.org/10.1016/S0167-4781(00)00032-4.

Li, Z., Le Roch, K., Geyer, J.a., Woodard, C. L., Prigge, S. T., Koh, J., et al. (2001). Influence of human p16(INK4) and p21(CIP1) on the in vitro activity of recombinant *Plasmodium falciparum* cyclin-dependent protein kinases. *Biochemical and Biophysical Research Communications*, *288*(5), 1207–1211. https://doi.org/10.1006/bbrc.2001.5920.

Li, J. L., Robson, K. J. H., Chen, J. L., Targett, G. A. T., & Baker, D. A. (1996). Pfmrk, a MO15-related protein kinase from *Plasmodium falciparum*. Gene cloning, sequence, stage-specific expression and chromosome localization. *European Journal of Biochemistry*, *241*(3), 805–813. https://doi.org/10.1111/j.1432-1033.1996.00805.x.

Lin, J. T., Juliano, J. J., & Wongsrichanalai, C. (2010). Drug-resistant malaria: The era of ACT. *Current Infectious Disease Reports*, *12*(3), 165–173. https://doi.org/10.1007/s11908-010-0099-y.

Lochner, A., & Moolman, J. A. (2006). The many faces of H89: A review. *Cardiovascular Drug Reviews*, *24*(3–4), 261–274. https://doi.org/10.1111/j.1527-3466.2006.00261.x.

Lourido, S., & Moreno, S. N. J. (2015). The calcium signaling toolkit of the Apicomplexan parasites Toxoplasma gondii and Plasmodium spp. *Cell Calcium*, *57*(3), 186–193. https://doi.org/10.1016/j.ceca.2014.12.010.

Low, H., Lye, Y. M., & Sim, T. S. (2007). Pfnek3 functions as an atypical MAPKK in *Plasmodium falciparum*. *Biochemical and Biophysical Research Communications*, *361*(2), 439–444. https://doi.org/10.1016/j.bbrc.2007.07.047.

Lucet, I. S., Tobin, A., Drewry, D., Wilks, A. F., & Doerig, C. (2012). Plasmodium kinases as targets for new-generation antimalarials. *Future Medicinal Chemistry*, *4*(18), 2295–2310. https://doi.org/10.4155/fmc.12.183.

Malumbres, M. (2014). Cyclin-dependent kinases. *Genome Biology*, *15*(6), 122. https://doi.org/10.1186/gb4184.

Manhani, K. K., Arcuri, H. A., da Silveira, N. J. F., Uchôa, H. B., de Azevedo, W. F., & Canduri, F. (2005). Molecular models of protein kinase 6 from *Plasmodium falciparum*. *Journal of Molecular Modeling*, *12*(1), 42–48. https://doi.org/10.1007/s00894-005-0002-1.

Manning, G., Whyte, D. B., Martinez, R., Hunter, T., & Sudarsanam, S. (2002). The protein kinase complement of the human genome. *Science*, *298*(5600), 1912–1934. https://doi.org/10.1126/science.1075762.

Masch, A., & Kunick, C. (2015). Selective inhibitors of *Plasmodium falciparum* glycogen synthase-3 (PfGSK-3): New antimalarial agents? *Biochimica et Biophysica Acta—Proteins and Proteomics*, *1854*(10), 1644–1649. https://doi.org/10.1016/j.bbapap.2015.03.013.

Masch, A., Nasereddin, A., Alder, A., Bird, M. J., Schweda, S. I., Preu, L., et al. (2019). Structure-activity relationships in a series of antiplasmodial thieno[2,3-b]pyridines. *Malaria Journal*, *18*(1), 1–10. https://doi.org/10.1186/s12936-019-2725-y.

McRobert, L., Taylor, C. J., Deng, W., Fivelman, Q. L., Cummings, R. M., Polley, S. D., et al. (2008). Gametogenesis in malaria parasites is mediated by the cGMP-dependent protein kinase. *PLoS Biology*, *6*(6), e139. https://doi.org/10.1371/journal.pbio.0060139.

Merckx, A., Echalier, A., Langford, K., Sicard, A., Langsley, G., Joore, J., et al. (2008). Structures of *P. falciparum* protein kinase 7 identify an activation motif and leads for inhibitor design. *Structure*, *16*(2), 228–238. https://doi.org/10.1016/j.str.2007.11.014.

Merckx, A., Le Roch, K., Nivez, M.-P., Dorin, D., Alano, P., Gutierrez, G. J., et al. (2003). Identification and initial characterization of three novel cyclin-related proteins of the human malaria parasite plasmodium falciparum. *Journal of Biological Chemistry*, *278*(41), 39839–39850.

Mockenhaupt, F. P. (1995). Mefloquine resistance in *Plasmodium falciparum*. *Parasitology Today*, *11*(7), 248–253. https://doi.org/10.1016/0169-4758(95)80201-0.

Moon, R. W., Taylor, C. J., Bex, C., Schepers, R., Goulding, D., Janse, C. J., et al. (2009). A cyclic GMP signalling module that regulates gliding motility in a malaria parasite. *PLoS Pathogens*, *5*(9), e1000599. https://doi.org/10.1371/journal.ppat.1000599.

Morton, L. C., Huber, C., Okoth, S. A., Griffing, S., Lucchi, N., Ljolje, D., et al. (2016). Plasmodium falciparum drug-resistant haplotypes and population structure in post-earthquake Haiti, 2010. *The American Journal of Tropical Medicine and Hygiene*, *95*(4), 811–816. https://doi.org/10.4269/ajtmh.16-0214.

Murphy, R. C., Ojo, K. K., Larson, E. T., Castellanos-Gonzalez, A., Perera, B. G. K., Keyloun, K. R., et al. (2010). Discovery of potent and selective inhibitors of CDPK1 from *C. parvum* and *T. gondii*. *ACS Medicinal Chemistry Letters*, *1*(7), 331–335. https://doi.org/10.1021/ml100096t.

Nagamune, K., Moreno, S. N., Chini, E. N., & Sibley, L. D. (2008). Calcium regulation and signaling in apicomplexan parasites. *Sub-Cellular Biochemistry*, *47*, 70–81. https://doi.org/10.1007/978-0-387-78267-6_5.

Ojo, K. K., Eastman, R. T., Vidadala, R., Zhang, Z., Rivas, K. L., Choi, R., et al. (2014). A specific inhibitor of pfcdpk4 blocks malaria transmission: Chemical-genetic validation. *Journal of Infectious Diseases*, *209*(2), 275–284. https://doi.org/10.1093/infdis/jit522.

Ojo, K. K., Pfander, C., Mueller, N. R., Burstroem, C., Larson, E. T., Bryan, C. M., et al. (2012). Transmission of malaria to mosquitoes blocked by bumped kinase inhibitors. *Journal of Clinical Investigation*, *122*(6), 2301–2305. https://doi.org/10.1172/JCI61822.

Ono, T., Cabrita-Santos, L., Leitao, R., Bettiol, E., Purcell, L. A., Diaz-Pulido, O., et al. (2008). Adenylyl cyclase a and cAMP signaling mediate Plasmodium sporozoite apical regulated exocytosis and hepatocyte infection. *PLoS Pathogens*, *4*(2), e1000008. https://doi.org/10.1371/journal.ppat.1000008.

Osolodkin, D. I., Zakharevich, N. V., Palyulin, V. A., Danilenko, V. N., & Zefirov, N. S. (2011). Bioinformatic analysis of glycogen synthase kinase 3: Human versus parasite kinases. *Parasitology*, *138*(06), 725–735. https://doi.org/10.1017/S0031182011000151.

Panchal, D., & Bhanot, P. (2010). Activity of a trisubstituted pyrrole in inhibiting sporozoite invasion and blocking malaria infection. *Antimicrobial Agents and Chemotherapy*, *54*(10), 4269–4274. https://doi.org/10.1128/AAC.00420-10.

Patel, G., Roncal, N. E., Lee, P. J., Leed, S. E., Erath, J., Rodriguez, A., et al. (2014). Repurposing human Aurora kinase inhibitors as leads for anti-protozoan drug discovery. *MedChemComm*, *5*(5), 655. https://doi.org/10.1039/c4md00045e.

Pease, B. N., Huttlin, E. L., Jedrychowski, M. P., Talevich, E., Harmon, J., Dillman, T., et al. (2013). Global analysis of protein expression and phosphorylation of three stages of *Plasmodium falciparum* intraerythrocytic development. *Journal of Proteome Research*, *12*(9), 4028–4045. https://doi.org/10.1021/pr400394g.

Plouffe, D., Brinker, A., McNamara, C., Henson, K., Kato, N., Kuhen, K., et al. (2008). In silico activity profiling reveals the mechanism of action of antimalarials discovered in a high-throughput screen. *Proceedings of the National Academy of Sciences of the United States of America*, *105*(26), 9059–9064. https://doi.org/10.1073/pnas.0802982105.

Popovici, J., & Ménard, D. (2015). Challenges in antimalarial drug treatment for vivax malaria control. *Trends in Molecular Medicine*, *21*(12), 776–788. https://doi.org/10.1016/j.molmed.2015.10.004.

Price, R. N., & Nosten, F. (2014). Single-dose radical cure of *Plasmodium vivax*: A step closer. *The Lancet*, *383*(9922), 1020–1021. https://doi.org/10.1016/S0140-6736(13)62672-0.

Price, R. N., von Seidlein, L., Valecha, N., Nosten, F., Baird, J. K., & White, N. J. (2014). Global extent of chloroquine-resistant *Plasmodium vivax*: A systematic review and meta-analysis. *The Lancet Infectious Diseases*, *14*(10), 982–991. https://doi.org/10.1016/S1473-3099(14)70855-2.

Radwan, M., Hanora, A., Khalifa, S., & Abou-el-ela, S. H. (2012). Manzamines. *Cell Cycle*, *11*(9), 1765–1772. https://doi.org/10.4161/cc.20135.

Ramdani, G., Naissant, B., Thompson, E., Breil, F., Lorthiois, A., Dupuy, F., et al. (2015). cAMP-signalling regulates gametocyte-infected erythrocyte deformability required for malaria parasite transmission. *PLoS Pathogens*, *11*(5), e1004815. https://doi.org/10.1371/journal.ppat.1004815.

Rangarajan, R., Bei, A., Henry, N., Madamet, M., Parzy, D., Nivez, M. P., et al. (2006). Pbcrk-1, the Plasmodium berghei orthologue of *P. falciparum* cdc-2 related kinase-1 (Pfcrk-1), is essential for completion of the intraerythrocytic asexual cycle. *Experimental Parasitology*, *112*(3), 202–207. https://doi.org/10.1016/j.exppara.2005.11.002.

Rangarajan, R., Bei, A. K., Jethwaney, D., Maldonado, P., Dorin, D., Sultan, A. A., et al. (2005). A mitogen-activated protein kinase regulates male gametogenesis and transmission of the malaria parasite Plasmodium berghei. *EMBO Reports*, *6*(5), 464–469. https://doi.org/10.1038/sj.embor.7400404.

Ranjan, R., Ahmed, A., Gourinath, S., & Sharma, P. (2009). Dissection of mechanisms involved in the regulation of *Plasmodium falciparum* calcium-dependent protein kinase 4. *The Journal of Biological Chemistry*, *284*(22), 15267–15276. https://doi.org/10.1074/jbc.M900656200.

Rao, K. V., Santarsiero, B. D., Mesecar, A. D., Schinazi, R. F., Tekwani, B. L., & Hamann, M. T. (2003). New manzamine alkaloids with activity against infectious and tropical parasitic diseases from an Indonesian sponge. *Journal of Natural Products*, *66*(6), 823–828. https://doi.org/10.1021/np020592u.

Read, L. K., & Mikkelsen, R. B. (1991). Comparison of adenylate cyclase and cAMP-dependent protein kinase in gametocytogenic and nongametocytogenic clones of *Plasmodium falciparum*. *The Journal of Parasitology*, 77(3), 346–352. http://www.ncbi.nlm.nih.gov/pubmed/2040946.

Reininger, L., Billker, O., Tewari, R., Mukhopadhyay, A., Fennell, C., Dorin-Semblat, D., et al. (2005). A NIMA-related protein kinase is essential for completion of the sexual cycle of malaria parasites. *Journal of Biological Chemistry*, *280*(36), 31957–31964. https://doi.org/10.1074/jbc.M504523200.

Reininger, L., Garcia, M., Tomlins, A., Müller, S., & Doerig, C. (2012). The *Plasmodium falciparum*, Nima-related kinase Pfnek-4: A marker for asexual parasites committed to sexual differentiation. *Malaria Journal*, *11*(1), 250. https://doi.org/10.1186/1475-2875-11-250.

Reininger, L., Tewari, R., Fennell, C., Holland, Z., Goldring, D., Ranford-cartwright, L., et al. (2009). An essential role for the Plasmodium Nek-2 Nima-related protein kinase in the sexual development of malaria parasites. *The Journal of Biological Chemistry*, *284*(31), 20858–20868. https://doi.org/10.1074/jbc.M109.017988.

Reininger, L., Wilkes, J. M., Bourgade, H., Miranda-Saavedra, D., & Doerig, C. (2011). An essential Aurora-related kinase transiently associates with spindle pole bodies during *Plasmodium falciparum* erythrocytic schizogony. *Molecular Microbiology*, *79*(1), 205–221. https://doi.org/10.1111/j.1365-2958.2010.07442.x.

Reker, D., Seet, M., Pillong, M., Koch, C. P., Schneider, P., Witschel, M. C., et al. (2014). Deorphaning pyrrolopyrazines as potent multi-target antimalarial agents. *Angewandte Chemie, International Edition*, *53*(27), 7079–7084. https://doi.org/10.1002/anie.201311162.

Rieckmann, K. H., Davis, D. R., & Hutton, D. C. (1989). *Plasmodium vivax* resistance to chloroquine? *The Lancet*, *334*(8673), 1183–1184. https://doi.org/10.1016/S0140-6736(89)91792-3.

Rogerio, K. R., Carvalho, L. J. M., Domingues, L. H. P., Neves, B. J., Moreira Filho, J. T., Castro, R. N., et al. (2018). Synthesis and molecular modelling studies of pyrimidinones and pyrrolo[3,4-d]-pyrimidinodiones as new antiplasmodial compounds. *Memórias do Instituto Oswaldo Cruz*, *113*(8), 1–19. https://doi.org/10.1590/0074-02760170452.

Roskoski, R. (2015). A historical overview of protein kinases and their targeted small molecule inhibitors. *Pharmacological Research*, *100*, 1–23. https://doi.org/10.1016/j.phrs.2015.07.010.

Ross-Macdonald, P. B., Graeser, R., Kappes, B., Franklin, R., & Williamson, D. H. (1994). Isolation and expression of a gene specifying a cdc2-like protein kinase from the human malaria parasite *Plasmodium falciparum*. *European Journal of Biochemistry*, *220*(3), 693–701.

Rotella, D. P. (2012). Recent results in protein kinase inhibition for tropical diseases. *Bioorganic & Medicinal Chemistry Letters*, *22*(22), 6788–6793. https://doi.org/10.1016/j.bmcl.2012.09.044.

Sahu, N. K., & Kohli, D. V. (2012). Structural insight for imidazopyridazines as malarial kinase PfPK7 inhibitors using QSAR techniques. *Medicinal Chemistry*, *8*(4), 636–648. http://www.ncbi.nlm.nih.gov/pubmed/22530846.

Schiffer, C. A. (2007). BCR-ABL tyrosine kinase inhibitors for chronic myelogenous leukemia. *The New England Journal of Medicine*, *357*, 258–265.

Sebastian, S., Brochet, M., Collins, M. O., Schwach, F., Jones, M. L., Goulding, D., et al. (2012). A Plasmodium calcium-dependent protein kinase controls zygote development and transmission by translationally activating repressed mRNAs. *Cell Host & Microbe*, *12*(1), 9–19. https://doi.org/10.1016/j.chom.2012.05.014.

Sibley, C. H. (2013). Tracking artemisinin resistance in Plasmodium falciparum. *The Lancet Infectious Diseases*, *13*(12), 999–1000. https://doi.org/10.1016/S1473-3099(13)70260-3.

Siden-Kiamos, I., Ecker, A., Nyb??ck, S., Louis, C., Sinden, R. E., & Billker, O. (2006). Plasmodium berghei calcium-dependent protein kinase 3 is required for ookinete gliding motility and mosquito midgut invasion. *Molecular Microbiology*, *60*(6), 1355–1363. https://doi.org/10.1111/j.1365-2958.2006.05189.x.

Sinha, S., Medhi, B., & Sehgal, R. (2014). Challenges of drug-resistant malaria. *Parasite*, *21*, 61. https://doi.org/10.1051/parasite/2014059.

Solyakov, L., Halbert, J., Alam, M. M., Semblat, J.-P., Dorin-Semblat, D., Reininger, L., et al. (2011). Global kinomic and phospho-proteomic analyses of the human malaria parasite *Plasmodium falciparum*. *Nature Communications*, *2*, 565. https://doi.org/10.1038/ncomms1558.

Spitzmüller, A., & Mestres, J. (2013). Prediction of the *P. falciparum* target space relevant to malaria drug discovery. *PLoS Computational Biology*, *9*(10), e1003257. https://doi.org/10.1371/journal.pcbi.1003257.

Sudo, A., Kato, K., Kobayashi, K., Tohya, Y., & Akashi, H. (2008). Susceptibility of *Plasmodium falciparum* cyclic AMP-dependent protein kinase and its mammalian homologue to the inhibitors. *Molecular and Biochemical Parasitology*, *160*(2), 138–142. https://doi.org/10.1016/j.molbiopara.2008.03.011.

Syin, C., Parzy, D., Traincard, F., Boccaccio, I., Joshi, M. B., Lin, D. T., et al. (2001). The H89 cAMP-dependent protein kinase inhibitor blocks *Plasmodium falciparum* development in infected erythrocytes. *European Journal of Biochemistry*, *268*(18), 4842–4849. https://doi.org/10.1046/j.1432-1327.2001.02403.x.

Talevich, E., Mirza, A., & Kannan, N. (2011). Structural and evolutionary divergence of eukaryotic protein kinases in Apicomplexa. *BMC Evolutionary Biology*, *11*(1), 321. https://doi.org/10.1186/1471-2148-11-321.

Talevich, E., Tobin, A. B., Kannan, N., & Doerig, C. (2012). An evolutionary perspective on the kinome of malaria parasites. *Philosophical Transactions of the Royal Society of London. Series B, Biological Sciences*, *367*(1602), 2607–2618. https://doi.org/10.1098/rstb.2012.0014.

Talisuna, A. O., Bloland, P., & D'Alessandro, U. (2004). History, dynamics, and public health importance of malaria parasite resistance. *Clinical Microbiology Reviews*, *17*(1), 235–254. https://doi.org/10.1128/CMR.17.1.235.

Taylor, H. M., McRobert, L., Grainger, M., Sicard, A., Dluzewski, A. R., Hopp, C. S., et al. (2010). The malaria parasite cyclic GMP-dependent protein kinase plays a central role in blood-stage schizogony. *Eukaryotic Cell*, *9*(1), 37–45. https://doi.org/10.1128/EC.00186-09.

Tewari, R., Dorin, D., Moon, R., Doerig, C., & Billker, O. (2005). An atypical mitogen-activated protein kinase controls cytokinesis and flagellar motility during male gamete formation in a malaria parasite. *Molecular Microbiology*, *58*(5), 1253–1263. https://doi.org/10.1111/j.1365-2958.2005.04793.x.

Tewari, R., Patzewitz, E. M., Poulin, B., Stewart, L., & Baker, D. A. (2014). Development of a transgenic Plasmodium berghei line (Pbpfpkg) expressing the *P. falciparum* cGMP-dependent protein kinase, a novel antimalarial drug target. *PLoS One*, *9*(5), e96923. https://doi.org/10.1371/journal.pone.0096923.

Tewari, R., Straschil, U., Bateman, A., Böhme, U., Cherevach, I., Gong, P., et al. (2010). The systematic functional analysis of Plasmodium protein kinases identifies essential regulators of mosquito transmission. *Cell Host & Microbe*, *8*(4), 377–387. https://doi.org/10.1016/j.chom.2010.09.006.

Tolle, N., & Kunick, C. (2011). Paullones as inhibitors of protein kinases. *Current Topics in Medicinal Chemistry*, *11*(11), 1320–1332. http://www.ncbi.nlm.nih.gov/pubmed/21513499.

Tong, L., Pav, S., White, D. M., Rogers, S., Crane, K. M., Cywin, C. L., et al. (1997). A highly specific inhibitor of human p38 MAP kinase binds in the ATP pocket. *Nature Structural Biology*, *4*(4), 311–316. http://www.ncbi.nlm.nih.gov/pubmed/9095200.

Tracer, W., & Gill, G. S. (1989). *Plasmodium falciparum* gametocyte formation in vitro: Its stimulation by phorbol diesters and by 8-bromo cyclic adenosine monophosphate. *The Journal of Protozoology*, *36*(5), 451–454.

Treeck, M., Sanders, J. L., Elias, J. E., & Boothroyd, J. C. (2011). The Phosphoproteomes of *Plasmodium falciparum* and Toxoplasma gondii reveal unusual adaptations within and beyond the parasites' boundaries. *Cell Host & Microbe*, *10*(4), 410–419. https://doi.org/10.1016/j.chom.2011.09.004.

Trivedi, V., & Nag, S. (2012). In silico characterization of atypical kinase PFD0975w from Plasmodium kinome: A suitable target for drug discovery. *Chemical Biology and Drug Design*, *79*(4), 600–609. https://doi.org/10.1111/j.1747-0285.2012.01321.x.

Vaid, A., Ranjan, R., Smythe, W. A., Hoppe, H. C., & Sharma, P. (2010). PfPI3K, a phosphatidylinositol-3 kinase from *Plasmodium falciparum*, is exported to the host erythrocyte and is involved in hemoglobin trafficking. *Blood*, *115*(12), 2500–2507. https://doi.org/10.1182/blood-2009-08-238972.

Vaid, A., & Sharma, P. (2006). PfPKB, a protein kinase B-like enzyme from *Plasmodium falciparum*: II. Identification of calcium/calmodulin as its upstream activator and dissection of a novel signaling pathway. *Journal of Biological Chemistry*, *281*(37), 27126–27133. https://doi.org/10.1074/jbc.M601914200.

Vaid, A., Thomas, D. C., & Sharma, P. (2008). Role of Ca2+/calmodulin-PfPKB signaling pathway in erythrocyte invasion by *Plasmodium falciparum*. *Journal of Biological Chemistry*, *283*(9), 5589–5597. https://doi.org/10.1074/jbc.M708465200.

Vaidya, A. B., Morrisey, J. M., Zhang, Z., Das, S., Daly, T. M., Otto, T. D., et al. (2014). Pyrazoleamide compounds are potent antimalarials that target Na+ homeostasis in intraerythrocytic Plasmodium falciparum. *Nature Communications*, *5*, 5521. https://doi.org/10.1038/ncomms6521.

Vanaerschot, M., Murithi, J. M., Pasaje, C. F. A., Ghidelli-Disse, S., Dwomoh, L., Bird, M., et al. (2020). Inhibition of resistance-refractory *P. falciparum* kinase PKG delivers prophylactic, blood stage, and transmission-blocking antiplasmodial activity. *Cell Chemical Biology*, *27*(7), 806–816.e8. https://doi.org/10.1016/j.chembiol.2020.04.001.

Ward, P., Equinet, L., Packer, J., & Doerig, C. (2004). Protein kinases of the human malaria parasite Plasmodium falciparum: The kinome of a divergent eukaryote. *BMC Genomics*, *5*(1), 79. No title https://doi.org/10.1186/1471-2164-5-79.

Waters, N. C., Woodard, C. L., & Prigge, S. T. (2000). Cyclin H activation and drug susceptibility of the Pfmrk cyclin dependent protein kinase from *Plasmodium falciparum*. *Molecular and Biochemical Parasitology*, *107*(1), 45–55. http://www.ncbi.nlm.nih.gov/pubmed/10717301.

Wernimont, A. K., Artz, J. D., Finerty, P., Lin, Y.-H., Amani, M., Allali-Hassani, A., et al. (2010). Structures of apicomplexan calcium-dependent protein kinases reveal mechanism of activation by calcium. *Nature Structural & Molecular Biology*, *17*(5), 596–601. https://doi.org/10.1038/nsmb.1795.

Wernsdorfer, W. H., & Payne, D. (1991). The dynamics of drug resistance in *Plasmodium falciparum*. *Pharmacology & Therapeutics*, *50*(1), 95–121. http://www.sciencedirect.com/science/article/pii/016372589190074V.

WHO. (2019). *WHO regional office for Africa*. WHO. https://www.who.int/news-room/fact-sheets/detail/malaria.

Widmann, C., Gibson, S., Jarpe, M. B., & Johnson, G. L. (1999). Mitogen-activated protein kinase: Conservation of a three-kinase module from yeast to human. *Physiological Reviews*, *79*(1), 143–180. http://www.ncbi.nlm.nih.gov/pubmed/9922370.

Wilde, M. L., Triglia, T., Marapana, D., Thompson, J. K., Kouzmitchev, A. A., Bullen, H. E., et al. (2019). Protein kinase A is essential for invasion of *Plasmodium falciparum* into human erythrocytes. *MBio*, *10*(5), 1–15. https://doi.org/10.1128/mBio.01972-19.

Woodard, C. L., Keenan, S. M., Gerena, L., Welsh, W. J., Geyer, J. A., & Waters, N. C. (2007). Evaluation of broad spectrum protein kinase inhibitors to probe the architecture of the malarial cyclin dependent protein kinase Pfmrk. *Bioorganic and Medicinal Chemistry Letters*, *17*(17), 4961–4966. https://doi.org/10.1016/j.bmcl.2007.06.032.

Woodard, C. L., Li, Z., Kathcart, A. K., Terrell, J., Gerena, L., Lopez-Sanchez, M., et al. (2003). Oxindole-based compounds are selective inhibitors of *Plasmodium falciparum* cyclin dependent protein kinases. *Journal of Medicinal Chemistry*, *46*(18), 3877–3882. https://doi.org/10.1021/jm0300983.

Wurtz, N., Chapus, C., Desplans, J., & Parzy, D. (2011). cAMP-dependent protein kinase from *Plasmodium falciparum*: An update. *Parasitology*, *138*(1), 1–25. https://doi.org/10.1017/S003118201000096X.

Xiao, Z., Waters, N. C., Woodard, C. L., Li, Z., & Li, P. K. (2001). Design and synthesis of Pfmrk inhibitors as potential antimalarial agents. *Bioorganic & Medicinal Chemistry Letters*, *11*(21), 2875–2878. https://doi.org/10.1016/S0960-894X(01)00578-9.

Yousaf, M., Hammond, N. L., Peng, J., Wahyuono, S., McIntosh, K. A., Charman, W. N., et al. (2004). New manzamine alkaloids from an indo-Pacific sponge. Pharmacokinetics, oral availability, and the significant activity of several manzamines against HIV-I, AIDS opportunistic infections, and inflammatory diseases. *Journal of Medicinal Chemistry*, *47*(14), 3512–3517. https://doi.org/10.1021/jm030475b.

Yuda, M., Yano, K., Tsuboi, T., Torii, M., & Chinzei, Y. (2001). Von Willebrand factor A domain-related protein, a novel microneme protein of the malaria ookinete highly conserved throughout Plasmodium parasites. *Molecular and Biochemical Parasitology*, *116*(1), 65–72. https://doi.org/10.1016/S0166-6851(01)00304-8.

Zhang, V. M., Chavchich, M., & Waters, N. C. (2012). Targeting protein kinases in the malaria parasite: Update of an antimalarial drug target. *Current Topics in Medicinal Chemistry*, *12*(5), 456–472. https://doi.org/10.2174/156802612799362922.

Zhao, Y., Kappes, B., & Franklin, R. M. (1993). Gene structure and expression of an unusual protein kinase from *Plasmodium falciparum* homologous at its carboxyl terminus with the EF hand calcium-binding proteins. *Journal of Biological Chemistry*, *268*(6), 4347–4354.

CHAPTER EIGHT

Yeast-based high-throughput screens for discovery of kinase inhibitors for neglected diseases

T.A. Tavella[a], G.C. Cassiano[a,b], Fabio Trindade Maranhão Costa[a], P. Sunnerhagen[c], and E. Bilsland[a,d,e,*]

[a]Laboratory of Tropical Diseases—Prof. Luiz Jacintho da Silva, Department of Genetics, Evolution and Bioagents, University of Campinas, Campinas, SP, Brazil
[b]Global Health and Tropical Medicine (GHTM), Instituto de Higiene e Medicina Tropical, Universidade Nova de Lisboa, Lisbon, Portugal
[c]Department of Chemistry and Molecular Biology, Lundberg Laboratory, University of Gothenburg, Gothenburg, Sweden
[d]Department of Structural and Functional Biology, Institute of Biology, UNICAMP, Campinas, Brazil
[e]Synthetic Biology Laboratory, Institute of Biology, UNICAMP, Campinas, Brazil
*Corresponding author: e-mail address: bilsland@unicamp.br

Contents

Abstract

The discovery and development of a new drug is a complex, time consuming and costly process that typically takes over 10 years and costs around 1 billion dollars from bench to market. This scenario makes the discovery of novel drugs targeting neglected tropical diseases (NTDs), which afflict in particular people in low-income countries, prohibitive.

Advances in Protein Chemistry and Structural Biology, Volume 124
ISSN 1876-1623
https://doi.org/10.1016/bs.apcsb.2020.09.007

Despite the intensive use of High-Throughput Screening (HTS) in the past decades, the speed with which new drugs come to the market has remained constant, generating doubts about the efficacy of this approach. Here we review a few of the yeast-based high-throughput approaches that can work synergistically with parasite-based, *in vitro*, or *in silico* methods to identify and optimize novel antiparasitic compounds. These yeast-based methods range from HTP screens to identify novel hits against promising parasite kinase targets to the identification of potential antiparasitic kinase inhibitors extracted from databases of yeast chemical genetic screens.

1. Introduction

Cellular life is regulated by protein phosphorylation. Perturbations in phosphorylation due to kinase mutations or overexpression can affect cellular differentiation and metabolism, contributing to many diseases such as cancer, autoimmune and inflammatory disorders, as well as cardiovascular problems, making kinases promising targets for pharmaceutical intervention. Rapamycin was the first specific protein kinase inhibitor to be approved by the FDA in 1999, revolutionizing organ transplants (Li, Kim, & Blenis, 2014). Later, in 2001, the FDA approved imatinib, a protein tyrosine kinase inhibitor, for the treatment of Philadelphia chromosome positive chronic myeloid leukemia (Jeon, Sparreboom, & Baker, 2018). Since then, innumerous kinase inhibitors have entered the clinical trials and are available on the market for the treatment of several diseases (Ferguson & Gray, 2018; Wu, Nielsen, & Clausen, 2015). In 2020, from the 29 new drugs approved by the FDA, six were human protein kinase inhibitors (www.fda.gov, 2020). Protein kinases represent the largest gene family many eukaryotic genomes, presumably as a result of numerous duplication events during evolution. All eukaryotic protein kinases, having a common evolutionary origin, are structurally related. Therefore, an inherent challenge when designing protein kinase inhibitors is the sequence conservation of the protein kinase catalytic domain. As a result, ATP-competitive inhibitors targeting the catalytic domain, the largest kinase inhibitor class, as a rule, display side activity against other protein kinases. Despite this, the success of drugs targeting human kinases has fostered prospects to target kinases from pathogenic organisms in a way to find new treatments for human infectious diseases.

The discovery of new drugs is a non-trivial process that requires the design of compounds with low toxicity, high specificity, activity and optimal physicochemical properties. The time required for a compound to progress from the research laboratory to the market is typically over

10 years, however, in conditions where there is low financial investment, such as for neglected tropical diseases, drug development can take longer (Pink, Hudson, Mouriès, & Bendig, 2005). High-Throughput Screening (HTS) technologies have been widely employed since the 1990s, as the possibility to test thousands of chemical compounds in a single test could potentially help to launch new drugs on the market in a shorter time-frame, with a lower investment (Pereira & Williams, 2007). Unfortunately, this did not occur. Few compounds against neglected infectious diseases identified in HTS have been successfully approved for clinical use, bringing the feeling of frustration about this approach (Ashburn & Thor, 2004; Paul et al., 2010; Springs & Janssen, 2003). Conventional high-throughput screening assays targeting parasitic diseases generally rely on the use of intact parasites for their assays (Baniecki, Wirth, & Clardy, 2007; Macgregor et al., 2014; Mandal et al., 2011). Unfortunately, the establishment of parasite-based HTS is not always possible (Bermúdez et al., 2018). Furthermore, the assays generate neither information regarding the toxicity to human cells, nor clues to the target affected by the hit compound. At the first stage of drug discovery, the majority of hit compounds identified in low- or high-throughput screens do not pass to the next stage of evaluation due to toxicity and low activity (Buckley & Dorato, 2006).

Parasite-based high-throughput screening (PHTS) for drug discovery on neglected diseases can be challenging due to the complex biological requirements of different parasites, which can be incompatible with an automated procedure (Visvesvara & Garcia, 2002). Probably the most studied tropical parasite, due to its medical importance and consolidated culturing methods, is *Plasmodium falciparum*, hence it is not surprising that this is one of the few parasites for which PHTS is possible (Gamo et al., 2010; Guiguemde et al., 2010). PHT screens have been employed by different groups to identify thousands of compounds with antiplasmodial activity, however these, as all parasite-based screens, provide no clue about the molecular target or mechanism of action of the hit compounds. Even if you use cheminformatics for compound-target validation, develop machine-learning models and virtual screenings to find molecules for a particular kinase target before PHTS, there is no guarantee that your molecule is in fact inhibiting that particular target (Réda, Kaufmann, & Delahaye-duriez, 2020).

A widely used technique to find small molecules interacting with a target protein with precision is high-throughput thermal shift assays (Differential Scanning Fluorimetry, DSF) (Fedorov, Niesen, & Knapp, 2012; Rudolf, Skovgaard, Knapp, Jensen, & Berthelsen, 2014). DSF relies on the employment of dyes that fluoresce when in contact with a protein in their denatured transition state. Potential ligands are found following the premise that most binding molecules are able to thermally stabilize their targets, that is, when there is a positive shift in the melting temperature (Tm) between the target protein incubated with a particular compound *vs* with vehicle control. The bottleneck for using DSF for finding kinase inhibitors is that usually only a part of the protein, the kinase domain, is expressed for these assays, leaving aside promising allosteric inhibitors, or molecules that inhibit kinase activity by binding other protein domains important for kinase functioning (Profeta et al., 2019). Moreover, protein assays might be good for detecting inhibitors but give us no clue regarding cellular permeability of these molecules. Cellular thermal shift assays were developed to fill this gap leaved by traditional DSF, by involving whole cells assays instead of using purified protein fragments (Jafari et al., 2014; Martinez et al., 2018). However, screening compounds for pathogenic organisms with complex or unavailable *in vitro* culturing methodology (like the case of *P. vivax*) remains a challenge (Bermúdez et al., 2018).

Here we will discuss how yeast-based screening assays can fill some of the gaps in the drug discovery pipeline for neglected diseases, with a focus on the search for compounds targeting the kinome from pathogenic agents. We also highlight the potential of exploring yeast-based chemical genome profiling database in a way to track novel potential kinase inhibitors, thus facilitating their optimization for clinical use. Hence, we aim to demonstrate that yeast-based assays can optimize kinase-based drug discovery and development.

2. Yeast-based drug screens

The first eukaryotic organism to be fully sequenced, in 1996, was the budding yeast *Saccharomyces cerevisiae*, which is now undoubtedly the best characterized eukaryote cell (Botstein, Chervitz, & Cherry, 1997). *S. cerevisiae* shares most biochemical pathways with higher eukaryotes and has been used for almost 40 years as a host for heterologous expression of genes, with the initial goal of better understanding their function (Zhang & Bilsland, 2011), and for drug screening against human targets (Couplan et al., 2011).

Denny and Steel (2015) reviewed in great detail different approaches in which the budding yeast *S. cerevisiae* and the fission yeast *Schizosaccharomyces pombe* can be employed in screens targeting neglected tropical diseases, using lethal expression, transactivation or substitution. They described how the expression of genes under the regulation of inducible promoters can be used in screens for inhibitors of lethal viral proteins (lethal expression); an approach that could potentially be employed in the search for antivirals targeting diseases such as Dengue, Zika or even COVID19. A similar approach using yeast lethal expression was developed by Fernández-Acero and colleagues, as a way to find inhibitors for human phosphatidylinositol-3 kinase (PI3K). The group took advantage of the fact that overproduction of PI3K is toxic in yeast, and tested the ability of commercial PI3K inhibitors to restore yeast growth, validating a platform using yeast-based drug screen for PI3K inhibitors (Fernández-Acero et al., 2012). The same rational was used to find inhibitors of human p38α, a protein kinase toxic to yeast, which plays a major role in oncogenesis (Lee et al., 2000).

Denny and Steel also provide examples on how yeast cells can be engineered to allow the activation of a reporter gene following the modulation of a heterologous protein target (transactivation) (Cerejo et al., 2012; Didem Demirbas, Shimizu-Albergine, Cakici, Beavo, & Hoffman, 2012; Douglas, Wang, Demirbas, Allain, & Hoffman, 2008). However, to date this approach has not been employed in the search of compounds targeting NTDs. A limitation of this approach is that the heterologously expressed protein has to be toxic to the yeast cell in the first place. Furthermore, by screening for suppression of this toxicity, inhibitors that target host factors affecting the toxicity may be found, rather than the heterologous protein itself. Hence, we will focus on the approach that has been successfully employed in NTD drug discovery: substitution or complementation, and further discuss how these techniques can be employed in the search for kinase inhibitors for different disease-causing organisms.

3. Yeast-based drug screens by substitution

Back in the 90s, Klein and co-workers pioneered the development of high-throughput screens for antiparasitic compounds in *S. cerevisiae*, through expression of potential *Haemonchus contortus* drug targets in yeast (Klein et al., 1997). In this work, 90,000 compounds were tested for the ability to inhibit the helminth's ornithine decarboxylase (ODC) expressed in yeast. In spite of the elegance of these screens, they generated only one single hit, which failed to inhibit the enzyme *in vitro*. Sibley and co-workers established yeast strains suitable for drug screens against *Plasmodium* dihydrofolate reductase (Sibley & Macreadie, 2001; Wooden, Hartwell, Vasquez, & Sibley, 1997). In spite of performing screens at a much lower throughput than the screens described for *H. contortus*, the compounds identified by Sibley's group appeared more promising, as they could be further validated in other systems. Since then, many groups have developed yeast-based assays to aid drug discovery (Smith, Ammar, Nislow, & Giaever, 2010; Wooden et al., 1997). In a series of related papers (Bilsland et al., 2011, 2013, 2018; Bilsland, Bean, Devaney, & Oliver, 2016; Chugh et al., 2015; Williams et al., 2015), Bilsland and co-workers described how yeast-based screens by substitution can be employed in the discovery of novel anti-parasitic compounds targeting malaria and neglected tropical diseases.

3.1 Fluorescent HTP antiparasitic drug screens

Genetically engineered yeast can be used to express antiparasitic targets and its human orthologues to facilitate quicker identification of potential therapeutic compounds by HTS, supplying composite activity and toxicity in human cells at once (Bilsland et al., 2013; Williams et al., 2015). Bilsland and co-workers cloned a series of potential drug targets of *P. falciparum*, *P. vivax*, *Schistosoma mansoni*, *Leishmania major*, *Trypanosoma brucei*, *T. cruzi* (Bilsland et al., 2011) and *Brugia malayi* (Bilsland et al., 2016) into yeast expression vectors under the control of a TetO2 regulatable promoter (Garí, Piedrafita, Aldea, & Herrero, 1997). In parallel, the authors cloned the human counterpart for each of the targets to be used as an internal control for specificity. They constructed yeast strains and designed experimental conditions making them dependent of the functional expression of each of the heterologous target, and then characterized the fitness of each engineered strain. These were then labeled with high-copy plasmids expressing one of four different fluorescent proteins (Bilsland et al., 2013): mCherry (Keppler-Ross, Noffz, & Dean, 2008), CFP, Venus or Sapphire (Sheff & Thorn, 2004). Such labeling of yeast cells with distinct fluorophores allows quantitation of growth of the two to four different strains in competition in presence of different library compounds, providing an internal control for general toxicity as well as plate positioning effects. The setup thus enables rapid identification of chemical inhibitors with selectivity for the parasite target protein over the human ortholog (Fig. 1).

These strains were successfully used in screens using chemical libraries with high diversity (Bilsland et al., 2013), libraries of FDA approved compounds (Williams et al., 2015) and the Malaria Box collection of 400 compounds with antiplasmodial activity (Bilsland et al., 2016). The hits identified by this approach have low general cytotoxicity and a higher

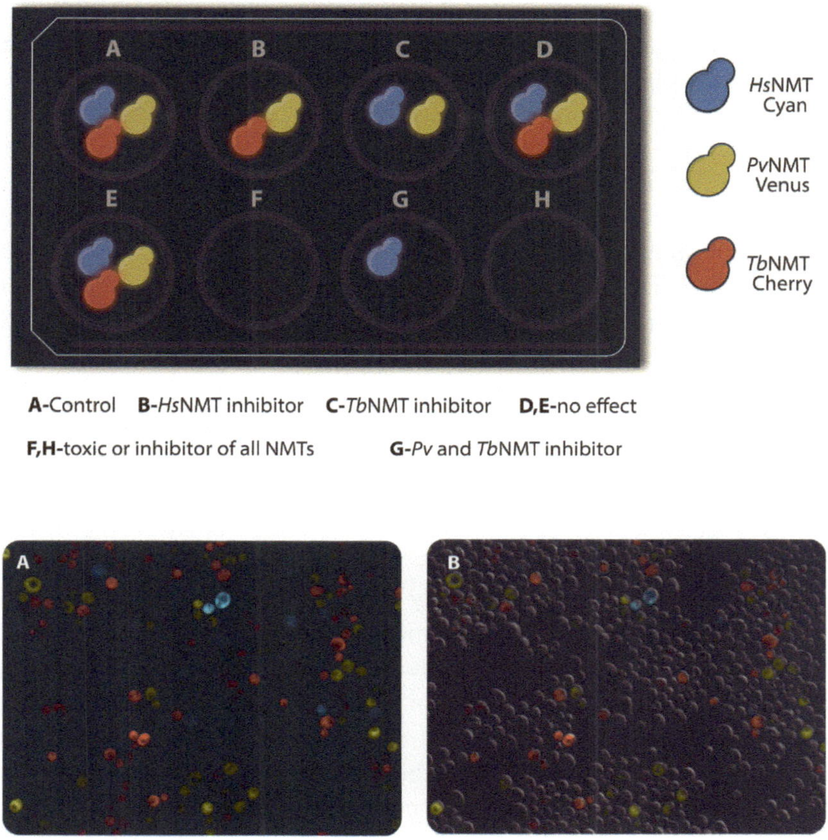

Fig. 1 Experimental setup of yeast-based screens with fluorescently labeled strains. Schematic representation of yeast strains expressing *P. vivax* (*Pv*NMT), *T. brucei* (*Tb*NMT) or human (*Hs*NMT) N-myristoyl transferases, and labeled with Venus (yellow), Cherry (red) or CFP (cyan) fluorescent proteins. (A–H) indicate different wells of a 384-well black plate with solvent control (A) or 10 μM of different test compounds (B–H).

specificity for the parasite target when compared to the human counterpart, with approximately 50% of the compounds identified in yeast-based screens do kill the parasites efficiently *in vitro*. This approach is particularly interesting for targets from parasites that cannot be cultivated *in vitro* (Balaña-Fouce & Reguera, 2016). Furthermore, it selects compounds able to cross the plasma membrane of eukaryotic cells and which are specific to the parasite target, not inhibiting the human counterpart, suggesting low general toxicity.

3.1.1 Optimization of fluorescent HTP screens

To ensure the quality control of yeast-based HTP screens, it is important to obtain an optimal expression of different heterologous proteins by the engineered yeast strains. There is a wide-range of variables able to tune yeast expression, such as the choice of particular promoters/terminators, selections markers and origin of replication of plasmid vectors or site of integration of (ORF) (Karim, Curran, & Alper, 2013). One study that characterized the use of yeast as a chassis for synthetic biology evaluated the impact of three different copy number evaluation systems on overall protein expression: single-copy chromosome integration, low-copy CEN6/ARS4 plasmids, and high-copy 2-μm plasmids (Lee, DeLoache, Cervantes, & Dueber, 2015). For this experiment, cassettes expressing mRuby2 or Venus fluorescent tags under weak, moderate and strong promoters were cloned for each copy number system. Unlike what one would expect, when fluorescence was measured for both tagged proteins, it was pointed a disagreement between the copy number systems, which not always represented gene expression accordingly.

Strains containing chromosomally integrated cassettes expressed under the different promoters displayed a more precise cell-to-cell expression of the genes, and filled the expected fluorescence grid. On the other hand, both, low-copy and high-copy 2 μm plasmids presented higher levels of fluorescence acquired by single-cell measurements, indicating that gene expression is not tightly regulated within these systems. Another experiment by the same group pointed out DNA copy number as the major contributor of this variation. Lee and colleagues showed that yeast strains expressing mRuby2 and Venus from different chromosomal loci presented nearly the same fluorescent distribution as the single chromosomal loci strain. This correlation was shown to be lost when two genes are expressed from different plasmids within the same cell (Lee et al., 2015).

Despite the homologous recombination machinery permissive to chromosome site-specific integration, plasmid transformation is more efficient than chromosomal integration (Lee et al., 2015). That explains the wide use of plasmid transformation in yeast synthetic biology. From this work, we can conclude that chromosomal integration should the preferred cloning method, as plasmids present high variation in copy number, and uncorrelated protein expression when a single cell contains more than one protein-expressing plasmid.

3.1.1.1 Genomic integration

One strategy adopted to increase the efficiency of homologous recombination, is the addition of endonucleases for the DNA integration process, which generates double-strand breaks within the target chromosome, facilitating genome integration. CRISPR-Cas9, a low cost technique that revolutionized genome-editing methodology, also serves a similar purpose. In CRISPR, a linear DNA molecule is combined with genome *via* the cotransformation of a plasmid containing the endonuclease (Cas9) with a guide RNA strand, together with endogenous DNA repair machinery. The high fidelity efficient system is suitable for HTS, and allows refined genomic alterations, as precise as a single-nucleotide change. Lee and colleagues actually compared two genome integration techniques examining their outputs on chromosomal library assembly: CRISPR-Cas9 and I-SceI (Lee et al., 2015). The group saw that chromosome integration of linear DNA through unassisted homologous recombination generates six times fewer colonies than that obtained through plasmid transformation. This yield could be improved in twofold and fivefold when chromosomal integration is assisted by CRISPR-Cas9 and I-Sel recognition site, respectively. The use of linearized expression plasmids containing both the Cas9 and I-Sel cutting systems, could increase transformation efficiency to similar levels of that of plasmid transformation for Cas9 strategy, with I-SceI system surpassing it by up to 1.8-fold (Lee et al., 2015).

3.1.2 Limitations of yeast HTP drug screens

Besides the problems with differential expression of heterologous proteins addressed previously, there are many other barriers to overcome for the use of yeast as a host for synthetic biology. Probably the major limitation of yeast-based screens for drug discovery is that not all mammalian or protozoan pathways and proteins are conserved in yeast, and heterologous proteins expressed in yeast might not be functional due to different post-translational modification machineries. Hence, functional complementation aiming at drug discovery is not applicable to all targets. Importantly, this technique relies on proteins that are essential for yeast viability, or at least become essential under specific culturing conditions. This is because the output of the screen is analyzed based on cell proliferation (fluorescence measurements). Thus, many kinases that might be important or essential for the parasite and non-essential or only conditionally essential in yeast might be left aside. Another limitation lies in the number of orthologs for the protein of interest present in humans or in the pathogen of interest. It is important that the gene to be substituted does not have another copy within the target organism, as that could bias protein inhibition or complementation.

One further concern in using *S. cerevisiae* assays in drug discovery is the high level of resistance of the yeast to many drugs due to the low permeability of the cell wall, as well as the high activity of multiple membrane-bound drug efflux pumps (Hoepfner et al., 2014). Therefore, drug screens in yeast frequently require higher concentration of active compounds than those used in assays employing mammalian cells. It is possible to circumvent this problem by using yeast genetic backgrounds with multiple mutations in genes encoding or regulating the expression of drug efflux pumps, or that increase membrane permeability (Hoepfner et al., 2014).

3.1.3 *Yeast-based HTP screens for kinase inhibitors*

Nearly the totality of protein phosphorylation in eukaryotes is conducted by the superfamily of protein kinases, which usually shares conserved catalytic domains within different organisms. The kinase superfamily is among the most abundant protein families, representing around 2% of all eukaryote genes (Manning et al., 2002). Despite the lack of some metazoan kinase families in yeast (Manning et al., 2002), most of yeast kinome has been characterized (Rubenstein & Schmidt, 2007). We can take advantage of the high rates of gene orthology within the kinase superfamily to use yeast as a chassis for studying the kinome of other eukaryotes (Fig. 2).

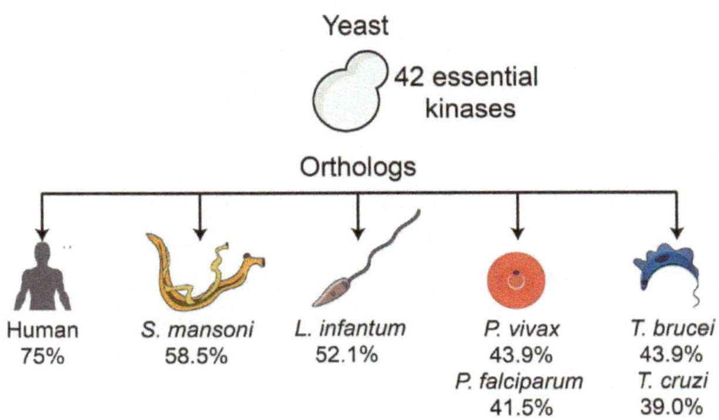

Fig. 2 Representation of the yeast essential kinome space and the percentage of coverage (orthology) among humans and different organisms causing neglected diseases.

To evaluate the potential of yeast-based HTP screens of kinase inhibitors toward pathogenic organisms, we identified the essential kinases that compose the yeast kinome and their orthologs in human and different pathogenic agents of interest (Fig. 2, Table 1). Out of the 42 essential kinases in *S. cerevisiae*, seven presented orthologs against humans and all species of pathogens evaluated (Table 1, highlighted in pink), comprising an interesting group of kinases as therapeutic targets. This approach relies on the substitution of an essential yeast gene of interest by orthologous proteins being present in the genomes of the parasite in question, and in the human host, for yeast functional cross-complementation. Not all protein kinase classes present in humans are conserved in yeast, notably the tyrosine kinase branch is absent, although mixed specificity (tyrosine/threonine) kinases are present (Manning et al., 2002). This limits the range of protein kinases approachable by yeast-based assays. Unicellular parasites such as *Plasmodium*, *Trypanosoma*, *Leishmania*, and *Toxoplasma* also lack the tyrosine kinases, so this does not present a problem for employing yeast as a surrogate for screens targeting protozoan kinases. Being specific to metazoans, the dedicated tyrosine kinases are present in parasitic worms such as the nematode *Brugia* or the trematode *Schistosoma*, however, and those protein targets are thus beyond the scope of this technology. Altogether, the versatility of fluorescent HTP drug screens allows the search for specific inhibitors toward different organisms at the same time, providing a strong tool for drug discovery targeting parasites that are challenging to culture for whole cellular drug assays.

4. Yeast-based methods for tracking the mode of action of compounds

Yeast chemical-genetic profiling is an elegant group of techniques used to track the mode of action of a molecule. The approach generates as an output, a hint of biochemical pathways affected by the compound of interest, also highlightening potential off target effects. The term "chemical-genetic profile" was first used in 2004 through the pioneering work of Parsons and co-workers (Parsons et al., 2004, 2006). Chemical-genetic profiling (CGP) in *S. cerevisiae* is particularly efficient due to the strategic introduction of unique 25 bp DNA marker sequences (barcodes) upstream and downstream of an antibiotic marker cassette (KanMX) used in the genome-wide gene deletion efforts (Lee et al., 2014; Pierce, Davis, Nislow, & Giaever, 2007). Each of these DNA barcodes is flanked by

Table 1 Essential *S. cerevisiae* kinases and their human and parasite orthologues.

Gene name	Entry name	Yeast essential kinases	Ortholog						
			Human	*P. vivax*	*P. falciparum*	*L. infantum*	*T. cruzi*	*T. brucei*	*S. mansoni*
YAL038W	KPYK1	P52489	P14618 P30613	A5K2A0	C6KTA4	A4IAQ1 A4IAQ4	K4DWR8	Q388I8	G4VAI0 G4VAH9
YAR019C	CDC15	P27636	–	–	–	–	–	–	–
YBL105C	KPC1	P24583	Q02156 P17252 P24723	–	–	–	–	–	G4VE95 G4M0Q1
YBR136W	ATR	P38111	Q13535	–	–	–	–	–	G4VRC4
YBR160W	CDK1	P00546	Q00526 P24941 P06493	A5K2H6 A5K4A7 A5KDI0	P61075 Q8IIR9	A4ICT0	K4DUA6 K4E8K7	Q38BA2 Q57TX3	G4VLX9
YCR012W	PGK	P00560	P00558 P07205	A5K6T9	P27362	E9AGT9 E9AGU0	K4E587	Q4GZE6 Q4GZG4 Q4GZG3	G4VG20
YDL017W	CDC7	P06243	O00311	P06243	–	–	–	–	G4VG92
YDL028C	MPS1	P54199	–	–	–	–	–	–	G4LUY3
YDL108W	KIN28	P06242	P50613	–	–	E9AGS0	–	–	G4VI20 G4VI21
YDL166C	KAD6	Q12055	–	A5K9R1	Q8I236	A4I5L5	–	Q584W4	G4VIC9
YDR196C	CAB5	Q03941	Q8WVC6	A5K3X3	Q8IL34	A4I012	K4E8Y9	Q583T2	–
YDR208W	MSS4	P38994	P78356 P48426 Q8TBX8	–	Q8I239	A4IA85	–	–	–

YDR226W	KAD2	P07170	P54819	A5K7B3	Q8IJV6	A4I9I0	K4E1Y3	Q38BZ2	G4VNE5 G4VNE6
YDR236C	RIFK	Q03778	Q969G6	–	–	A4IBK7	K4DYC9	Q38DG4	G4M0L1
YDR454C	KGUA	P15454	Q16774	A5K709	Q8I2M1	A4I8X4	K4E0J2	Q389H1	G4VGU4
YDR531W	PANK	Q04430	Q9H999 Q9BZ23 Q8TE04	A5K130	–	–	–	–	G4VQ31 G4VQ30
YFL029C	CAK1	P43568	–	–	–	–	–	–	–
YIL147C	SLN1	P39928	–	–	–	–	–	–	–
YJL087C	TRNL	P09880	–	–	–	–	–	–	–
YJR049C	UTR1	P21373 P32622	O95544	–	–	–	–	–	–
YJR057W	KTHY	P00572	P23919	A5K915	Q8I4S1	–	–	Q581B1 Q57YW2	G4VDI4 C4PXA6
YKL024C	KCY	P15700	P30085	A5K9Q6	–	A4HS93	–	Q38EC4	G4V9S0
YKL139W	CTK1	Q03957	P50750	A5K626	Q8I1Q9	–	–	–	G4VTU1 G4VJ86
YKL203C	TOR2	P32600 P35169	P42345	–	–	A4IE36	K4E111	Q38AC2	–
YLL035W	GRC3	Q07845	–	–	–	–	–	–	–

Continued

Table 1 Essential *S. cerevisiae* kinases and their human and parasite orthologues.—cont'd

Gene name	Entry name	Yeast essential kinases	Ortholog						
			Human	*P. vivax*	*P. falciparum*	*L. infantum*	*T. cruzi*	*T. brucei*	*S. mansoni*
YLR240W	VPS34	P22543	Q8NEB9	A5K9K5	–	A4I109	K4EB11	Q57YM2	G4VD93
YLR305C	STT4	P37297	P42356	–	–	–	–	–	–
YMR001C	CDC5	P32562	P53350 Q9H4B4 Q9NYY3	–	–	A4HXE5	K4DPB8	Q57VI0	G4V5L2
YMR013C	SEC59	P20048	Q9UPQ8	–	B9ZSI8	–	–	–	–
YMR208W	KIME	P07277	Q03426	–	–	A4I6L2	K4E5Y8	Q57Y11	G4VIV3
YMR220W	ERG8	P24521	–	–	–	–	–	–	–
YNL161W	CBK1	P53894	Q9Y2H1	–	–	E9AG71	–	Q38BA7	G4VBN3
YNL207W	RIO2	P40160	Q9BVS4	A5K649	Q8I1N8	A4I5G7	K4E6C1	Q583Q0	G4VIX6
YNL256W	FOL1	P53848	–	A5JZS1	–	–	–	–	–
YNL267W	PIK1	P39104	Q9UBF8	–	Q8I406	–	–	–	G4VGB3
YOL100W	PKH2	Q12236 Q03407	Q9Y2H9 Q6P0Q8 O15021 O60307	–	–	A4I3S8	–	–	–

YOR119C	RIO1	Q12196	Q9BRS2	A5K0B0	Q8I5B1	A4I477	K4DJT1	Q580T5	G4LWE1
YOR143C	THI80	P35202	Q9H3S4	–	C0H556	–	K4E637	–	–
YPL153C	RAD53	P22216 P39009	O96017	–	–	–	–	–	–
YPL204W	HRR25	P29295	P48730 P49674 P48729	A5K525	Q8IHZ9	A4IAZ8	K4DZI9 K4E849 K4DNP2 K4E0T8 K4E070 K4ECB7	–	G4VHW4
YPL209C	IPL1	P38991	O14965 Q96GD4 Q9UQB9	A5KBA7	C6KSQ1	A4I3C6	K4E6X3	Q384J7	G4VQ85
YPR161C	BUR1	P23293	–	–	–	–	–	–	–

DNA sequences (universal primer binding sites) common to all deletion cassettes, allowing the simultaneous amplification of all DNA barcodes from all strains grown in competition. These amplified barcodes can be sequenced or hybridized to barcode specific DNA chips for quantification of each strain in the pool (Pierce et al., 2007). CGP can be subdivided into several approaches: haplo-insufficiency profiling (HIP), homozygous profiling (HOP) or haploid deletion chemical genomic profiling, multi-copy suppression profiling (MSP) (Ericson, Hoon, St Onge, Giaever, & Nislow, 2010), comparative expression profiling, and high-content screening, among others (Hillenmeyer et al., 2009; Pries & Hoepfner, 2016). HIP is based on the principle that strains with a reduced dosage of a given target will be more sensitive to the test compound than strains with full expression of that gene.

HIP profiling can be performed in liquid cultures, with 6000 strains growing in competition, or on agar plates, with different strains arrayed on defined spots on plates The colonies are arranged in a high-density grid (up to 1536 μL format), and the entire surface is optically scanned every 20 min (Zackrisson et al., 2016). The number of cells in each colony at each time-point is estimated from a function of the distribution of pixel intensities, and the resulting data is compiled into length of lag phase, maximal growth rate, and yield. Thus, effects on each of these growth parameters can be measured separately and with high accuracy, on a massively parallel scale. Growing pools in liquid is faster, cheaper, and consumes less compound than the agar-based colony screening method, as the entire mutant pool can be assayed in a few 100 μL. By contrast, the agar-based method requires an entire agar slab of ~100 mL. On the other hand, agar-based screening provides data with higher accuracy, since time-resolved growth curves with hundreds of data points are obtained for each mutant, whereas only 2–3 timepoints are typically harvested for experiments with mutant pools in liquid. Moreover, agar-based screening obviates complications from cross-feeding and other interfering influences between different mutants mixed in the same pool (Zackrisson et al., 2016).

Homozygous or haploid deletion chemical-genetic profiling rely on growing ~4500 homozygous or haploid deletion mutants (non-essential genes) in the presence or absence of the test compound (Pries & Hoepfner, 2016; Silberberg, Kupiec, & Sharan, 2016). Unlike HIP, where the target protein is present in half the normal amount in the mutant compared to wild-type cells and is thus titrated out by a lower amount of compound, HOP does not give a direct indication of the target inhibited by the compound. This is because the target protein is completely absent from the mutant cells since all copies of the gene have been deleted from the genome. All effects on the mutant cells of the compound thus have to result from compensatory changes in other pathways (Wong et al., 2019). These can be, e.g., parallel pathways activated by the compound of interest, transport proteins, or drug-modifying proteins. Multi-copy suppression profiling relies on utilizing overexpression libraries to identify what genes can compensate for the inhibitory effects of the test compound (Hernández-López, García-Marqués, Randez-Gil, & Prieto, 2011). Hence, MSP frequently identifies pleiotropic drug resistance genes (such as the gene encoding the multidrug efflux pump Pdr5), or the direct target of the compound of interest (Jaime et al., 2012).

4.1 Yeast-based chemical genomic database: A hidden source of kinase inhibitors

The discovery of potential pathways affected by compounds through yeast-based chemogenomics is important in drug discovery, especially if you don't have a hint of what is the target of your molecule. Finding potential inhibitors for a particular target is also possible thanks to databases of yeast chemical genetic screens, which store data from thousands of compounds tested in pools of heterozygous or homozygous mutant yeast libraries. A public HIP/HOP yeast database created and administered by researchers from Giaever and Nislow labs at the Terrence Donnelly Centre for Cellular and Biomedical Research, from the University of Toronto (http://chemogenomics.pharmacy.ubc.ca), provides data from more than 3000 bioactive molecules. This Canadian databank not only allows the search for genetic profiles from chemicals of interest, but also offers a list of possible inhibitors from a particular target, in case you use a specific gene as an input (Fig. 3).

The database also allows the search for compounds affecting major cellular responses such as the unfolded protein response, or protein kinase C cell wall signaling. Thus, this platform is helpful and probably underestimated or underused by the scientific community, as having a list of chemicals with genetic validation of potential target inhibition might be a good starting point in the search for new bioactive inhibitors.

As we are particularly interested in kinase inhibitors for parasitic diseases, we can explore this database for the search of possible inhibitors of the kinases listed in Table 1, highlighted in pink, that is, yeast kinases with orthologs distributed among different pathogens. To simplify, we arbitrarily selected only potential inhibitors with fitness defect (FD) score above five, in

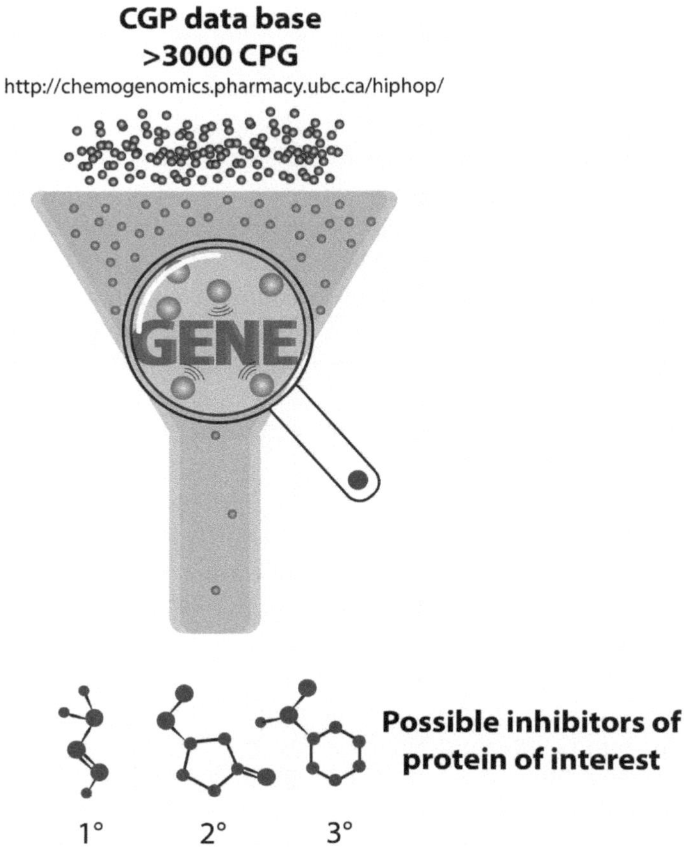

Fig. 3 Scheme showing use of the HIP/HOP chemical genomic database as a strategy to find potential inhibitors.

the HIP database. Out of the seven kinases from the table, six had potential inhibitors pointed by the CGP yeast database: Pyruvate kinase 1 (CDC19), Cyclin-dependent kinase 1 (CDC28), Adenylate kinase (ADK1), RIO1 protein kinase, RIO2 protein kinase and Spindle assembly checkpoint protein kinase (IPL1).

4.2 Potential Pyruvate kinase 1 inhibitors

Pyruvate kinase is an enzyme that catalyzes the last stage of glycolysis, transferring the phospho group from PEP (phosphoenolpyruvate) to ADP, forming ATP and pyruvate, which is crucial for cell metabolism (Schormann, Hayden, Lee, Banerjee, & Chattopadhyay, 2019). There are four isoforms in humans, which are differentially expressed according to cell type (Israelsen & Vander Heiden, 2015). In yeasts, two isoforms are known, Cdc19 and Pyk2. Cdc19 is expressed in the presence of glucose, and when yeast is on glucose deprivation, this enzyme is rapidly suppressed (Xu et al., 2012). Its deletion causes arrest at the G1/S transition and is lethal on glucose, while suppression of Pyk2 does not cause phenotypic changes (Boles, De Jong-Gubbels, & Pronk, 1998). Bioinformatics analyzes identified two pyruvate kinase in *P. falciparum* (PfPyKI and PfPyKII), which are phylogenetically related to the two pyruvate kinase of another Apicomplexa, *Toxoplasma gondii*. It has been suggested that these enzymes in the Apicomplexa are related to the dephosphorylation of PEP in the apicoplast, providing pyruvate for the synthesis of fatty acids in this organelle (Ralph et al., 2004), and are synthesized predominantly during the asexual blood stage.

In the absence of the Kreb's cycle, trypanosomatids depend exclusively on glycolysis to generate ATP. In *T. brucei*, pyruvate kinase is allosterically activated by fructose 2,6-biphosphate (Van Schaftingen, Opperdoes, & Hers, 1985) and is considered a validated target, since its silencing by RNAi affected the growth of the parasite (Albert et al., 2005). Below are putative Pyruvate kinase 1 inhibitors, predicted by HIP results (Fig. 4).

4.3 Potential Cyclin-dependent protein kinase 1 inhibitors

The first Cyclin-dependent kinase (CDK), called *cdc2* (now CDK1, ortholog of budding yeast CDC28), was discovered in fission yeast genetic studies in the 80s (Lee & Nurse, 1987; Malumbres, 2014). As implied by the name, the serine/threonine Cyclin-dependent kinase activation relies on their interaction with proteins named cyclins, which compose a complex with CDK to form the active holoenzyme (Malumbres, 2014). CDK1 is ubiquitous to all eukaryotes and was the first member of the CDK family to be characterized, being shown to play an essential role in mitosis (Diril et al., 2012). In *Plasmodium*, PfPK5 was the first CDK to be characterized (Graeser, Wernli, Franklin, & Kappes, 1996; Holton et al., 2003). It shows overall 60% amino acid sequence similarity with human CDK1 and CDK2, and the catalytic domain 82% and 92% of sequence identity with the human counterparts, respectively (Holton et al., 2003). A study showed that the *T. brucei* CRK3, probably the main CDK controlling cell cycle, has homologs identified in *T. cruzi* and *L. major* that shares around 75% amino acid sequence identity between themselves and around 50% between human CDK1 (Tu & Wang, 2004). Inhibitors targeting CDKs belonging to

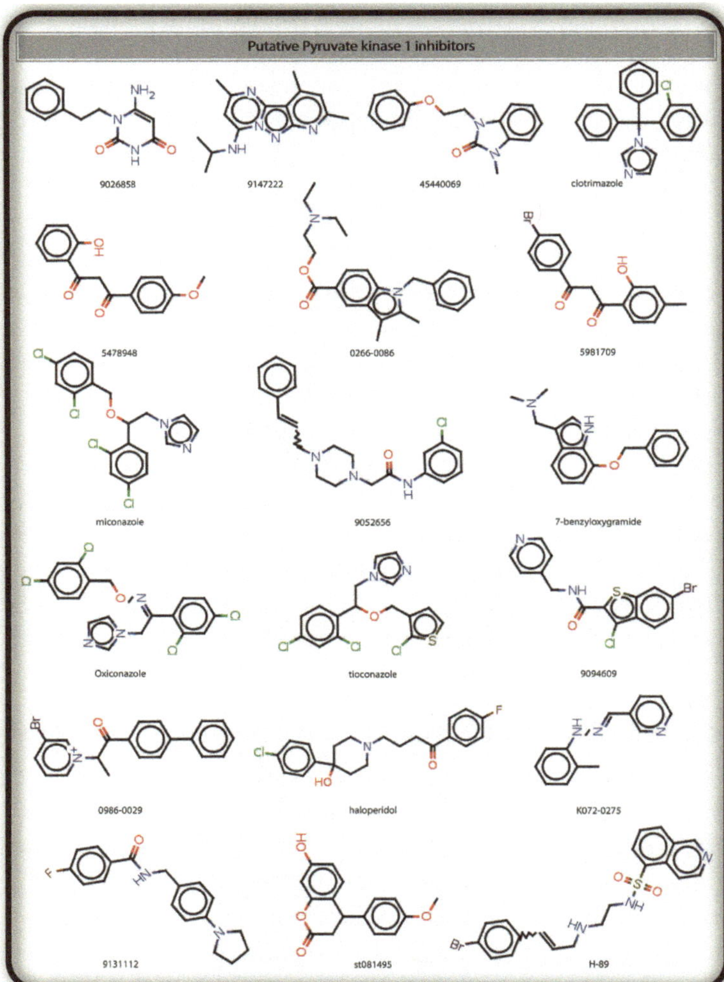

Fig. 4 Potential Pyruvate kinase 1 inhibitors proposed by HIP results.

parasites have been extensively explored and pursuit (Geyer et al., 2009; Grant et al., 2004; Woodard et al., 2003). Inhibitors of CDK1 proposed by HIP results are presented below (Fig. 5).

4.4 Potential Adenylate kinase 1 inhibitors

The Adenylate kinase (AK) is a housekeeping enzyme involved in energy metabolism *via* catalysis of the nucleotide exchange reaction 2ADP ↔ ATP + AMP. The unique delivery of γ- and β-phosphoryl groups to ATP, doubling its energetic potential, promoted by AK enzyme, plays a major role

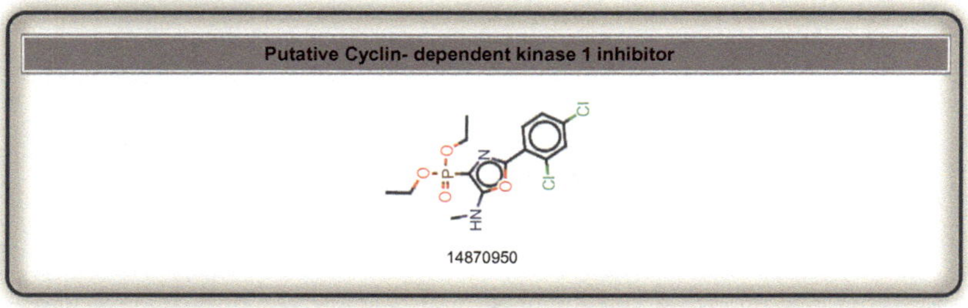

Fig. 5 Potential Cyclin-dependent protein kinase 1 inhibitors proposed by HIP results.

in several cellular processes like hormone secretion, endocytosis, cellular motility and nuclear transport, besides ubiquitous participation in energetic metabolic signaling pathways (Dzeja & Terzic, 2009). The search for AK inhibitors ended as fast as it started, with a collection of synthetic inhibitors arising in the 70s (Feldhaus, Frohlich, Goody, Isakov, & Schirmer, 1975). The knockdown of AK in tumor cells overexpressing AK2 can impair multiple processes intrinsic for cancer cells proliferation and metastasis, like migration, invasion, and apoptosis (Klepinin et al., 2020). Although there are nine members of AK family expressed in humans (K1–K9), four of them involved in cancer metabolism, this number varies depending on the organism. While *Plasmodium* and *Schistosoma* have two known variants of the gene and additional putative AKs (Ma et al., 2012), Trypanosomatids have an impressive high number of genes coding Adenylate kinases (Pereira, Bouvier, De Los Milagros Cámara, & Miranda, 2011). Here we present the potential AK1 inhibitors proposed by HIP results (Fig. 6).

4.5 Potential RIO1/RIO2 kinase inhibitors

The RIO protein kinase family was first identified in a screening with *Aspergillus nidulans* in 1996 (LaRonde-LeBlanc & Wlodawer, 2005). Because mutations in RIO1 conferred fitness defects, abnormal chromatin organization and chromosome loss, it was though that this kinase played a role in kinetochore assembly or participated of nuclear processes like chromosome condensation/segregation (Iacovella et al., 2015; LaRonde, 2014; LaRonde-LeBlanc & Wlodawer, 2005). This ancient protein family is essential in all eukaryotes and comparative protein structure studies suggest that it might have conserved functions between different organisms. So far we know little about the RIO families, most of the current knowledge being around RIO1 (LaRonde, 2014). RIO1 helps to promote pre-40S ribosomal subunit maturation, and controls the entrance of nascent Small Ribosomal Subunit (SSU) into translation in a range of organisms from

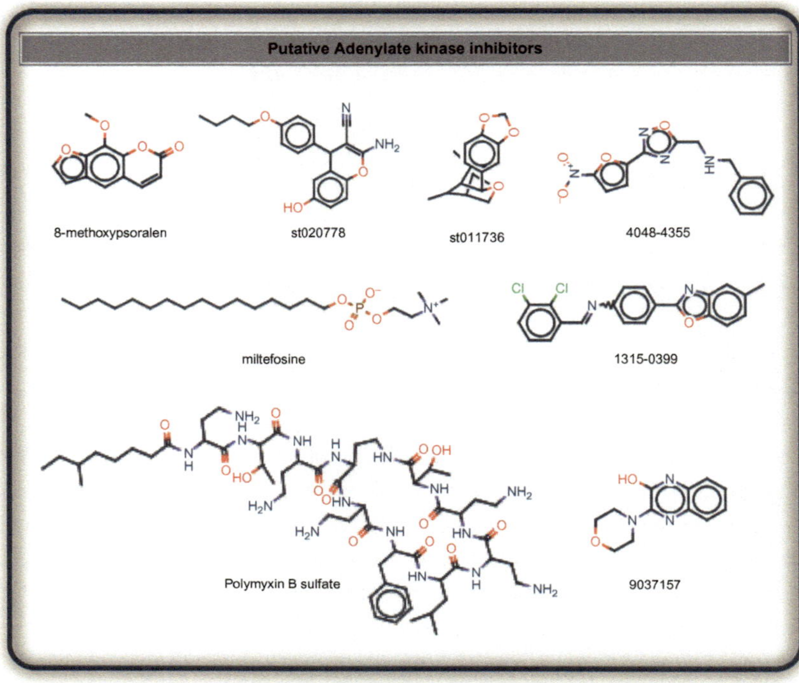

Fig. 6 Potential Adenylate kinase 1 inhibitors proposed by HIP results.

Archea to higher eukaryotes (Iacovella et al., 2015). A recent study from Iacovella and colleagues showed that the role of RIO kinase goes far beyond ribosome maturation, as around 66% of all genes transcripts are directly or indirectly regulated by RIO1 (Iacovella et al., 2018). Proteasome expression seems to be negatively regulated by RIO1, indicating also a role in proteostasis (Iacovella et al., 2018). RIO1 is overexpressed in 8% of primary cancer cells, being quoted as a potential tumor biomarker and an attractive target for therapeutics. In general, there is not much research focusing on targeting the RIO family for drug discovery. Despite available crystal structures for both, RIO1 and RIO2 kinases (LaRonde-LeBlanc, Guszczynski, Copeland, & Wlodawer, 2005; Wang, Varin, Vieth, & Elkins, 2019), so far, specific and potent inhibitors are still needed, especially for RIO1 (Kiburu & LaRonde-LeBlanc, 2012). Another study used virtual screening to propose putative inhibitors for RIO2 in *Plasmodium*, and the compounds tested showed micromolar-range antiplasmodial activity (Balaji, Ahsan, Jadav, & Trivedi, 2019). Putative inhibitors for both kinases, RIO1 and RIO2, proposed by the yeast chemical genetic HIP database are shown in Figs. 7 and 8.

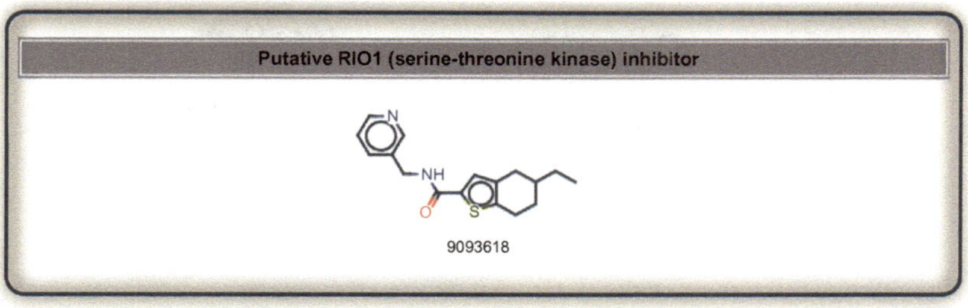

Fig. 7 Potential RIO1 kinase inhibitors proposed by HIP results.

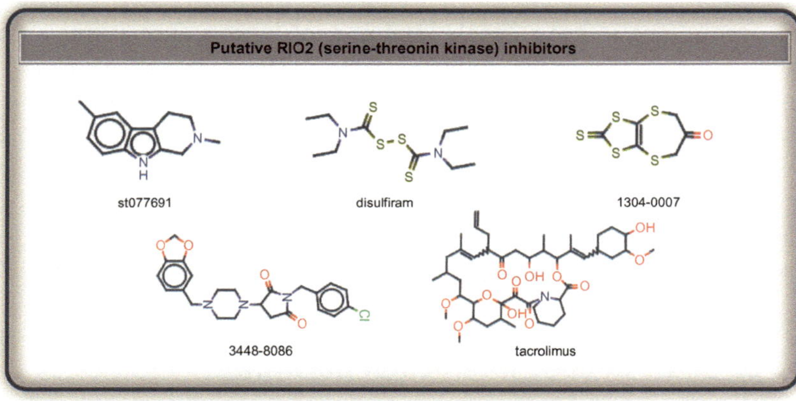

Fig. 8 Potential RIO2 kinase inhibitors proposed by HIP results.

4.6 Potential Spindle assembly checkpoint kinase inhibitors

The serine/threonine Spindle assembly checkpoint kinase (encoded by the *IPL1* gene in yeast) was discovered in 1993 during a yeast genetic screening performed to identify mutants able to promote an increase in ploidy (Chan & Botstein, 1993). In 2000, a group of scientists found that IPL1 kinase, the only member of aurora kinases in yeast, and their interacting phosphatase, Glc7/PP1, were responsible for the regulation of histone-3 (H3) phosphorylation levels in the cell cycle (Chan & Botstein, 1993). Mutations in H3 were known to cause abnormal chromosomal segregation and chromosome loss during the cell cycle (Guo & Kemphues, 1995), and when H3 phosphorylation was inhibited, chromosome condensation was impaired, preventing cells for entering mitosis (Guo & Kemphues, 1995). Because of its major role in cell cycle regulation, aurora kinases are found overexpressed in cancer cells, consisting in promising targets for tumor therapy, which lead to the discovery of several kinase inhibitors filling

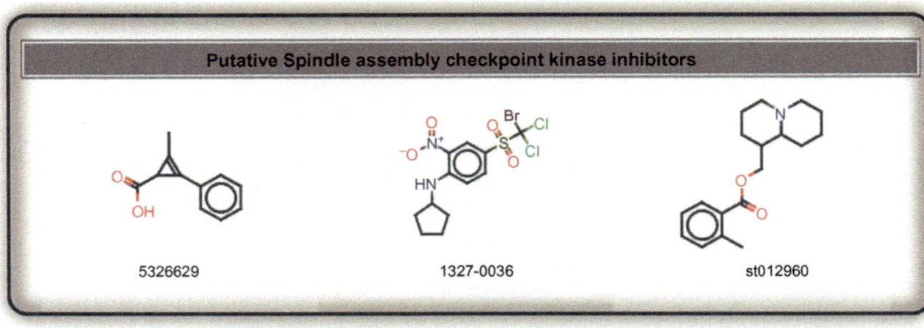

Fig. 9 Potential Spindle assembly checkpoint kinase inhibitors proposed by HIP results.

this purpose (Patel et al., 2014). Inhibitors of human aurora kinase were also shown to antiparasitic activity against *Trypanosoma*, *Leishmania* and *Plasmodium* (Giuliani et al., 2018; Ochiana et al., 2013; Patel et al., 2014). Aurora kinases were also found to be promising druggable targets for the *S. mansoni* drug discovery pipeline. A group performed *in silico* computational analysis to propose potential inhibitors for a wide range of *Schistosoma* kinases, including aurora kinase (Giuliani et al., 2018). Potential inhibitors for aurora kinases identified by this work using HIP database are found in Fig. 9.

5. Yeast genetic interaction network

One can also take advantage of genetic interplays, or the ability to generate negative or synthetic lethal interactions or positive genetic interactions, as another arm for the drug discovery pipeline (Costanzo et al., 2016; Tong et al., 2004). By "negative genetic interaction" we mean that the phenotype of a combination of two or more mutations is more severe than what would be expected from the phenotypes of the respective single mutants. Synthetic lethality, the most extreme form of negative genetic interaction, occurs when two mutations that are not lethal individually produce a lethal phenotype when combined. Following the same line, "positive" genetic interactions happen when two combined mutations alleviate or produce a phenotype less severe than what one would anticipate. Costanzo and colleagues constructed a global genetic interaction network in yeast, finding ~550,000 negative and ~350,000 positive interactions covering nearly all the yeast genome (Costanzo et al., 2016). The map of genetic interplays generated from this work contains not only data from interactions between non-essential yeast genes, but also covers interactions with some essential genes. For the analysis of essential genes, the authors developed a temperature-sensitive (TS) mutant yeast library composed by more than 800 yeast strains carrying conditional deletion for essential genes. TS mutants we screened at a semipermissive temperature, where cells were partially weakened by compromised protein function.

The genetic negative interaction maps generate information regarding central bioprocesses, and pleiotropic genetics, usually connecting genes with similar functions. By contrast, positive genetic interaction maps often allow the identification of regulatory connections linked with problems in housekeeping processes such as cell cycle management and proteostasis (Costanzo et al., 2016).

We can access The Cell Map website (Usaj et al., 2017), a website that makes available quantitative genomic data and provides free access tools for data visualization and data mining from The Global Yeast Genetic Interaction Network, to gain information that can be used for finding new kinase inhibitors. The yeast Synthetic Genetic Array analysis (SGA) scores produced by Costanzo and colleagues are based on the difference between a double mutant colony size and their expected colony size. We can explore the SGA platform to look for genes that confer synthetic lethal interactions with protein kinases (one kinase is not essential alone, but can be essential when combined with a different mutation). The output of SGA data mining (list of genes that make negative genetic interactions with protein kinases) can be used together with HOP database for the search of putative kinase inhibitors. In this case, this workflow can be used as a screen for kinase inhibitors for both non-essential and essential kinases (Fig. 10).

We identified the genetic negative interactions for our seven essential kinase genes that have orthologs in different pathogens (Table 1) and picked one among the most promising interacting genes based on SGA score and *P*-value (Fig. 11). Out of the seven kinases, we could extract genetic interaction data for six of them: Pyruvate kinase 1 (CDC19), Cyclin-dependent kinase 1 (CDC28), Adenylate kinase (ADK1), RIO1 kinase, RIO2 kinase and Spindle assembly checkpoint kinase (IPL1). We next sought to investigate if the kinase partner gene, contributor to the negative interaction, had hits suggested by the HOP database. Just as an example, we proceeded with hits from HOP found for four out of the six interactor genes, considering the SGA score of the negative interaction and the FD score of the inhibitor in the HOP database. To simplify, we selected just one possible inhibitor for each member of the different kinase families of interest, which presented FD score above five in the HOP database (Fig. 11). As expected, all the pair of genes composing the negative interactions presented related

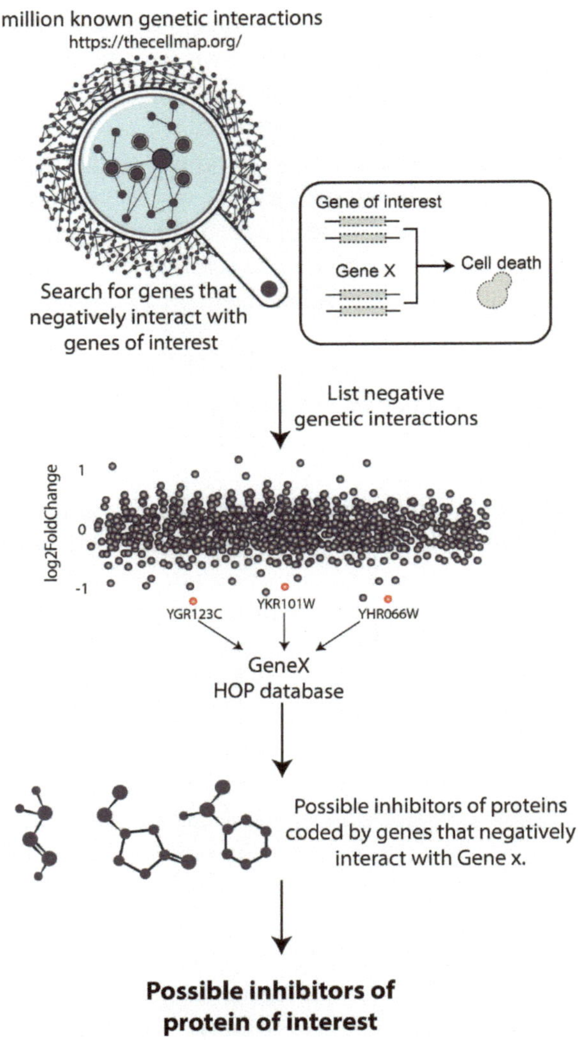

Fig. 10 Scheme showing how Global Yeast Genetic Interaction Network can be explored together with HOP database for preliminary screening of novel inhibitors for protein of interest.

cellular functions: the gene that interacts with Pyruvate kinase 1 is also involved in pyruvate metabolism; the gene paired with Cyclin-dependent protein kinase 1 is important for cell cycle regulation (Fig. 11).

Predicting inhibitors through the combination of yeast negative genetic interactions followed by HOP database analysis has many limitations.

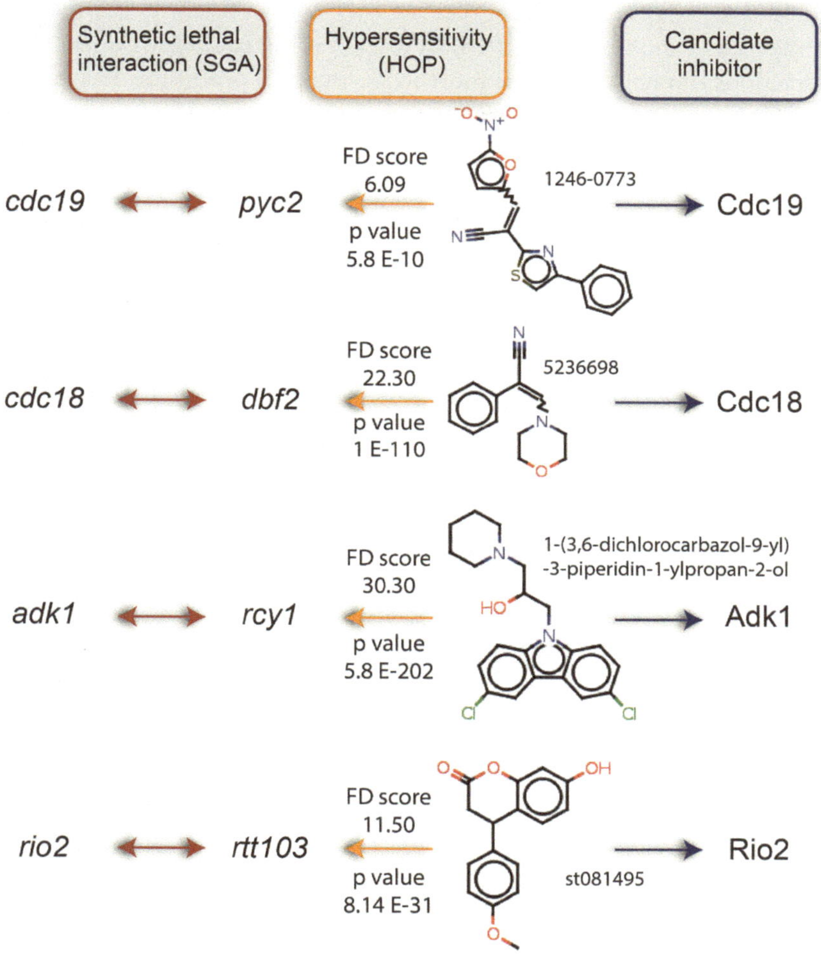

Fig. 11 Potential inhibitors for kinases of interest proposed by Global Yeast Genetic Network and HOP results.

It is important to have in mind that this strategy allows one to select potential inhibitors for any gene that might interact negatively with the gene that interacts negatively with your gene of interest, thus, this is quite an indirect search. However, it might be useful as a starting point, if combined with other predicting methods, such as structure-based computational approaches (Docking, Machine Learning, Shape-based assays).

To investigate not only the essential kinases themselves, but also the proteins they interact with genetically, considerably extends the reach of the search. Negative chemical-genetic interactions are more informative as predictors of the biological processes they are involved in than positive interactions (Simpkins et al., 2018). It is reasonable that negative gene-gene interactions also will contribute more to discovery of promising drug targets, given that most drugs inhibit, rather than enhance, activity of a gene product. Functional large-scale screens, such as genetic interaction mapping in yeast, are orthogonal to structure-based searches; cheminformatics tools to find small molecules with similar structural properties, and virtual ligand screens attempting to fit particular target macromolecules. Therefore, functional and structure-based methods should ideally be combined. Artificial intelligence approaches including machine learning have been used extensively in cheminformatics and structural bioinformatics. So far this has been less explored for genetic information, but for the future, similar ideas could be applied to analyze large-scale genetic interaction networks to predict favorable drug targets.

6. Conclusion

The advance of molecular and synthetic biology tools in the past decades has made it possible to bioengineer screening platforms to improve drug discovery and development. In this review we provided a flavor for how these advances have been translated into the optimization of drug discovery for the search of kinase inhibitors against pathogenic targets. The genetic tractability of the budding yeast *S. cerevisiae* provides a limitless resource that should be further explored to optimize the drug discovery process. We are convinced on the potential of greatly expanding the range of applications for yeast-based drug discovery both in academia as well as in industrial settings.

Acknowledgments

This work was supported by Fundação de Amparo à Pesquisa do Estado de São Paulo (FAPESP, Grants #2012/16525-2 and #2015/03553-6), the Conselho Nacional de Desenvolvimento Científico e Tecnológico (CNPq), and the Swedish Research Council (2016-05627). TAT was sponsored by CAPES and FAPESP (2019/27626-3) fellowships. FTMC is a CNPq research fellow. GCC was sponsored by FAPESP fellowship (2015/20774-6).

Conflict of interest

The authors declare no conflict of interest.

References

Albert, M. A., Haanstra, J. R., Hannaert, V., Van Roy, J., Opperdoes, F. R., Bakker, B. M., et al. (2005). Experimental and in silico analyses of glycolytic flux control in bloodstream form Trypanosoma brucei. *Journal of Biological Chemistry*, *280*(31), 28306–28315. https://doi.org/10.1074/jbc.M502403200.

Ashburn, T. T., & Thor, K. B. (2004). Drug repositioning: Identifying and developing new uses for existing drugs. *Nature Reviews. Drug Discovery*, *3*, 673–683. https://doi.org/10.1038/nrd1468.

Balaji, S. N., Ahsan, M. J., Jadav, S. S., & Trivedi, V. (2019). Molecular modelling, synthesis, and antimalarial potentials of curcumin analogues containing heterocyclic ring. *Arabian Journal of Chemistry*, *12*(8), 2492–2500. https://doi.org/10.1016/j.arabjc.2015.04.011.

Balaña-Fouce, R., & Reguera, R. M. (2016). Yeast-based systems for tropical disease drug discovery. *Expert Opinion on Drug Discovery*, *11*(5), 429–432. https://doi.org/10.1517/17460441.2016.1160052.

Baniecki, M. L., Wirth, D. F., & Clardy, J. (2007). High-throughput Plasmodium falciparum growth assay for malaria drug discovery. *Antimicrobial Agents and Chemotherapy*, *51*(2), 716–723. https://doi.org/10.1128/AAC.01144-06.

Bermúdez, M., Andrés, D., Pérez, M., Pinzón, G. A., Curtidor, H., & Patarroyo, M. A. (2018). Plasmodium vivax in vitro continuous culture: The spoke in the wheel. *Malaria Journal*, *17*, 1–12. https://doi.org/10.1186/s12936-018-2456-5.

Bilsland, E., Bean, D. M., Devaney, E., & Oliver, S. G. (2016). Yeast-based high-throughput screens to identify novel compounds active against Brugia malayi. *PLoS Neglected Tropical Diseases*, *10*(1), 1–15. https://doi.org/10.1371/journal.pntd.0004401.

Bilsland, E., Pir, P., Gutteridge, A., Johns, A., King, R. D., & Oliver, S. G. (2011). Functional expression of parasite drug targets and their human orthologs in yeast. *PLoS Neglected Tropical Diseases*, *5*(10), 1–9. https://doi.org/10.1371/journal.pntd.0001320.

Bilsland, E., Sparkes, A., Williams, K., Moss, H. J., de Clare, M., Pir, P., et al. (2013). Yeast-based automated high-throughput screens to identify anti-parasitic lead compounds. *Open Biology*, *3*, 120158. https://doi.org/10.1098/rsob.120158.

Bilsland, E., Van Vliet, L., Williams, K., Feltham, J., Carrasco, M. P., Fotoran, W. L., et al. (2018). Plasmodium dihydrofolate reductase is a second enzyme target for the antimalarial action of triclosan. *Scientific Reports*, *8*(1), 1–8. https://doi.org/10.1038/s41598-018-19549-x.

Boles, E., De Jong-Gubbels, P., & Pronk, J. T. (1998). Identification and characterization of MAE1, the Saccharomyces cerevisiae structural gene encoding mitochondrial malic enzyme. *Journal of Bacteriology*, *180*(11), 2875–2882. https://doi.org/10.1128/jb.180.11.2875-2882.1998.

Botstein, D., Chervitz, S. A., & Cherry, J. M. (1997). Yeast as a model organism. *Science*, *277*(5330), 1259–1260.

Buckley, L. A., & Dorato, M. A. (2006). Toxicology in the drug discovery and development process. *Current Protocols in Pharmacology*, *32*, 1–35. https://doi.org/10.1002/0471141755.ph1003s32.

Cerejo, M., Andrade, G., Roca, C., Sousa, J., Rodrigues, C., Pinheiro, R., et al. (2012). A powerful yeast-based screening assay for the identification of inhibitors of indoleamine 2,3-dioxygenase. *Journal of Biomolecular Screening*, *17*, 1362–1371. https://doi.org/10.1177/1087057112452595.

Chan, C. S. M., & Botstein, D. (1993). Isolation and characterization of chromosome-gain and increase-in-ploidy mutants in yeast. *Genetics*, *135*(3), 677–691.

Chugh, M., Scheurer, C., Sax, S., Bilsland, E., Van Schalkwyk, D. A., Wicht, K. J., et al. (2015). Identification and deconvolution of cross-resistance signals from antimalarial compounds using multidrug-resistant plasmodium falciparum strains. *Antimicrobial Agents and Chemotherapy*, *59*(2), 1110–1118. https://doi.org/10.1128/AAC.03265-14.

Costanzo, M., VanderSluis, B., Koch, E. N., Baryshnikova, A., Pons, C., Tan, G., et al. (2016). A global genetic interaction network maps a wiring diagram of cellular function. *Science*, *353*(6306), 1–16. https://doi.org/10.1126/science.aaf1420.

Couplan, E., Aiyar, R. S., Kucharczyk, R., Kabala, A., Ezkurdia, N., & Gagneur, J. (2011). A yeast-based assay identifies drugs active against human mitochondrial disorders. *Proceedings of the National Academy of Sciences of the United States of America*, *108*(29), 11989–11994. https://doi.org/10.1073/pnas.1101478108.

Denny, P. W., & Steel, P. G. (2015). Yeast as a potential vehicle for neglected tropical disease drug discovery. *Journal of Biomolecular Screening*, *20*, 56–63. https://doi.org/10.1177/1087057114546552.

Didem Demirbas, A. R. W., Shimizu-Albergine, M., Cakici, O., Beavo, J. A., & Hoffman, C. S. (2012). A yeast-based chemical screen identifies a PDE inhibitor that elevates steroidogenesis in mouse Leydig cells via PDE8 and PDE4 inhibition. *PLoS One*, *23*(3), 594–601. https://doi.org/10.1371/journal.pone.0071279.

Diril, M. K., Ratnacaram, C. K., Padmakumar, V. C., Du, T., Wasser, M., Coppola, V., et al. (2012). Cyclin-dependent kinase 1 (Cdk1) is essential for cell division and suppression of DNA re-replication but not for liver regeneration. *Proceedings of the National Academy of Sciences of the United States of America*, *109*(10), 3826–3831. https://doi.org/10.1073/pnas.1115201109.

Douglas, I. F., Wang, L., Demirbas, D., Allain, C., & Hoffman, C. S. (2008). Development of a fission yeast-based high throughput screen to identify chemical regulators of cAMP phosphodiesterases. *Journal of Biomolecular Screening*, *13*(1), 62–71. https://doi.org/10.1177/1087057107312127.

Dzeja, P., & Terzic, A. (2009). Adenylate kinase and AMP signaling networks: Metabolic monitoring, signal communication and body energy sensing. *International Journal of Molecular Sciences*, *10*(4), 1729–1772. https://doi.org/10.3390/ijms10041729.

Ericson, E., Hoon, S., St Onge, R. P., Giaever, G., & Nislow, C. (2010). Exploring gene function and drug action using chemogenomic dosage assays. In *Vol. 470. Methods in enzymology* (2nd ed.). Elsevier Inc. https://doi.org/10.1016/S0076-6879(10)70010-0.

Fedorov, O., Niesen, F. H., & Knapp, S. (2012). Kinase inhibitor selectivity profiling using differential scanning fluorimetry. *Methods in Molecular Biology*, *795*, 109–118. https://doi.org/10.1007/978-1-61779-337-0_7.

Feldhaus, P., Frohlich, T., Goody, R. S., Isakov, M., & Schirmer, R. H. (1975). Synthetic inhibitors of adenylate kinases in the assays for ATPases and phosphokinases. *European Journal of Biochemistry*, *57*(1), 197–204. https://doi.org/10.1111/j.1432-1033.1975.tb02291.x.

Ferguson, F. M., & Gray, N. S. (2018). *Kinase inhibitors: The road ahead*. Nature Publishing Group. https://doi.org/10.1038/nrd.2018.21.

Fernández-Acero, T., Rodríguez-escudero, I., Vicente, F., Monteiro, M. C., Tormo, J. R., Cantizani, J., et al. (2012). A yeast-based in vivo bioassay to screen for class I phosphatidylinositol 3-kinase specific inhibitors. *Journal of Biomolecular Screening*, *8*, 1018–1029. https://doi.org/10.1177/1087057112450051.

Gamo, F.-J., Sanz, L. M., Vidal, J., de Cozar, C., Alvarez, E., Lavandera, J.-L., et al. (2010). Thousands of chemical starting points for antimalarial lead identification. *Nature*, *465*(7296), 305–310. https://doi.org/10.1038/nature09107.

Garí, E., Piedrafita, L., Aldea, M., & Herrero, E. (1997). A set of vectors with a tetracycline-regulatable promoter system for modulated gene expression in Saccharomyces cervisiae. *Yeast*, *13*(9), 837–848. https://doi.org/10.1002/(SICI)1097-0061(199707)13:9<837::AID-YEA145>3.0.CO;2-T.

Geyer, J. A., Keenan, S. M., Woodard, C. L., Thompson, P. A., Gerena, L., Nichols, D. A., et al. (2009). Selective inhibition of Pfmrk, a Plasmodium falciparum CDK, by antimalarial 1,3-diaryl-2-propenones. *Bioorganic and Medicinal Chemistry Letters*, *19*(7), 1982–1985. https://doi.org/10.1016/j.bmcl.2009.02.042.

Giuliani, S., Silva, A. C., Borba, J. V. V. B., Ramos, P. I. P., Paveley, R. A., Muratov, E. N., et al. (2018). Computationally-guided drug repurposing enables the discovery of kinase targets and inhibitors as new schistosomicidal agents. *PLoS Computational Biology*, *14*(10), 1–21. https://doi.org/10.1371/journal.pcbi.1006515.

Graeser, R., Wernli, B., Franklin, R. M., & Kappes, B. (1996). Plasmodium falciparum protein kinase 5 and the malarial nuclear division cycles. *Molecular and Biochemical Parasitology*, *82*(1), 37–49. https://doi.org/10.1016/0166-6851(96)02716-8.

Grant, K. M., Dunion, M. H., Yardley, V., Skaltsounis, A., Marko, D., Eisenbrand, G., et al. (2004). Inhibitors of Leishmania mexicana CRK3 cyclin-dependent kinase: Chemical library screen and antileishmanial activity. *Antimicrobial Agents and Chemotherapy*, *48*(8), 3033–3042. https://doi.org/10.1128/AAC.48.8.3033.

Guiguemde, W. A., Shelat, A. A., Bouck, D., Duffy, S., Gregory, J., Davis, P. H., et al. (2010). Chemical genetics of Plasmodium falciparum. *Nature*, *465*(7296), 311–315. https://doi.org/10.1038/nature09099.

Guo, S., & Kemphues, K. J. (1995). par-1, a gene required for establishing polarity in C. elegans embryos, encodes a putative Ser/Thr kinase that is asymmetrically distributed. *Cell*, *81*(4), 611–620. https://doi.org/10.1016/0092-8674(95)90082-9.

Hernández-López, M. J., García-Marqués, S., Randez-Gil, F., & Prieto, J. A. (2011). Multicopy suppression screening of Saccharomyces cerevisiae identifies the ubiquitination machinery as a main target for improving growth at low temperatures. *Applied and Environmental Microbiology*, 77(21), 7517–7525. https://doi.org/10.1128/AEM.00404-11.

Hillenmeyer, M. E., Fung, E., Wildenhain, J., Pierce, S. E., Lee, W., Proctor, M., et al. (2009). The chemical genomic portrait of yeast: Uncovering a phenotype for all genes. *Science*, *320*(5874), 362–365. https://doi.org/10.1126/science.1150021.

Hoepfner, D., Helliwell, S. B., Sadlish, H., Schuierer, S., Filipuzzi, I., Brachat, S., et al. (2014). High-resolution chemical dissection of a model eukaryote reveals targets, pathways and gene functions. *Microbiological Research*, *169*(2–3), 107–120. https://doi.org/10.1016/j.micres.2013.11.004.

Holton, S., Merckx, A., Burgess, D., Doerig, C., Noble, M., & Endicott, J. (2003). Structures of P. falciparum PfPK5 test the CDK regulation paradigm and suggest mechanisms of small molecule inhibition. *Structure*, *11*(11), 1329–1337. https://doi.org/10.1016/j.str.2003.09.020.

Iacovella, M. G., Bremang, M., Basha, O., Giacò, L., Carotenuto, W., Golfieri, C., et al. (2018). Integrating Rio1 activities discloses its nutrient-activated network in Saccharomyces cerevisiae. *Nucleic Acids Research*, *46*(15), 7586–7611. https://doi.org/10.1093/nar/gky618.

Iacovella, M. G., Golfieri, C., Massari, L. F., Busnelli, S., Pagliuca, C., Dal Maschio, M., et al. (2015). Rio1 promotes rDNA stability and downregulates RNA polymerase I to ensure rDNA segregation. *Nature Communications*, *6*, 6643. https://doi.org/10.1038/ncomms7643.

Israelsen, W. J., & Vander Heiden, M. G. (2015). Pyruvate kinase: Function, regulation and role in cancer. *Seminars in Cell & Developmental Biology*, *43*(1), 43–51. https://doi.org/10.1016/j.semcdb.2015.08.004.

Jafari, R., Almqvist, H., Axelsson, H., Ignatushchenko, M., Lundbäck, T., Nordlund, P., et al. (2014). The cellular thermal shift assay for evaluating drug target interactions in cells. *Nature Protocols*, *9*, 2100–2122. https://doi.org/10.1038/nprot.2014.138.

Jaime, M. D. L. A., Lopez-Llorca, L. V., Conesa, A., Lee, A. Y., Proctor, M., Heisler, L. E., et al. (2012). Identification of yeast genes that confer resistance to chitosan oligosaccharide (COS) using chemogenomics. *BMC Genomics*, *13*(1), 267. https://doi.org/10.1186/1471-2164-13-267.

Jeon, J. Y., Sparreboom, A., & Baker, S. D. (2018). Kinase inhibitors: The reality behind the success. *Clinical Pharmacology and Therapeutics*, *102*(5), 726–730. https://doi.org/10.1002/cpt.815.

Karim, A. S., Curran, K. A., & Alper, H. S. (2013). Characterization of plasmid burden and copy number in Saccharomyces cerevisiae for optimization of metabolic engineering applications. *FEMS Yeast Research*, *13*(1), 107–116. https://doi.org/10.1111/1567-1364.12016.

Keppler-Ross, S., Noffz, C., & Dean, N. (2008). A new purple fluorescent color marker for genetic studies in Saccharomyces cerevisiae and Candida albicans. *Genetics*, *179*(1), 705–710. https://doi.org/10.1534/genetics.108.087080.

Kiburu, I. N., & LaRonde-LeBlanc, N. (2012). Interaction of Rio1 kinase with toyocamycin reveals a conformational switch that controls oligomeric state and catalytic activity. *PLoS One*, 7(5), 1–13. https://doi.org/10.1371/journal.pone.0037371.

Klein, R. D., Favreau, M. A., Alexander-Bowman, S. J., Nulf, S. C., Vanover, L., Winterrowd, C. A., et al. (1997). Haemonchus contortus: Cloning and functional expression of a cDNA encoding ornithine decarboxylase and development of a screen for inhibitors. *Experimental Parasitology*, *87*(3), 171–184. https://doi.org/10.1006/expr.1997.4213.

Klepinin, A., Zhang, S., Klepinina, L., Rebane-Klemm, E., Terzic, A., Kaambre, T., et al. (2020). Adenylate kinase and metabolic signaling in cancer cells. *Frontiers in Oncology*, *10*, 660. https://doi.org/10.3389/fonc.2020.00660.

LaRonde, N. A. (2014). The ancient microbial RIO kinases. *Journal of Biological Chemistry*, *289*(14), 9488–9492. https://doi.org/10.1074/jbc.R113.538090.

LaRonde-LeBlanc, N., Guszczynski, T., Copeland, T., & Wlodawer, A. (2005). Structure and activity of the atypical serine kinase Rio1. *FEBS Journal*, *272*(14), 3698–3713. https://doi.org/10.1111/j.1742-4658.2005.04796.x.

LaRonde-LeBlanc, N., & Wlodawer, A. (2005). The RIO kinases: An atypical protein kinase family required for ribosome biogenesis and cell cycle progression. *Biochimica et Biophysica Acta, Proteins and Proteomics*, *1754*(1–2), 14–24. https://doi.org/10.1016/j.bbapap.2005.07.037.

Lee, M. E., DeLoache, W. C., Cervantes, B., & Dueber, J. E. (2015). A highly characterized yeast toolkit for modular, multipart assembly. *ACS Synthetic Biology*, *4*(9), 975–986. https://doi.org/10.1021/sb500366v.

Lee, J. C., Kumar, S., Griswold, D. E., Underwood, D. C., Votta, B. J., & Adams, J. L. (2000). Inhibition of p38 MAP kinase as a therapeutic strategy. *Immunopharmacology*, *47*, 185–201.

Lee, M. G., & Nurse, P. (1987). Complementation used to clone a human homologue of the fission yeast cell cycle control gene cdc2. *Nature*, *327*(6117), 31–35. https://doi.org/10.1038/327031a0.

Lee, A. Y., St Onge, R. P., Proctor, M. J., Wallace, I. M., Nile, A. H., Spagnuolo, P. A., et al. (2014). Mapping the cellular response chemogenomic fitness signatures. *Science*, *344*, 208–212. https://doi.org/10.1126/science.1250217.

Li, J., Kim, S. G., & Blenis, J. (2014). Perspective rapamycin: One drug, many effects. *Cell Metabolism*, *19*(3), 373–379. https://doi.org/10.1016/j.cmet.2014.01.001.

Ma, J., Rahlfs, S., Jortzik, E., Heiner Schirmer, R., Przyborski, J. M., & Becker, K. (2012). Subcellular localization of adenylate kinases in Plasmodium falciparum. *FEBS Letters*, *586*(19), 3037–3043. https://doi.org/10.1016/j.febslet.2012.07.013.

Macgregor, P., Ivens, A., Shave, S., Collie, I., Gray, D., Auer, M., et al. (2014). High-throughput chemical screening for antivirulence developmental phenotypes in Trypanosoma brucei. *Eukaryotic Cell*, *13*(3), 412–426. https://doi.org/10.1128/EC.00335-13.

Malumbres, M. (2014). Cyclin-dependent kinases. *Genome Biology*, *15*(6), 122. https://doi.org/10.1186/gb4184.

Mandal, S., Maharjan, M., Ganguly, S., Chatterjee, M., Singh, S., Buckner, F. S., et al. (2011). High-throughput screening of amastigotes of Leishmania donovani clinical isolates against drugs using a colorimetric beta-lactamase assay. *Indian Journal of Experimental Biology*, *47*(6), 475–479.

Manning, G., Plowman, G. D., Hunter, T., Sudarsanam, S., Manning, G., & Plowman, G. D. (2002). Evolution of protein kinase signaling from yeast to man. *Trends in Biochemical Sciences*, *27*(10), 514–520. https://doi.org/10.1016/s0968-0004(02)02179-5.

Martinez, N. J., Asawa, R. R., Cyr, M. G., Zakharov, A., Urban, D. J., Roth, S., et al. (2018). A widely-applicable high-throughput cellular thermal shift assay (CETSA) using split Nano Luciferase. *Scientific Reports*, *8*, 1–16. https://doi.org/10.1038/s41598-018-27834-y.

Ochiana, S. O., Pandarinath, V., Wang, Z., Kapoor, R., Ondrechen, M. J., Ruben, L., et al. (2013). The human Aurora kinase inhibitor danusertib is a lead compound for anti-trypanosomal drug discovery via target repurposing. *European Journal of Medicinal Chemistry*, *62*, 777–784. https://doi.org/10.1016/j.ejmech.2012.07.038.

Parsons, A. B., Brost, R. L., Ding, H., Li, Z., Zhang, C., Sheikh, B., et al. (2004). Integration of chemical-genetic and genetic interaction data links bioactive compounds to cellular target pathways. *Nature Biotechnology*, *22*(1), 62–69. https://doi.org/10.1038/nbt919.

Parsons, A. B., Lopez, A., Givoni, I. E., Williams, D. E., Gray, C. A., Porter, J., et al. (2006). Exploring the mode-of-action of bioactive compounds by chemical-genetic profiling in yeast. *Cell*, *126*, 611–625. https://doi.org/10.1016/j.cell.2006.06.040.

Patel, G., Roncal, N. E., Lee, P. J., Leed, S. E., Erath, J., Rodriguez, A., et al. (2014). Repurposing human Aurora kinase inhibitors as leads for anti-protozoan drug discovery. *MedChemComm*, *5*(5), 655. https://doi.org/10.1039/c4md00045e.

Paul, S. M., Mytelka, D. S., Dunwiddie, C. T., Persinger, C. C., Munos, B. H., Lindborg, S. R., et al. (2010). How to improve R&D productivity: The pharmaceutical industry's grand challenge. *Nature Reviews. Drug Discovery*, *9*, 203–214. https://doi.org/10.1038/nrd3078.

Pereira, C. A., Bouvier, L. A., De Los Milagros Cámara, M., & Miranda, M. R. (2011). Singular features of trypanosomatids' phosphotransferases involved in cell energy management. *Enzyme Research*, *2011*(1), 576483. https://doi.org/10.4061/2011/576483.

Pereira, D. A., & Williams, J. A. (2007). Review: Historical perspectives in pharmacology origin and evolution of high throughput screening. *152*, 53–61. https://doi.org/10.1038/sj.bjp.0707373.

Pierce, S. E., Davis, R. W., Nislow, C., & Giaever, G. (2007). Genome-wide analysis of barcoded Saccharomyces cerevisiae gene-deletion mutants in pooled cultures. *Nature Protocols*, *2*, 2958–2974. https://doi.org/10.1038/nprot.2007.427.

Pink, R., Hudson, A., Mouriès, M., & Bendig, M. (2005). Opportunities and challenges in antiparasitic drug discovery. *Nature Reviews. Drug Discovery*, *4*, 727–740. https://doi.org/10.1038/nrd1824.

Pries, V., & Hoepfner, D. (2016). Chemogenomic profiling: Past, present and beyond. *Cellular & Molecular Medicine: Open Access*, *02*(02), 6–9. https://doi.org/10.21767/2573-5365.100017.

Profeta, G. S., Reis, C. V., Santiago, A. S., Godoi, P. H. C., Fala, A. M., Wells, C. I., et al. (2019). Binding and structural analyses of potent inhibitors of the human Ca^{2+}/calmodulin dependent protein kinase kinase 2 (CAMKK2) identified from a collection of commercially-available kinase inhibitors. *Scientific Reports*, *9*, 16452. https://doi.org/10.1038/s41598-019-52795-1.

Ralph, S. A., van Dooren, G. G., Waller, R. F., Crawford, M. J., Fraunholz, M. J., Foth, B. J., et al. (2004). Metabolic maps and functions of the Plasmodium falciparum apicoplast. *Nature Reviews Microbiology*, *2*(3), 203–216. https://doi.org/10.1038/nrmicro843.

Réda, C., Kaufmann, E., & Delahaye-duriez, A. (2020). Machine learning applications in drug development. *Computational and Structural Biotechnology Journal*, *18*, 241–252. https://doi.org/10.1016/j.csbj.2019.12.006.

Rubenstein, E. M., & Schmidt, M. C. (2007). Mechanisms regulating the protein kinases of Saccharomyces cerevisiae. *Eukaryotic Cell*, *6*(4), 571–583. https://doi.org/10.1128/EC.00026-07.

Rudolf, A. F., Skovgaard, T., Knapp, S., Jensen, L. J., & Berthelsen, J. (2014). A comparison of protein kinases inhibitor screening methods using both enzymatic activity and binding affinity determination. *PLoS One*, *9*(6), 2–6. https://doi.org/10.1371/journal.pone.0098800.

Schormann, N., Hayden, K. L., Lee, P., Banerjee, S., & Chattopadhyay, D. (2019). An overview of structure, function, and regulation of pyruvate kinases. *Protein Science*, *28*(10), 1771–1784. https://doi.org/10.1002/pro.3691.

Sheff, M. A., & Thorn, K. S. (2004). Optimized cassettes for fluorescent protein tagging in Saccharomyces cerevisiae. *Yeast*, *21*(8), 661–670. https://doi.org/10.1002/yea.1130.

Sibley, C. H., & Macreadie, I. (2001). Novel approaches to tackling malarial drug resistance using yeast. *IUBMB Life*, *52*(6), 285–289. https://doi.org/10.1080/15216540131729 1138.

Silberberg, Y., Kupiec, M., & Sharan, R. (2016). Utilizing yeast chemogenomic profiles for the prediction of pharmacogenomic associations in humans. *Scientific Reports*, *6*, 1–12. https://doi.org/10.1038/srep23703.

Simpkins, S. W., Nelson, J., Deshpande, R., Li, S. C., Piotrowski, J. S., Wilson, E. H., et al. (2018). Predicting bioprocess targets of chemical compounds through integration of chemical-genetic and genetic interactions. *PLoS Computational Biology*, *14*, e1006532. https://doi.org/10.1371/journal.pcbi.1006532.

Smith, A. M., Ammar, R., Nislow, C., & Giaever, G. (2010). A survey of yeast genomic assays for drug and target discovery. *Pharmacology and Therapeutics*, *127*(2), 156–164. https://doi.org/10.1016/j.pharmthera.2010.04.012.

Springs, P., & Janssen, P. (2003). Who wants to be irrational? *Drug Discovery Today*, *8*(15), 655–656. https://doi.org/10.1016/s1359-6446(03)02734-x.

Tong, A. H. Y., Lesage, G., Bader, G. D., Ding, H., Xu, H., Xin, X., et al. (2004). Global mapping of the yeast genetic interaction network. *Science*, *303*(5659), 808–813. https://doi.org/10.1126/science.1091317.

Tu, X., & Wang, C. C. (2004). The involvement of two cdc2-related kinases (CRKs) in Trypanosoma brucei cell cycle regulation and the distinctive stage-specific phenotypes caused by CRK3 depletion. *Journal of Biological Chemistry*, *279*(19), 20519–20528. https://doi.org/10.1074/jbc.M312862200.

Usaj, M., Tan, Y., Wang, W., VanderSluis, B., Zou, A., Myers, C. L., et al. (2017). TheCellMap.org: A web-accessible database for visualizing and mining the global yeast genetic interaction network. *G3: Genes, Genomes, Genetics*, 7(5), 1539–1549. https://doi.org/10.1534/g3.117.040220.

Van Schaftingen, E., Opperdoes, F. R., & Hers, H.-G. (1985). Stimulation of Trypanosoma brucei pyruvate kinase by fructose 2,6-bisphosphate. *European Journal of Biochemistry*, *153*(2), 403–406. https://doi.org/10.1111/j.1432-1033.1985.tb09316.x.

Visvesvara, G. S., & Garcia, L. S. (2002). Culture of protozoan parasites. *Clinical Microbiology Reviews*, *15*(3), 327–328. https://doi.org/10.1128/CMR.15.3.327.

Wang, J., Varin, T., Vieth, M., & Elkins, J. M. (2019). Crystal structure of human RIOK2 bound to a specific inhibitor. *Open Biology*, *9*(4), 190037. https://doi.org/10.1098/rsob.190037.

Williams, K., Bilsland, E., Sparkes, A., Aubrey, W., Young, M., Soldatova, L. N., et al. (2015). Cheaper faster drug development validated by the repositioning of drugs against neglected tropical diseases. *Journal of the Royal Society Interface*, *12*(104), 20141289. https://doi.org/10.1098/rsif.2014.1289.

Wong, J. H., Alfatah, M., Kong, K. W., Hoon, S., Yeo, W. L., Ching, K. C., et al. (2019). Chemogenomic profiling in yeast reveals antifungal mode-of-action of polyene macrolactam auroramycin. *PLoS One*, *14*(6), 1–22. https://doi.org/10.1371/journal.pone.0218189.

Woodard, C. L., Li, Z., Kathcart, A. K., Terrell, J., Gerena, L., Lopez-Sanchez, M., et al. (2003). Oxindole-based compounds are selective inhibitors of Plasmodium falciparum cyclin dependent protein kinases. *Journal of Medicinal Chemistry*, *46*(18), 3877–3882. https://doi.org/10.1021/jm0300983.

Wooden, J. M., Hartwell, L. H., Vasquez, B., & Sibley, C. H. (1997). Analysis in yeast of antimalaria drugs that target the dihydrofolate reductase of Plasmodium falciparum. *Molecular and Biochemical Parasitology*, *85*(1), 25–40. https://doi.org/10.1016/S0166-6851(96)02808-3.

Wu, P., Nielsen, T. E., & Clausen, M. H. (2015). Small-molecule kinase inhibitors: An analysis of FDA-approved drugs. *Drug Discovery Today*, *21*, 5–10. https://doi.org/10.1016/j.drudis.2015.07.008.

www.fda.gov. (2020). *Novel drug approvals for 2020*. FDA. https://www.fda.gov/drugs/new-drugs-fda-cders-new-molecular-entities-and-new-therapeutic-biological-products/novel-drug-approvals-2020.

Xu, Y. F., Zhao, X., Glass, D. S., Absalan, F., Perlman, D. H., Broach, J. R., et al. (2012). Regulation of yeast pyruvate kinase by ultrasensitive allostery independent of phosphorylation. *Molecular Cell*, *48*(1), 52–62. https://doi.org/10.1016/j.molcel.2012.07.013.

Zackrisson, M., Hallin, J., Ottosson, L. G., Dahl, P., Fernandez-Parada, E., Ländström, E., et al. (2016). Scan-o-matic: High-resolution microbial phenomics at a massive scale. *G3: Genes, Genomes, Genetics*, *6*(9), 3003–3014. https://doi.org/10.1534/g3.116.032342.

Zhang, N., & Bilsland, E. (2011). Contributions of *Saccharomyces cerevisiae* to understanding mammalian gene function and therapy. *Methods in Molecular Biology*, *759*, 501–523. https://doi.org/10.1007/978-1-61779-173-4_28.

CHAPTER NINE

Structural studies of full-length receptor tyrosine kinases and their implications for drug design

Adam Bajinting[a] and Ho Leung Ng[b,*]
[a]St. Louis University School of Medicine, St. Louis, MO, United States
[b]Department of Biochemistry and Molecular Biophysics, Kansas State University, Manhattan, KS, United States
*Corresponding author: e-mail address: hng@ksu.edu

Contents

Abstract

Receptor tyrosine kinases (RTKs) are important drug targets for cancer and immunological disorders. Crystal structures of individual RTK domains have contributed greatly to the structure-based drug design of clinically used drugs. Low-resolution structures from electron microscopy are now available for the RTKs, EGFR, PDGFR, and Kit. However, there are still no high-resolution structures of full-length RTKs due to the technical challenges of working with these complex, membrane proteins. Here, we review what has

Advances in Protein Chemistry and Structural Biology, Volume 124
ISSN 1876-1623
https://doi.org/10.1016/bs.apcsb.2020.10.007

been learned from structural studies of these three RTKs regarding their mechanisms of ligand binding, activation, oligomerization, and inhibition. We discuss the implications for drug design. More structural data on full-length RTKs may facilitate the discovery of druggable sites and drugs with improved specificity and effectiveness against resistant mutants.

1. Introduction

The characteristic structure of the receptor tyrosine kinases (RTKs) family of transmembrane proteins consists of four distinct, linked regions: an extracellular domain (ECD), a transmembrane (TM) region, a juxtamembrane (JM) region, and a tyrosine kinase domain (TKD) (Lemmon & Schlessinger, 2010). According to the conventional activation model, growth factors specifically bind to the RTK ECD to cause dimerization/oligomerization with another RTK:ligand complexes. In a complex process that is still not well understood, oligomerization transduces a signal across the cell membrane to the TKDs, resulting in transphosphorylation of the cytoplasmic regions. The phosphorylated receptor intracellular regions then function as docking sites for other signal transduction proteins. RTKs activate signal cascades integral to the growth, differentiation, development, and survival of cells (Lemmon & Schlessinger, 2010). As such, they are drivers of oncogenesis and important drug targets.

RTKs are drivers of many cancers, including adenocarcinomas, sarcomas, leukemias, and lymphomas. Mutant RTKs have been found to lead to constitutively activated angiogenesis and growth pathways regardless of ligand-binding. As a result, these mutations are linked to increased tumor mass and subsequent invasion and metastasis (Regad, 2015). Researchers have relied on structural data of EGFR to design effective cancer drugs. This includes cetuximab to target hyperactive EGFR in colorectal cancer. An important challenge for research is developing expression systems that can produce functional full-length RTKs for structural studies (Bajinting & Ng, 2017; Mi et al., 2008).

Of the identified RTKs, studies on EGFR, a type I RTK, and Kit and PDGFR, Type III RTKs, have advanced the furthest in discerning the conformational mechanism of signal transduction. Though this review centers primarily on two subfamilies, there are 50 known subfamilies of RTKs, each classified according to variations of the ECD-TM-JM-TKD arrangement and function (Lemmon & Schlessinger, 2010). Type I RTKs, the prototypical RTK, are classified by their dimerization via the receptor backbone

(Lemmon & Schlessinger, 2010). Their specific roles range from intestinal adaptation to neurodegenerative diseases, and mutations have been attributed to breast, gliomas, ovarian carcinomas, and colorectal cancers (Bublil & Yarden, 2007; Shin et al., 1998). Type III RTKs are classified by their dimerization that is activated and "bridged" by a growth factor dimer. Their roles have been attributed to and mutations have been associated with multiple cancers including astrocytoma and glioma (Paul & Mukhopadhyay, 2004). Structural determination for RTKs has progressed from crystal structures of individual domains, to electron microscopy of fully intact receptors (Devan, Thaker, & Jura, 2019). EM structures have provided us with information on symmetry, orientation, and relative energetics of the RTK domains in relation to each other. However, crystal structures of full length RTKs have not been determined to provide atomic resolution insights. The effective resolution of the EM structures is >20 Å. By determining the mechanism of signal transduction, researchers will be able to design more effective chemotherapeutics that can selectively inhibit active conformations of the protein and prevent cancer cells from progressing.

2. Mechanistic questions about intact RTKs

Comparing Type I and III RTKs reveals broad similarities on their mechanisms. The majority of RTK structural research have focused on characterizing domains as separate structures as opposed to being a part of the full-length structure due to the difficulty in expressing full length (or nearly full length) RTKs. Understanding the nature of separate RTK domains gives limited insight into RTK behavior and activation because isolated domains behave differently than when part of a functional intact receptor.

The first question RTK comparison can answer is the role that homotypic ECD interactions play in the activation of RTKs. They are likely essential in downstream coupling of the TM region and TKD. Though the crystal structures of the ECD domains of Type I and III RTKs have been deciphered, they do not tell us the true behavior of the ECDs' terminal regions. Crystal structures of the EGFR ECD have shown terminal domains not homotypically dimerizing and yet, electron microscopy (EM) of the intact receptor revealed homotypic dimerization (Endres, Engel, Das, Kovacs, & Kuriyan, 2011, Endres et al., 2013, Bessman, Freed, & Lemmon, 2014, Mi et al., 2011).

The second question is the role of the transmembrane domain in dimerization and intramembrane coupling. Conventionally, researchers have focused on analyzing the sequence of the transmembrane domain (TMD) to discern any structural characteristic that may give clues as to its role in dimerization and coupling. However, there has been an increasing focus on the lipid membrane composition and its impact on TMD conformation (Bocharov, Sharonov, Bocharova, & Pavlov, 2017; Maeda, Sato, Okamoto, Yanagawa, & Sako, 2018). Perhaps lipid composition of the surrounding and embedding membrane may play a greater role in determining the active conformation of RTKs at these regions.

The third question is the importance of asymmetric tyrosine kinase dimerization. Several types of active RTK dimers have been shown to exhibit asymmetric tyrosine kinase association (Chen, Unger, & He, 2015; Kovacs, Zorn, Huang, Barros, & Kuriyan, 2015). This indicates that both monomers interact in complementary ways as opposed to identical ways. This gives rise to the question: what makes one monomer behaves as the acceptor while the other behave as the receiver?

The fourth question is the role of inactive dimers. Inactive dimers have been shown to exist naturally as evidenced by EM of nearly fully intact EGFR and FRET tests of FGFR, it is still under investigation as to why inactive dimers form (Sarabipour & Hristova, 2016; Yu, Sharma, Takahashi, Iwamoto, & Mekada, 2002). Yu et al., Sako, Mekada and Yanagida conducted crosslinking experiments that demonstrated ligand independent dimers in B effector and CHO cells. Moreover, Yu reports that the existence of indirect EGFR dimers and demonstrates concentration of these dimers in BE cells were similar despite being EGF treated or not (Yu et al., 2002). This suggests that receptor dimerization and activation are independent mechanisms. Originally, it was hypothesized that inactive dimers occur as a safeguard to protect the cell from unwanted overactive signaling. It was believed to be a way for the cell to sequester RTKs when expression levels are excessive (Arteaga, Ramsey, Shawver, & Guyer, 1997). However, more recent experiments have led researchers to argue otherwise: that the inactive dimer is a stage necessary for EGF binding and subsequently activation (Endres et al., 2011). Moreover, it is interesting to note that that only a few classes of RTKs have been identified to exhibit inactive dimers. For instance, while class I and IV RTKs have been shown to exhibit this phenomenon, class III's have not demonstrated this feature. Reviewing relevant studies may lead to clues and avenues for investigation. Unlocking the reason for inactive dimerization will also be important in understanding conditions that lead to an activation scheme of fully intact in vivo RTKs. Perhaps RTK activation schemes, at least for some classes, require more than growth factors such as other environmental factors.

3. RTKs of interest

Before comparing the structural and biophysical studies of these proteins, it is essential to summarize what is structurally known about these proteins, particularly structural studies of full and nearly full length RTKs.

3.1 Class I prototypical RTK: EGFR

The human epidermal growth factor receptor, Erb1, is the first and most structurally studied RTK. It is a member of the ErbB RTK family. Mutations of EGFR have been linked to 60% of small cell lung cancer cases, 40% of glioblastoma cases, and breast cancer, and it was the first receptor to be identified as an oncogene (da Cunha Santos et al., 2011, Kovacs et al., 2015). Crystal structures of the extracellular portion (ECD) of the monomer and dimer, with and without EGF ligand, have been obtained revealing the role of EGF in EGFR ECD's activation. Dimerization of EGFRs occurs because of interaction of EGF at ECD Ig domain II, where EGF interacts away from the dimer interface. The current mechanism of EGF-induced dimerization at the ECD is known. The EGF:EGFR interaction removes autoinhibitory intramolecular interactions resulting in dimerization (Ferguson et al., 2003). The transmembrane of EGFR is an alpha helix composed of 24 amino acids (Kovacs et al., 2015). TOXCAT and NMR studies on lipid bicelles with EGFR homo and heterodimers reveal dimer interaction at the TMD's N-terminal GxxxG motifs and juxtamembrane region's LRRLL motifs (Kovacs et al., 2015; Mendrola, Berger, King, & Lemmon, 2002). Transmembrane coupling is essential to the activity of EGFR as oncogenic mutations within the motif have been associated with stronger stabilization of the dimer (Endres et al., 2013; Purba, Saita, & Maruyama, 2017; Shan et al., 2012; Sinclair, Walker, Doerner, & Schepartz, 2018). *trans*-Autophosphorylation is not dependent upon TKD dimerization but rather on the allosteric mechanism of the tyrosine kinase domain dimer (TKD) where the C-lobe of one TKD acts as the acceptor that interacts with the N-lobe of the other termed the receiver (Kovacs et al., 2015; Zhang, Gureasko, Shen, Cole, & Kuriyan, 2006).

However, studying separate domains does not unveil the true nature of EGFR mechanisms. Studying the fully intact EGFR is the ultimate goal.

However, there are major technical obstacles associated with expression and preserving functionality of the intact receptor. Four studies by the lab of Timothy Springer have successfully visualized nearly full-length EGFR receptors via low resolution negative staining EM. The only portion of the EGFR missing was the auto phosphorylation tail (residues 999–1186) which is not believed to affect dimerization or structure. Significant biophysical characteristics have been identified regarding structure stability, an equilibrium of inactive and active states, dimer structures, and the likely direction of activation.

One of the earliest publications involving nearly full-length EGFR centered on the stability and tyrosine kinase of the protein in various lipid environments. Springer and colleagues concluded that using phospholipid nanodiscs rather than the detergents DDM and Triton X-100 increased the stability and activity of EGFR. This put into question the degree of importance of transmembrane domain conformation and whether it was influenced by its environment rather than its sequence (Mi et al., 2008).

This question became increasingly pressing after another Springer study in 2010 used EM of a fully intact EGFR and X-ray crystallography of EGFR's ECD to examine the signaling mechanism of EGFR activation, a study that is a first of its kind. It was found that activation of EGFR was mediated by loose coupling between the ECD and TKD which leads to more questions about how flexibility of the dimer especially at the transmembrane domain can connect ligand binding extracellularly to allosteric activation intracellularly (Bessman et al., 2014; Lu et al., 2010). This suggests that flexibility most likely contributes to the weak dimerization association between the transmembrane domains, a feature that is most likely responsible for the coupling strength of ECD to TKD.

Another study in 2011 by Springer utilized EM of intact EGFRs to examine the structural changes upon ligand binding to further characterize the mechanism of activation. They concluded that the ECD dimerizes in a characteristic heart shape conformation upon EGF binding, and the resulting coupling leads to one of two TKD dimer conformations, active asymmetrical or inactive symmetrical, as shown by the doughnut and rod electron density shapes, respectively (Mi et al., 2011) (Fig. 1). Most of all, this phenomenon emphasizes the complexity of coupling between the ECD and TM. What causes EGFR to dimerize into the asymmetric or symmetric form is still in question.

Springer also examined the signaling direction of EGFR to lend better insight into the coupling relationship between all three domains as a

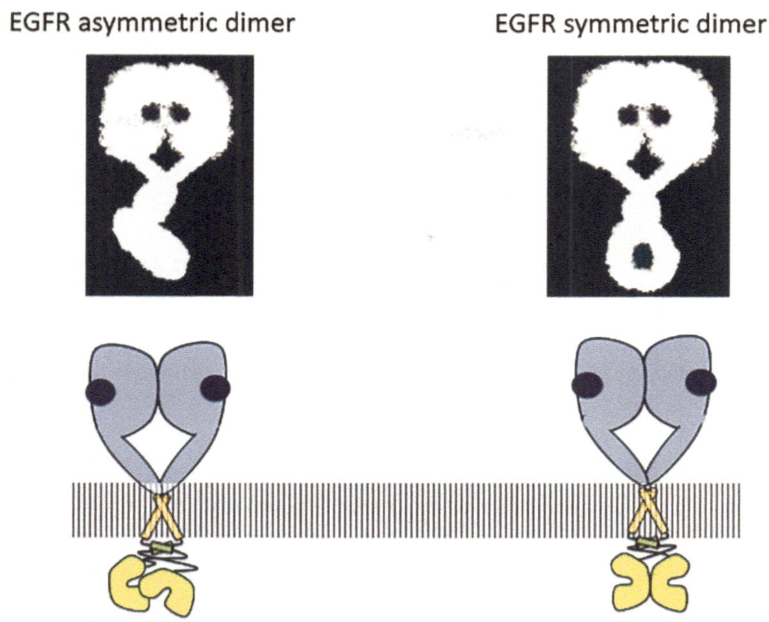

Fig. 1 Nearly full-length EGFR dimers identified by electron microscopy, asymmetric vs symmetric dimers. Top: Schematic of EM images. Bottom: Organization of domains in the receptor: ECD (blue), TMD (orange), JM region (green), TKD (yellow).

monomer and as a dimer. This is currently the most in depth study of a nearly full-intact EGFR. Springer first produced nearly full-length EGFR lacking only the autophosphorylated tail, using the detergents Triton X-100 and later, n-dodecyl β-D-maltoside (DDM), for isolation and purification (Mi et al., 2008). First, by cross-correlating EM images with crystal structures of dimeric and monomeric EGFR ectodomains, it confirmed two hypotheses about full length receptors: (1) unbound EGFR exhibited a monomeric tethered structure and (2) ligand-bound dimers exhibited two conformations with symmetric and asymmetric TKDs. Second, Springer studied the effects of TKD mutations on the nearly full-length construct. The effects of the V924R mutation, which was previously shown to prevent the activation of the tyrosine kinase domain and asymmetric dimerization in the absence of the other receptor domains, was found to still allow for the dimerization of the ectodomain and transmembrane domain. However, the TKDs were either symmetric dimers or monomers that were unable to be autophosphorylated. This attests to the non-functionality of nearly full length symmetric EGFR dimers. Third, by using kinase inhibitors, he

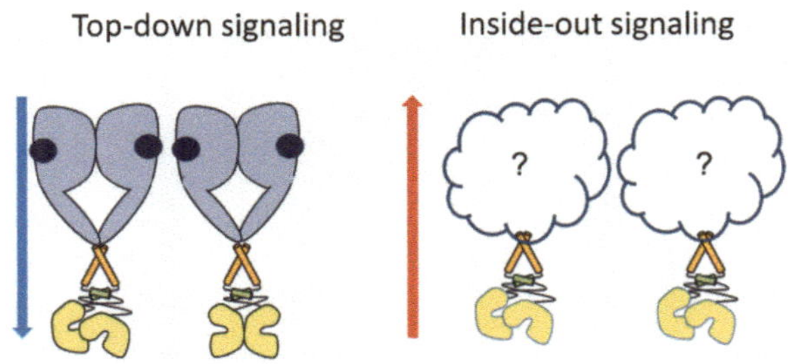

Fig. 2 Coupling schemes of EGFR. EM was used to visualize the coupling between the ECD and TKD within EGFR. The left column demonstrates strong top-down coupling. The right column demonstrates weak inside-out coupling. The electron density for the ECD is shown as clouds with a "?" because they are poorly defined in the EM images.

demonstrated support for a top down signaling mechanism. Upon stabilizing the asymmetric kinase dimer conformation using the inhibitors, gefitinib, erlotinib, or PD168393, the transmembrane and ecto-TM linker domains subsequently dimerized but not the ectodomain, indicating that inside-out signaling was not favorable (Fig. 2 right). In addition, the coupling between the TKD and TM region demonstrates that the conformation of the TM region plays an essential role. Rather than just embedding the RTK in the membrane, it also positions the TKD for either symmetric or asymmetric dimerization. In contrast, binding of EGF to the ECD of EGFR leads to spontaneous dimerization into both active or inactive states indicating outside-in signaling and coupling initiated from the ECD down to the TKD (Lu, Mi, Schurpf, Walz, & Springer, 2012).

Furthermore, mutations that prevented asymmetric dimerization of kinases showed that EGF still induced dimerization from the ECD to the TMD indicating that coupling between these two domains are energetically favorable from top down. Overall, the study supports the hypothesis that ECD and TKD are not coupled strongly as concluded by Springer's 2010 and 2011 studies. Asymmetric dimerization of the TKD dimer leads to multiple unresolved ECD EM densities and dimerization of the ECD leads to both symmetrical and asymmetrical TKD dimers (Mi et al., 2011). This study also supports Springer and Lu's conclusion in 2010 that no specific interfaces in ECD Ig Domain IV and TM were required for EGFR signaling (Lu et al., 2010). Overall, top down signaling from the ECD to TKD may be more energetically favorable than inside out signaling since the former yields a correct active dimer at the terminal portion (TKD) while the latter yields no active conformation of its extracellular region (ECD). The EGF ligand is essential to facilitate an energy favorable top down signaling scheme (Lu et al., 2012).

Aside from Springer's studies and the EM data, there also have been efforts to understand multimerization of full-length EGFRs. As demonstrated in a recent study by Kuriyan, mutations to ECD's Ig Domain IV prevented multimerization of full length EGFRs (Huang, Bharill, Karandur, Peterson, & Marita, 2016). This lends insight into the importance of Extracellular Ig Domain IV in an intact receptor, a domain that has often been challenging to study because crystal structures of this ECD's region lacks the adjacent transmembrane domain. Therefore, the actual relative orientation of the domain IV as a dimer as part of a full-length receptor or an ECD + TM construct has been difficult to determine.

3.2 Class III prototypical RTKs: PDGFR and Kit

Class III RTKs are defined by their presence of five Ig like domains within their extracellular domains. Both PDGFR and Kit are in this class and are often discussed together due to their structural similarities. Aside from the nearly full-length EGFR, these RTKs are the only ones that have been currently structurally studied in their entirety using negative stain electron microscopy (EM).

3.3 PDGFR

The Platelet Derived Growth Factor Receptor (PDGFR) plays a critical role in human developmental and immune systems and has been associated with multiple cancers such as dermatofibrosarcoma protuberance, gastrointestinal stromal tumors, glioblastoma, rhabdomyosarcoma, osteosarcoma, leukemias, lymphomas, prostate cancer, cholangiocarcinoma, and colorectal carcinoma (Heldin, 2013). Full-length PDGFR-β has been visualized by negative stain electron microscopy (EM) (Chen et al., 2015). Studies on the ECD reveal activation and dimerization by homologous and heterologous growth factor dimers with a cysteine-knot fold (Omura, Heldin, & Ostman, 1997). The PDGF ligand binds to the D2 and D3 domains of both PDGFR monomers which are subsequently stabilized by homotypic interactions at ECD Ig domains 4 and 5 (Chen, Chen, & He, 2013; Omura et al., 1997; Yang, Yuzawa, & Schlessinger, 2008). As for the transmembrane domain, which is an alpha helical structure composed of 25 amino acids, studies on just this domain have revealed TM oligomerization via the leucine-like zipper motif. These studies of separate domains have not been able to elucidate how oligomerization of the TM domains can occur when part of a fully intact receptor; one hypothesis is that the helixes bend or tilt within the membrane in order to adjust for tyrosine kinases dimerization, but this was believed to be energetically expensive and unstable (Chen et al., 2013; Muhle-Goll et al., 2012). It is likely that dimerization occurs at the N-terminal end rather than throughout the TMD to allow for steric freedom at the C-terminal so the TKDs can dimerize in their active form. The 40 amino acid juxtamembrane region adjacent to the TMD participates in an auto-inhibitory role on the TKDs (Chen et al., 2013). Studies on the TKD indicate activation via trans-autophosphorylation and asymmetric dimerization (Chen et al., 2013; Chen et al., 2015; Kazlauskas & Cooper, 1989).

EM and molecular dynamics simulations of full length PDGFR-β revealed multiple interactions within the full length dimer: interaction of PDGF with PDGFR ECD's Ig D2 and D3, and homotypic interactions at Ig D3 and D4, loose dimerization of TMDs, and asymmetric dimerization of TKDs (Chen et al., 2013, Chen et al., 2015). EM revealed inconsistencies between the fully intact PDGFR structure compared to crystal structures of separate domains. Originally, crystal structure of the D1-D3 ECD domains suggested steric hindrance at D3 would prevent homotypic interactions at the D4 and D5 domains. However, the EM showed homotypic interactions at D4 and D5, suggesting their roles in dimer stabilization. The study also indicated flexibility at the D2 and D3 hinge that may contribute to ligand affinity. Chen et al. also suggested that ECD's D4 and D5 may play a role in positioning the hinge so that affinity to the PDGF ligand dimer was optimized; this supports the hypothesis that intra-coupling existed within the ECD. Moreover, the EM confirms a prominent D5 homotypic interface which up until then was not known. As cancer associated mutations have been found in PDGFRβ's D5, EM data supports D5's importance for coupling the ECD to the TM and indirectly to the TKD. EM also revealed a characteristic funnel in the middle of the receptor suggesting a high degree of interaction at the N-terminus of the TMDs. The widening region below the funnel is presumed to be the majority of the TMD domain that is not dimerized. This lends further support to the hypothesis that aside from N-terminal interaction, the majority of the TMD remains flexible. He, Unger, and Chen suggest that this N-terminus association permits a large range of motion at the C-termini of the TMD so that the terminal TKDs can dimerize correctly. EM also supports TKD dimer asymmetry via an allosteric mechanism due to bending of the TKD as well as the bulkiness of the densities in the region of interest. The domain believed to be the TKD appears to be rigid and well-defined, suggesting stable interaction between both TKDs rather than the previous hypothesis suggesting transience (Chen et al., 2015).

3.4 Kit

Kit, also known as stem cell factor receptor (SCFR), is another class III receptor that has also been studied structurally in its entirety. Kit, unlike PDGFR, binds to growth factors with helical bundles as opposed to cysteine knot folds (Chen et al., 2015). Kit plays a critical role in hematopoiesis, gametogenesis, and melanogenesis (Chen et al., 2015; Opatowsky et al., 2014). Like PDGFR, it is activated and dimerizes upon binding to a growth factor dimer, in this case SCF. Like PDGFR, the protomers of the dimer are bridged by the growth factor (Liu, Chen, Focia, & He, 2007). Most structural research on kit domains focuses on studying the ECD, JM, and TKD individually. The ECD dimer has been resolved and like PDGFR, is composed of similar elements, including five Ig like domains, D1-D5 (Yuzawa et al., 2007, Opatowski et al., 2014). Based on this crystal structure, scientists have elucidated a mechanism of activation at the ECD. The ECD is divided into three functional parts. The first is composed of D1-D3 and interacts with the SCF dimer at three binding sites. The second region comprises D3-D4, and the third region comprises the D5 ECD domain, which stabilize the correct conformation of the ECD dimer via homotypic interactions. The juxtamembrane region plays a regulatory role in the activation of the TKD (Chan, Ilangumaran, La Rose, Chakrabartty, & Rottapel, 2003). There have been mutations identified at the D5 extracellular domain that are associated with cancer (Yuzawa et al., 2007). It is currently believed that these mutations reinforce homotypic interactions of the monomers, enabling constitutive activation. While some suggest that these mutations play a role in activation in a ligand independent manner, others propose that they strengthen a dimer already induced by the SCF ligand (Opatowsky et al., 2014; Reshetnyak et al., 2015). Perhaps both hypotheses are correct. That suggests that cancer mutations involving RTKs may not act in black or white ways. Depending on environmental conditions or the nature of these receptors, these mutations may contribute to different mechanisms of activation rather than just one approach, especially if other mutations in the region or related regions are present. Combinations of mutations may elicit different approaches to tumorigenesis. This scenario would certainly have implications for drug design and pharmacogenomics.

There has been one study that has characterized the structure of the intact Kit receptor using negative stain electron microscopy (Opatowsky et al., 2014). This study builds on findings of a previous study that determined the crystal structure of Kit receptor's extracellular domain before and after activation by SCF (Yuzawa et al., 2007). The ECDs (as part of an intact receptor) of the ligand-free monomers were found to be structurally similar to the ligand-bound dimers. EM 3D-reconstruction with docking analysis showed that the TKD regions assemble asymmetrically; this was based on the "kidney" and "pear" shaped forms of the kinase domains (Opatowsky et al., 2014). EM also confirmed the presence of a characteristic cavity at the D2-D3 ECD domains. EM 3D-reconstruction showed homotypic interactions at the D4 and D5 domains, affirming previous studies centered on the ECD. However, while interactions at the D4 domain were consistent with the crystal structure of the ECD domain, the D5 homotypic interactions disagreed with the crystal structure (Opatowsky et al., 2014). EM of the full-length receptor revealed a cavity at the D4-D5 sites which formed by interaction of the D5 domains. This D5 interaction may be essential to activation and correct ECD-TKD coupling (Opatowsky et al., 2014). This is also supported by the fact that D5 is a membrane proximal region and therefore is likely involved in the dimerization mechanism. Like PDGFR, D5's importance is further supported by the identification of cancer associated mutations at this domain (Reshetnyak et al., 2015). As for the transmembrane domain, EM did not reveal any significant data other than the characteristic funnel also seen in PDGFR which could indicate N-terminal association (Opatowsky et al., 2014).

4. Structural features of full-length type III RTKs identified by EM

PDGFR and Kit have important structural similarities. First, both exhibit homotypic interactions at D5 that could not be identified by x-ray crystallography of their extracellular domains alone. The D5 interactions were only observed by EM of the full-length receptors (Opatowsky et al., 2014) (Chen et al., 2015). This can be explained by a common elbow/hinged shaped conformation at the D2-D3 junction which favors the observed dimer conformations of D4-D4 and D5-D5. Moreover, both exhibit asymmetric dimerization at the tyrosine kinase domains. Both also exhibit a funnel at the TM region suggesting N-terminal interaction of the transmembrane domains, which would allow for more steric freedom in order for TKDs to asymmetrically dimerize (Chen et al., 2015; Opatowsky et al., 2014).

5. Mechanistic insights from comparisons between class I and III RTKs

Studies of nearly full-length type I and full-length type III RTKs have revealed structural similarities as well as differences that are crucial in understanding the mechanism of activation. Studies on nearly fully intact or fully intact receptors have confirmed as well as contradicted findings found from studying individual domains. This is significant as it reveals that behavior of domains may differ when studied separately than when part of the fully intact receptor. Therefore, pharmacological agents may have different effects on isolated TKDs than intact receptors: this may translate to in vivo effects.

6. Comparing extracellular domains (ECD)

Class I and III RTKs share similarities in the effects of ligand binding on the ECD. When EGF binds to ECD Ig D1 in the tethered monomer, the hydrogen bonds between D2 and 4 are broken allowing D2 to dimerize with D2 of another extended monomer. Dimerzation at D2 may bring the D4 domains together to homotypically dimerize by pointing the D3 domains inward (Ferguson et al., 2003). For PDGFR, the ECD also has structural components that alter ligand affinity, but dimerization does not occur in the "back to back" fashion as in EGFR. The D2-D3 linker is flexible. Therefore, homotypic interactions at D4 and D5, which result from coupling with movements in the D2-D3 hinge, are believed to affect ligand affinity at D2 and D3. Ligand binding prevents D3 from bending inward (Chen et al., 2015). This is supported by the partial crystal structure of the D1-D3 PDGFR dimer revealing a more bent D3 than the intact receptor D3 (Chen et al., 2015). While the binding sites are in fixed conformations, what remains unclear is whether the ligand affinities of PDGFR change because of shifts in the conformations of D4 and D5. Shim and He found that for PDGF-A, the ligand binding affinities decreased despite the activation-favored dimer conformation (Shim et al., 2010). However, there is no research on the relationship between ligand affinities for PDGF-B and conformational changes. As for Kit, data suggests that unlike

EGFR and PDGFR, there is no effect of the ECD conformation on ligand binding affinity (Lemmon, Pinchasi, Zhou, Lax, & Schlessinger, 1997).

The second major similarity is the significance of ECD homotypic interactions specifically at D4. It is believed that they define the asymmetric arrangement of TKD by positioning the TM domains due to their membrane proximal location (Arkhipov et al., 2013). This hypothesis, in addition to oncogenic mutations at these positions, suggests the significance of these homotypic interactions for activation. Most ECD homotypic interactions for both RTK classes occur adjacent to the linker, JM region, and TMD which underlies these interactions' importance in determining downstream coupling. For EGFR, EM studies of the nearly full-length receptor shows dimerization at the ECD D4 which contradicts previous conclusions based on the EGFR ECD dimer crystal structure. The crystal structure showed D4 to be disordered, not tightly interacting with the other regions of the ECD and not homotypically dimerizing. Reasons for the crystal structure's discrepancy could be due to the lack of the TMD which is needed to anchor, impart stability, and provide a well-defined conformation for D4; there was also a terminal segment of D4 missing from the crystal. The significance of the homotypic interactions of EGFR's ECD D4 is clinically relevant as missense mutations within this region are associated with gliomas, specifically C620Y, C624F, C628Y, and C646Y (Greenall, Donoghue, Gottardo, Johns, & Adams, 2015). As for PDGFR and Kit, homotypic interaction at the D4 and D5 domains of the ECD has been confirmed by crystallography and are consistent with negative stain EM and molecular dynamics simulations. Moreover, its significance to dimerization has been demonstrated via D4 mutations Arg-385 and Glu-390, which resulted in resulting in compromised PDGFR-induced activation by disrupting its salt bridge mediated homotypic interactions (Yang et al., 2008). As for Kit, homotypic interactions also occur at the D4 domains via salt bridges Arg 381 and Glu386 as well as at D5 via T418 hydroxyl groups and influenced by additional interactions of Tyr418 and Asn505 (Yuzawa et al., 2007). However, only the type III RTKs PDGFR-Beta and Kit's homotypic interactions in the membrane proximal regions have been structurally characterized. Those for EGFR, which have only been studied by EM, have not yet been structurally characterized at high resolution.

Third, both classes of receptors exhibit rigid ECDs dimers. ECD dimers exhibit rigidity in a fully or nearly fully intact receptor as shown by the consistent density arrangements in EM (Fig. 3). For EGFR, all active ECD dimers have exhibited symmetrical two ear shaped densities aligned

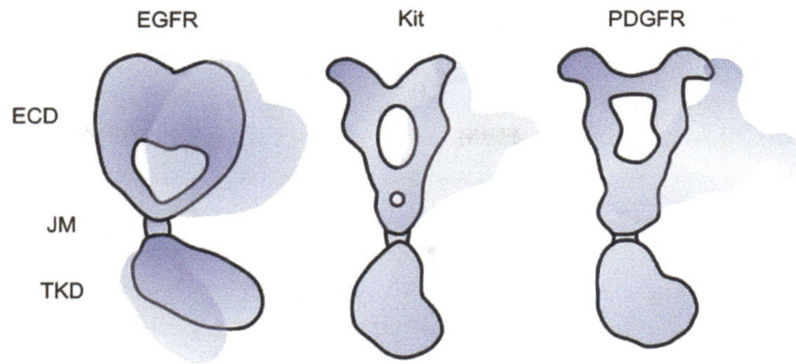

Fig. 3 The linker region of the TM region imparts flexibility between the ECD and TKD for both type I (EGFR) and type III (Kit and PDGFR) RTKs. The flexibility of nearly full-length EGFR (type I), and full-length Kit (type III), and PDGFR-B (type III) as shown by negative staining electron microscopy appears to occur at the transmembrane domain. The shaded regions show alternate conformations observed.

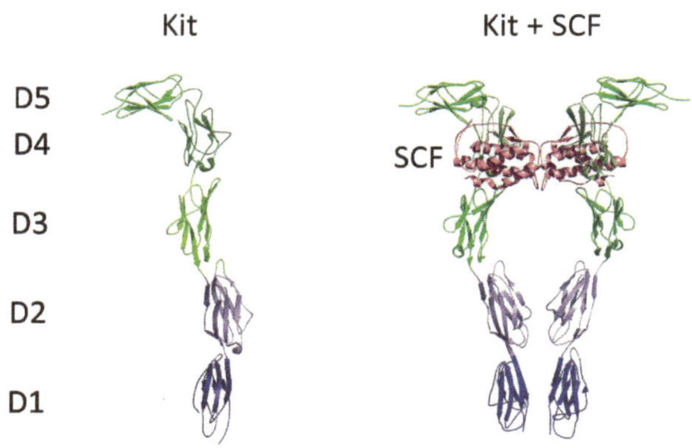

Fig. 4 Kit dimerization upon binding SCF, based on crystal structures of unbound monomer (PDB 2EC8) and SCF-bound dimer (PDB 2E9W). SCF is shown on the right in pink.

next to each other with the concave regions facing each other. All EM images indicate that the ECDs all share a characteristic upside down triangular gap at the center of the domain due to no homotypic interaction at D3 but homotypic interaction at D4 (Lu et al., 2012; Mi et al., 2008; Mi et al., 2011). The type III RTKs, PDGFR and Kit, also have consistent density arrangements. D1 of the ECD protrudes as "bunny ears" while the growth factor homodimer bridges the two receptor monomers at D2-D3 and show a characteristic diamond shaped gap at D3. Homotypic interactions at D4 and D5 also follow (Fig. 4). Both type III receptors share a characteristic elbow conformation due to inward bending of the D4 domains (Chen et al., 2015). The consistency of ECD class averages for both types of receptors suggest rigidity determines the conformation of downstream elements for activation. Mutations in the ECD that negatively affects the dimer's rigidity would negatively impact activation. Mutations that positively affects the rigidity of the dimer may result in hyperactivation. On the other hand, non-dimerized monomers are suggested to be flexible based on their class averages as by molecular modeling. Flexibility in both types of receptor dimers also occurs at the linker and transmembrane domains. Challenges with understanding the linker and transmembrane region is primarily its short sequence- current EM cannot resolve the full detail of this region and x-racy crystallography has not yet been able to characterize the linker and transmembrane domain in conjunction with the ECD. The similarities and differences identified at the ECD of these RTK classes highlight an important concept: RTKs have evolved to function, behave, and couple according to their structure.

7. Comparing transmembrane domains

Structural similarities are also found at the TM domains. N-terminal localized dimerization is the common paradigm. More recently, the lab of Kristina Hristova has extensively studied the biophysical characteristics of RTK TMDs. Hristova has focused on (1) studying TMD dimers by NMR and crystallography, (2) interactions of the TM domains and the lipid bilayer as well as with its conjoined domains, and (3) studying the role of TM motifs in RTK dimerization (Li, Wimley, & Hristova, 2012). Hristova and others have made progress deciphering biophysical properties of these TM dimers that may contribute to the dimerization and activation of these receptors.

Instead of a TM dimer interface that spans the entire domain, the interface of the TMD dimer occurs only at the ECD-TM linker and N-terminal region. EM studies showed a characteristic narrowing "funnel" between the ECD and TM between the ECD and TKD, suggesting that the TKDs cross at the N-terminal to dimerize (Chen et al., 2015). This allows greater range of movement at the C-terminal region so that TKDs can position themselves asymmetrically in the active conformation. The TMDs for type I and III RTKs differ greatly in sequence and structural motifs. The three receptors in this review share no common structural or sequence motifs. EGFR has

GxxxG like sequence motifs near the N and C terminus which have been found to participate in alpha helix dimerization among other transmembrane proteins (Lu et al., 2010). PDGFR-β contains a SxxxA motif near the N-terminus, which is similar to the GxxxG motif, as well as a leucine zipper in the last two thirds of the TMD (Li et al., 2012; Oates, King, & Dixon, 2010). Kit TMDs have no sequence and structural motifs. However, full-length receptor studies as well as studies on other prototypical RTKs suggest that these sequence and structural motifs may not play an important role in dimerization (Doura & Fleming, 2004). Studies have concluded that RTK TMD dimers do not require a specific interface for signaling to occur. This is significant because it has steered researchers away from sequence specific studies to biophysical membrane studies. Future approaches to studying TMD dimerization in RTKs should address the interaction of the RTK dimer with the membrane, including issues such as membrane dimensions, curvature, and lipid composition. A study conducted by Springer has suggested that EGFR TMD stability depends on the membrane lipid composition (Lu et al., 2010; Mi et al., 2008).

8. An emerging area of RTK structural study: JM regions

Studies centered on the JM-TKD region have revealed the significance of the JM region's role in activation. Though both type I and III RTKs have this region, its role differs across subfamilies. For type III PDGFR and Kit, the JM region plays an autoinhibitory role (Chan et al., 2003; Irusta et al., 2002). Tyrosine phosphorylation activates the TKD. In contrast, alanine scanning mutagenesis showed that EGFR's JM region functions as an activation domain that enhanced asymmetric dimerization of the TKD, which was necessary for autophosphorylation. This was also supported by the crystal structure of the JM region with the TKD that showed JM association with the C-lobe of the donor kinase, supporting the asymmetric arrangement of the TKD dimers (Red Brewer et al., 2009). A similar feature both subfamilies share is that the JM regions have oncogenic mutations, supporting their significance in activation. In EGFR, JM region mutations associated with non-small cell lung cancer were studied in full length constructs by measuring their tyrosine phosphorylation. The mutation V665M led to ligand independent autophosphorylation at the TKD (Red Brewer et al., 2009).

9. Tyrosine kinase domains

Of the three domains, the TKD is the most similar between both classes of RTKs, with two residues conserved, P675 and Y920 (EGFR numbering). Their significance is unknown. P675 is located at the end of the interface. Proline is a well-known disruptor of secondary structure: mutations would likely have major functional consequences. EM images and crystal structures support asymmetric dimerization as the active conformation for all three receptors. Studies of the EGFR TKD crystal structure reveals an asymmetry that is similar to that of CDK-cyclin 2 (Zhang et al., 2006) at the helix C dimer interface (Zhang et al., 2006). Moreover, EM results of almost fully intact EGFR show asymmetry of the tyrosine kinase domain dimer by its rod-like density arrangements (see Fig. 1A). Strong electron density indicates well-defined conformations of the interacting TKDs as opposed to the once widely believed transient interactions. Likewise, the PDGFR-β and Kit TKDs, when studied as part of full-length receptors by EM 3D-reconstruction, show kidney bean-like structures, further supporting asymmetric dimers.

TKD dimer asymmetry supports the idea that one receptor acts as the activator while the other acts as the receiver. This leads to a new question: what determines which TKD will be the activator or receiver? This is an important question because further research may reveal new mechanisms of regulation and avenues for drug discovery. Only EGFR has been shown to exhibit both asymmetric and symmetric arrangements. The functional significance of symmetrical arrangements of EGFR is still being explored.

10. Comparison to other classes of RTKs

Class I and III RTKs not only share structural characteristics between themselves but also with other less studied classes. Two relationships that have been identified from previous research are similarities between class I EGFR and class VI fibroblast growth factor receptors (FGFRs) (Bae & Schlessinger, 2010), and between type III Kit and PDGFRs and class V vascular endothelial growth factor receptors (VEGFRs) (Yang, Xie, Opatowsky, & Schlessinger, 2010). What is known about classes I and III can guide further research for classes V and VI RTKs.

The similar mechanisms of EGFR and FGFR may be due to similarities in their transmembrane motifs, specifically the GxxxG motif (Bocharov et al., 2013; Sarabipour & Hristova, 2016). They mediate interactions with the plasma membrane which could potentially affect the overall integrity and functionality of the EGFR and FGFR dimers. Moreover, like EGFR, FGFR exhibits back to back dimerization at their ECDs and their ligands do not bind directly at the ECD dimer interface. Also, like EGFR, FGFR has been found to exhibit inactive dimers and oligomers, a characteristic not found in other RTK classes. However, studies of the FGFR inactive dimers have employed the Forster resonance energy transfer based experiments rather than negative stain EM since the full length receptor has not yet been recombinantly expressed (Bajinting & Ng, 2017; Sarabipour & Hristova, 2016).

Likewise, understanding the characteristics of type III receptors, Kit and PDGFR, may help with understanding the class III RTK, VEGFR, an RTK that has not been structurally studied in its full-length form. Kit, PDGFR, and VEGFR all share the ECD domain D4-D4 dimerization motif and interface, suggesting they share dimerization mechanisms. Yuzawa et al. suggests that similar protomer interactions may take place (Yuzawa et al., 2007). This hypothesis is further supported by another feature these RTKs share: that Kit and VEGFR's ECD domains are Ig-like and that both classes dimerize at the ECD by a bridged dimer of growth factors rather than the back to back conformation of Type I and Type VI RTKs.

Tyrosine kinase asymmetry seems to be a nearly universal quality that indicates activity among RTKs. It is likely that important structural details of asymmetric dimerization differ between RTK classes. Additional structural research on intact kinases classes beyond I and III is sorely needed.

11. EGFR structure and pharmacological applications

In EGFR, structural data has driven the development of EGFR inhibitors in the form of monoclonal antibodies that target the ECD and tyrosine kinase inhibitors that target the TKD. These drugs are used to treat small cell lung cancer, colorectal cancer, breast cancer, and head and neck cancer (Xu, Johnson, & Grandis, 2017). Cetuximab is a monoclonal antibody that binds to the ECD D3 domain of a tethered EGFR monomer and blocks the EGF binding region. Structural research that contributed to the development of cetuximab include mechanistic studies of ECD activation and crystal structures of the ECD (Li et al., 2005). Small molecule drugs such as erlotinib, gefitinib, lapatinib, osimertinib, and afatinib target the TKD (Singh, Attri, Gill, & Nariwal, 2016). The invention of these and other kinase inhibitors make extensive use of structural data (Ferguson & Gray, 2018; Gagic, Ruzic, Djokovic, Djikic, & Nikolic, 2019). Structural studies are especially valuable in engineering drugs with high target specificity. For example, gefitinib has been found to bind EGFR with 200-fold higher affinity than other related receptors (Seshacharyulu et al., 2012).

An important issue in EGFR therapy is matching specific EGFR tumor mutations with the most effective drug. Certain mutations confer resistance to some drugs (Lynch et al., 2004; Paez et al., 2004; Singh & Jadhav, 2018; Sullivan & Planchard, 2017). Crystal structures aid interpretation of the structural compatibility between mutations and drug binding. Much current research is focused on developing next-generation EGFR inhibitors that are active against commonly occurring resistance mutations such as T790M in the ATP-binding site (Jia et al., 2016; Russo, Franchina, & Ricciardi, 2017; Wang, Song, Yan, & Liu, 2016) and exon 20 insertions in the C-helix and adjacent loop in the ECD (Ikemura et al., 2019; Ruan & Kannan, 2018; Vyse & Huang, 2019). As some resistance mutations are found outside of the ECD, structural studies of the intact receptor will facilitate the discovery of new drugs for these cases (Morgillo et al., 2016, Orellana et al., 2019, Zhong et al., 2017).

12. Kit and PDGFR structures and pharmacological applications

For Kit and PDGFR, structural data on the mutated and native structures have provided blueprints for drug design of inhibitors. Inhibitors including imatinib, sunitinib, and sorafenib are now in clinical use (Heldin, 2013). However, while PDGFR-B has been extensively studied, structural studies have been lacking for other PDGFR isoforms including AA, BB, CC, DD, and AB. This is especially important since some isoforms are tissue specific and are associated with different cancers (Kazlauskas, 2017, Papadopoulis & Lennartsson, 2018). For instance, PDGFR-A is associated with astrocytomas and glioblastomas (Heldin, 2013). Moreover, PDGFR contributes to maintenance of the tumor microenvironment (Belli et al., 2018; Östman, 2017). Structural biological studies of intact type III receptors provide opportunities for discovery of drugs with new mechanisms to address important medical needs.

13. New approaches to RTK structure-based drug discovery

Most RTK inhibitor drugs in clinical use are small molecules that target the TKDs. However, due to the high conservation of KDs across the ~500 protein kinases in the human genome, all kinase inhibitors have significant off-target effects (Anastassiadis, Deacon, Devarajan, Ma, & Peterson, 2011; Hanson et al., 2019; Hu, Kunimoto, & Bajorath, 2016). Moreover, mutations in the TKD that block drug binding are a common source of drug resistance. Antibody drugs targeting the ECD such as cetuximab are an effective alternative to TKD inhibitors. More structural data on fully intact receptors will facilitate discovery of alternative druggable sites.

13.1 EGFR

Full-length receptor studies have confirmed significant coupling between the ECD and TKD. Different ligands binding to the ECD lead to different conformational changes in the TM and JM regions (Sinclair et al., 2018). The JM region has also been shown to bind Traf4 (TNF receptor-associated factor 4) to activate the TKD (Cai et al., 2018). Thus, the JM region is a potential druggable site. Moreover, JM amino acid sequences are not conserved across RTK classes, improving prospects for drug specificity. Preliminary reports show JM peptide mimics can function as inhibitors (Boran et al., 2012; Gerhart, Thévenin, Bloch, King, & Thévenin, 2018). Further studies are needed to develop more drug-like inhibitors to explore the therapeutic potential of the JM site.

13.2 PDGFR and Kit

PDGFR and Kit drug discovery efforts have been more limited compared to EGFR. Full length EM studies of PDGFR and Kit reveal the importance of the homotypic interactions at the ECD D4 and D5 domains, features that are not seen in the crystal structure of the isolated ECD. This is an attractive target for drug design as these homotypic interactions facilitate dimerization and activation. Antibodies targeting the D4 of Kit have been shown to block the activation of both the wild type receptor and an oncogenic mutant, making this an especially promising approach (Reschetnyak et al., 2013). Corresponding antibodies for PDGFR have not yet been reported.

14. Conclusions

Currently, three low-resolution EM structures are available for full-length EGFR, PDGFR, and Kit. Immediate needs are to improve the resolution of these structures by utilizing cryo-EM and to expand structural details to the many other pharmacologically important RTKs. Already, the low-resolution structures have provided new insights into the relationships between the ECD, TM, JM, and TKD regions and their roles in activation and regulation. They have illuminated alternative druggable sites in addition to established targets in the TKD and ECD. These alternative sites may be especially important in developing drugs with improved specificity and the ability to overcome drug resistant mutations. Atomic resolution studies will be critical for enabling structure-based drug discovery. With the recent, rapid developments in cryo-EM technologies, we are confident that there will be many important breakthroughs in the near future.

References

Anastassiadis, T., Deacon, S. W., Devarajan, K., Ma, H., & Peterson, J. R. (2011). Comprehensive assay of kinase catalytic activity reveals features of kinase inhibitor selectivity. *Nature Biotechnology*, *29*, 1039–1045.

Arkhipov, A., Shan, Y., Das, R., Endres, N. F., Eastwood, M. P., Wemmer, D. E., et al. (2013). Architecture and membrane interactions of the EGF receptor. *Cell*, *152*, 557–569.

Arteaga, C. L., Ramsey, T. T., Shawver, L. K., & Guyer, C. A. (1997). Unliganded epidermal growth factor receptor dimerization induced by direct interaction of quinazolines with the ATP binding site. *The Journal of Biological Chemistry*, *272*, 23247–23254.

Bae, J. H., & Schlessinger, J. (2010). Asymmetric tyrosine kinase arrangements in activation or autophosphorylation of receptor tyrosine kinases. *Molecules and Cells*, *29*, 443–448.

Bajinting, A., & Ng, H. L. (2017). Recombinant expression in E. coli of human FGFR2 with its transmembrane and extracellular domains. *PeerJ*, *5*, e3512.

Belli, C., Trapani, D., Viale, G., D'Amico, P., Duso, B. A., Vigna, P. D., et al. (2018). Targeting the microenvironment in solid tumors. *Cancer Treatment Reviews*, *65*, 22–32.

Bessman, N. J., Freed, D. M., & Lemmon, M. A. (2014). Putting together structures of epidermal growth factor receptors. *Current Opinion in Structural Biology*, *29*, 95–101.

Bocharov, E. V., Leoyov, D. M., Goncharuk, S. A., Goncharuk, M. V., Hristova, K., & Arseniev, A. S. (2013). Structure of FGFR3 transmembrane domain dimer: Implications for Signaling and human pathologies. *Structure*, *21*, 2087–2093.

Bocharov, E. V., Sharonov, G. V., Bocharova, O. V., & Pavlov, K. V. (2017). Conformational transitions and interactions underlying the function of membrane embedded receptor protein kinases. *Biochimica et Biophysica Acta*, *1859*, 1417–1429.

Boran, A. D. W., Seco, J., Jayaraman, V., Jayaraman, G., Zhao, S., Reddy, S., et al. (2012). A potential peptide therapeutic derived from the juxtamembrane domain of the epidermal growth factor receptor. *PLoS One*, 7, e49702.

Bublil, E. M., & Yarden, Y. (2007). The EGF receptor family: Spearheading a merger of signaling and therapeutics. *Current Opinion in Cell Biology*, *19*, 124–134.

Cai, G., Zhu, L., Chen, X., Sun, K., Liu, C., Sen, G. C., et al. (2018). TRAF4 binds to the juxtamembrane region of EGFR directly and promotes kinase activation. *Proceedings of the National Academy of Sciences of the United States of America*, *115*, 11531–11536.

Chan, P. M., Ilangumaran, S., La Rose, J., Chakrabartty, A., & Rottapel, R. (2003). Autoinhibition of the kit receptor tyrosine kinase by the cytosolic juxtamembrane region. *Molecular and Cellular Biology*, *23*, 3067–3078.

Chen, P. H., Chen, X., & He, X. (2013). Platelet-derived growth factors and their receptors: Structural and functional perspectives. *Biochimica et Biophysica Acta*, *1834*, 2176–2186.

Chen, P. H., Unger, V., & He, X. (2015). Structure of full-length human PDGFRbeta bound to its activating ligand PDGF-B as determined by negative-stain electron microscopy. *Journal of Molecular Biology*, *427*, 3921–3934.

Da Cunha Santos, G., Shepherd, F. A., & Tsao, M. S. (2011). EGFR mutations and lung cancer. *Annual Review of Pathology*, *6*, 49–69.

Devan, D., Thaker, T., & Jura, N. (2019). More than the sum of the parts: Toward full-length receptor tyrosine kinase structures. *IUBMB Life*, *71*, 706–720.

Doura, A. K., & Fleming, K. G. (2004). Complex interactions at the helix-helix interface stabilize the glycophorin A transmembrane dimer. *Journal of Molecular Biology*, *343*, 1487–1497.

Endres, N. F., Das, R., Smith, A. W., Arkhipov, A., Kovacs, E., Huang, Y., et al. (2013). Conformational coupling across the plasma membrane in activation of the EGF receptor. *Cell*, *152*, 543–556.

Endres, N. F., Engel, K., Das, R., Kovacs, E., & Kuriyan, J. (2011). Regulation of the catalytic activity of the EGF receptor. *Current Opinion in Structural Biology*, *21*, 777–784.

Ferguson, K. M., Berger, M. B., Mendrola, J. M., Cho, H. S., Leahy, D. J., & Lemmon, M. A. (2003). EGF activates its receptor by removing interactions that autoinhibit ectodomain dimerization. *Molecular Cell*, *11*, 507–517.

Ferguson, F. M., & Gray, N. S. (2018). Kinase inhibitors: The road ahead. *Nature Reviews. Drug Discovery*, *17*, 353–376.

Gagic, Z., Ruzic, D., Djokovic, N., Djikic, T., & Nikolic, K. (2019). In silico methods for design of kinase inhibitors as anticancer drugs. *Frontiers in Chemistry*, *7*, 873.

Gerhart, J., Thévenin, A. F., Bloch, E., King, K. E., & Thévenin, D. (2018). Inhibiting epidermal growth factor receptor dimerization and signaling through targeted delivery of a juxtamembrane domain peptide mimic. *ACS Chemical Biology*, *13*, 2623–2632.

Greenall, S. A., Donoghue, J. F., Gottardo, N. G., Johns, T. G., & Adams, T. E. (2015). Glioma-specific domain IV EGFR cysteine mutations promote ligand-induced covalent receptor dimerization and display enhanced sensitivity to dacomitinib in vivo. *Oncogene*, *34*, 1658–1666.

Hanson, S. M., Georghiou, G., Thakur, M. K., Miller, W. T., Rest, J. S., Chodera, J. D., et al. (2019). What makes a kinase promiscuous for inhibitors? *Cell Chemical Biology*, *26*, 390–399.

Heldin, C. H. (2013). Targeting the PDGF signaling pathway in tumor treatment. *Cell Communication and Signaling: CCS*, *11*, 97.

Hu, Y., Kunimoto, R., & Bajorath, J. (2016). Mapping of inhibitors and activity data to the human kinome and exploring promiscuity from a ligand and target perspective. *Chemical Biology & Drug Design*, *89*, 834–845.

Huang, Y., Bharill, S., Karandur, D., Peterson, S. M., Marita, M., Shi, X., et al. (2016). Molecular basis for multimerization in the activation of the epidermal growth factor receptor. *eLife*, *5*, e14107.

Ikemura, S., Yasuda, H., Matsumoto, S., Kamada, M., Hamamoto, J., Masuzawa, K., et al. (2019). Molecular dynamics simulation-guided drug sensitivity prediction for lung cancer with rare EGFR mutations. *Proceedings of the National Academy of Sciences of the United States of America*, *116*, 10025–10030.

Irusta, P. M., Luo, Y., Bakht, O., Lai, C. C., Smith, S. O., & Dimaio, D. (2002). Definition of an inhibitory juxtamembrane WW-like domain in the platelet-derived growth factor beta receptor. *The Journal of Biological Chemistry*, *277*, 38627–38634.

Jia, Y., Yun, C.-H., Park, E., Ercan, D., Manuia, M., Juarez, J., et al. (2016). Overcoming EGFR(T790M) and EGFR(C797S) resistance with mutant-selective allosteric inhibitors. *Nature*, *534*, 129–132.

Kazlauskas, A. (2017). PDGFs and their receptors. *Gene*, *614*, 1–7.

Kazlauskas, A., & Cooper, J. A. (1989). Autophosphorylation of the PDGF receptor in the kinase insert region regulates interactions with cell proteins. *Cell*, *58*, 1121–1133.

Kovacs, E., Zorn, J. A., Huang, Y., Barros, T., & Kuriyan, J. (2015). A structural perspective on the regulation of the epidermal growth factor receptor. *Annual Review of Biochemistry*, *84*, 739–764.

Lemmon, M. A., Pinchasi, D., Zhou, M., Lax, I., & Schlessinger, J. (1997). Kit receptor dimerization is driven by bivalent binding of stem cell factor. *The Journal of Biological Chemistry*, *272*, 6311–6317.

Lemmon, M. A., & Schlessinger, J. (2010). Cell signaling by receptor tyrosine kinases. *Cell*, *141*, 1117–1134.

Li, S., Schmitz, K. R., Jeffrey, P. D., Wiltzius, J. J., Kussie, P., & Ferguson, K. M. (2005). Structural basis for inhibition of the epidermal growth factor receptor by cetuximab. *Cancer Cell*, *7*, 301–311.

Li, E., Wimley, W. C., & Hristova, K. (2012). Transmembrane helix dimerization: Beyond the search for sequence motifs. *Biochimica et Biophysica Acta*, *1818*, 183–193.

Liu, H., Chen, X., Focia, P. J., & He, X. (2007). Structural basis for stem cell factor-KIT signaling and activation of class III receptor tyrosine kinases. *The EMBO Journal*, *26*, 891–901.

Lu, C., Mi, L. Z., Grey, M. J., Zhu, J., Graef, E., Yokoyama, S., et al. (2010). Structural evidence for loose linkage between ligand binding and kinase activation in the epidermal growth factor receptor. *Molecular and Cellular Biology*, *30*, 5432–5443.

Lu, C., Mi, L. Z., Schurpf, T., Walz, T., & Springer, T. A. (2012). Mechanisms for kinase-mediated dimerization of the epidermal growth factor receptor. *The Journal of Biological Chemistry*, *287*, 38244–38253.

Lynch, T. J., Bell, D. W., Sordella, R., Gurubhagavatula, S., Okimoto, R. A., Brannigan, B. W., et al. (2004). Activating mutations in the epidermal growth factor receptor underlying responsiveness of non-small-cell lung cancer to gefitinib. *The New England Journal of Medicine*, *350*, 2129–2139.

Maeda, R., Sato, T., Okamoto, K., Yanagawa, M., & Sako, Y. (2018). Lipid-protein interplay in dimerization of juxtamembrane domains of epidermal growth factor receptor. *Biophysical Journal*, *114*, 893–903.

Mendrola, J. M., Berger, M. B., King, M. C., & Lemmon, M. A. (2002). The single transmembrane domains of ErbB receptors self-associate in cell membranes. *The Journal of Biological Chemistry*, *277*, 4704–4712.

Mi, L. Z., Grey, M. J., Nishida, N., Walz, T., Lu, C., & Springer, T. A. (2008). Functional and structural stability of the epidermal growth factor receptor in detergent micelles and phospholipid nanodiscs. *Biochemistry*, *47*, 10314–10323.

Mi, L. Z., Lu, C., Li, Z., Nishida, N., Walz, T., & Springer, T. A. (2011). Simultaneous visualization of the extracellular and cytoplasmic domains of the epidermal growth factor receptor. *Nature Structural & Molecular Biology*, *18*, 984–989.

Morgillo, F., Della Corte, C. M., Fasano, M., & Ciardiello, F. (2016). Mechanisms of resistance to EGFR-targeted drugs: Lung cancer. *ESMO Open*, *1*, e000060.

Muhle-Goll, C., Hoffmann, S., Afonin, S., Grage, S. L., Polyansky, A. A., Windisch, D., et al. (2012). Hydrophobic matching controls the tilt and stability of the dimeric platelet-derived growth factor receptor (PDGFR) beta transmembrane segment. *The Journal of Biological Chemistry*, *287*, 26178–26186.

Oates, J., King, G., & Dixon, A. M. (2010). Strong oligomerization behavior of PDGFbeta receptor transmembrane domain and its regulation by the juxtamembrane regions. *Biochimica et Biophysica Acta*, *1798*, 605–615.

Omura, T., Heldin, C. H., & Ostman, A. (1997). Immunoglobulin-like domain 4-mediated receptor-receptor interactions contribute to platelet-derived growth factor-induced receptor dimerization. *The Journal of Biological Chemistry*, *272*, 12676–12682.

Opatowsky, Y., Lax, I., Tome, F., Bleichert, F., Unger, V. M., & Schlessinger, J. (2014). Structure, domain organization, and different conformational states of stem cell factor-induced intact KIT dimers. *Proceedings of the National Academy of Sciences of the United States of America*, *111*, 1772–1777.

Orellana, L., Thorne, A. H., Lema, R., Gustavsson, J., Parisian, A. D., Hospital, A., et al. (2019). Oncogenic mutations at the EGFR ectodomain structurally converge to remove a steric hindrance on a kinase-coupled cryptic epitope. *Proceedings of the National Academy of Sciences of the United States of America*, *116*, 10009–10018.

Östman, A. (2017). PDGF receptors in tumor stroma: Biological effects and associations with prognosis and response to treatment. *Advanced Drug Delivery Reviews*, *121*, 117–123.

Paez, J. G., Janne, P. A., Lee, J. C., Tracy, S., Greulich, H., Gabriel, S., et al. (2004). EGFR mutations in lung cancer: Correlation with clinical response to gefitinib therapy. *Science*, *304*, 1497–1500.

Papadopoulis, N., & Lennartsson, J. (2018). The PDGF/PDGFR pathway as a drug target. *Molecular Aspects of Medicine*, *62*, 75–88.

Paul, M. K., & Mukhopadhyay, A. K. (2004). Tyrosine kinase—Role and significance in cancer. *International Journal of Medical Sciences*, *1*, 101–115.

Purba, E. R., Saita, E., & Maruyama, I. N. (2017). Activation of the EGF receptor by ligand binding and oncogenic mutations: The "rotation model". *Cell*, *6*, 13.

Red Brewer, M., Choi, S. H., Alvarado, D., Moravcevic, K., Pozzi, A., Lemmon, M. A., et al. (2009). The juxtamembrane region of the EGF receptor functions as an activation domain. *Molecular Cell*, *34*, 641–651.

Regad, T. (2015). Targeting RTK signaling pathways in cancer. *Cancers*, *7*, 1758–1784.

Reshetnyak, A. V., Nelson, B., Shi, X., Boggon, T. J., Pavlenco, A., Mandel-Bausch, E. M., et al. (2013). Structural basis for KIT receptor tyrosine kinase inhibition by antibodies targeting the D4 membrane-proximal region. *Proceedings of the National Academy of Sciences of the United States of America*, *110*, 17832–17837.

Reshetnyak, A. V., Opatowsky, Y., Boggon, T. J., Folta-Stogniew, E., Tome, F., Lax, I., et al. (2015). The strength and cooperativity of KIT ectodomain contacts determine normal ligand-dependent stimulation or oncogenic activation in cancer. *Molecular Cell*, *57*, 191–201.

Ruan, Z., & Kannan, N. (2018). Altered conformational landscape and dimerization dependency underpins the activation of EGFR by αC–β4 loop insertion mutations. *Proceedings of the National Academy of Sciences of the United States of America*, *115*, E8162–E8171.

Russo, A., Franchina, T., & Ricciardi, G. R. R. (2017). Third generation EGFR TKIs in EGFR-mutated NSCLC: Where are we now and where are we going. *Critical Reviews in Oncology/Hematology*, *117*, 38–47.

Sarabipour, S., & Hristova, K. (2016). Mechanism of FGF receptor dimerization and activation. *Nature Communications*, *7*, 10262.

Seshacharyulu, P., Ponnusamy, M. P., Haridas, D., Jain, M., Ganti, A. K., & Batra, S. K. (2012). Targeting the EGFR signaling pathway in cancer therapy. *Expert Opinion on Therapeutic Targets*, *16*, 15–31.

Shan, Y., Eastwood, M. P., Zhang, X., Kim, E. T., Arkhipov, A., Dror, R. O., et al. (2012). Oncogenic mutations counteract intrinsic disorder in the EGFR kinase and promote receptor dimerization. *Cell*, *149*, 860–870.

Shim, A. H., Liu, H., Focia, P. J., Chen, X., Lin, P. C., & He, X. (2010). Structures of a platelet-derived growth factor/propeptide complex and a platelet-derived growth factor/receptor complex. *Proceedings of the National Academy of Sciences of the United States of America, 107*, 11307–11312.

Shin, C. E., Helmrath, M. A., Falcone, R. A., Jr., Fox, J. W., Duane, K. R., Erwin, C. R., et al. (1998). Epidermal growth factor augments adaptation following small bowel resection: Optimal dosage, route, and timing of administration. *The Journal of Surgical Research, 77*, 11–16.

Sinclair, J. K. L., Walker, A. S., Doerner, A. E., & Schepartz, A. (2018). Mechanism of allosteric coupling into and through the plasma membrane by EGFR. *Cell Chemical Biology, 25*, 857–870.

Singh, D., Attri, B. K., Gill, R. K., & Nariwal, J. (2016). Review on EGFR inhibitors: Critical updates. *Mini-Reviews in Medicinal Chemistry, 16*, 1134–1166.

Singh, M., & Jadhav, H. M. (2018). Targeting non-small cell lung cancer with small-molecule EGFR tyrosine kinase inhibitors. *Drug Discovery Today, 23*, 745–753.

Sullivan, I., & Planchard, D. (2017). Next-generation EGFR tyrosine kinase inhibitors for treating EGFR-mutant lung cancer beyond first line. *Frontiers in Medicine, 3*, 76.

Vyse, S., & Huang, P. H. (2019). Targeting EGFR exon 20 insertion mutations in non-small cell lung cancer. *Signal Transduction and Targeted Therapy, 4*, 5.

Wang, S., Song, Y., Yan, F., & Liu, D. (2016). Mechanisms of resistance to third-generation EGFR tyrosine kinase inhibitors. *Frontiers in Medicine, 10*, 383–388.

Xu, M. J., Johnson, D. E., & Grandis, J. R. (2017). EGFR-targeted therapies in the post-genomic era. *Cancer Metastasis Reviews, 36*, 463–473.

Yang, Y., Xie, P., Opatowsky, Y., & Schlessinger, J. (2010). Direct contacts between extracellular membrane-proximal domains are required for VEGF receptor activation and cell signaling. *Proceedings of the National Academy of Sciences of the United States of America, 107*, 1906–1911.

Yang, Y., Yuzawa, S., & Schlessinger, J. (2008). Contacts between membrane proximal regions of the PDGF receptor ectodomain are required for receptor activation but not for receptor dimerization. *Proceedings of the National Academy of Sciences of the United States of America, 105*, 7681–7686.

Yu, X., Sharma, K. D., Takahashi, T., Iwamoto, R., & Mekada, E. (2002). Ligand-independent dimer formation of epidermal growth factor receptor (EGFR) is a step separable from ligand-induced EGFR signaling. *Molecular Biology of the Cell, 13*, 2547–2557.

Yuzawa, S., Opatowsky, Y., Zhang, Z., Mandiyan, V., Lax, I., & Schlessinger, J. (2007). Structural basis for activation of the receptor tyrosine kinase KIT by stem cell factor. *Cell, 130*, 323–334.

Zhang, X., Gureasko, J., Shen, K., Cole, P. A., & Kuriyan, J. (2006). An allosteric mechanism for activation of the kinase domain of epidermal growth factor receptor. *Cell, 125*, 1137–1149.

Zhong, J., Li, L., Wang, Z., Bai, H., Gai, F., Duan, J., et al. (2017). Potential resistance mechanisms revealed by targeted sequencing from lung adenocarcinoma patients with primary resistance to epidermal growth factor receptor (EGFR) tyrosine kinase inhibitors (TKIs). *Journal of Thoracic Oncology, 12*, 1766–1778.

Kolibri 360 GmbH
Ferdinand-Jühlke-Straße 7
99095 Erfurt, Deutschland
produktsicherheit@kolibri360.de